A First Course in

FUZZY

LOGIC

SECOND EDITION

A First Course in

FUZZY

LOGIC

SECOND EDITION

Hung T. Nguyen
Elbert A. Walker

Department of Mathematical Sciences
New Mexico State University
Las Cruces, New Mexico

CHAPMAN & HALL/CRC

Boca Raton   London   New York   Washington, D.C.

Library of Congress Cataloging-in-Publication Data

Nguyen, Hung T., 1944–
    A first course in fuzzy logic / Hung T. Nguyen and Elbert A.
Walker. — 2nd ed.
      p. cm.
    ISBN 0-8493-1659-6 (alk. paper)
    1. Neural networks (Computer science). 2. Fuzzy logic.
I. Walker, E. (Elbert), 1930– . II. Title. III. Title: Fuzzy
logic.
QA76.87.N497 1999
511.3—dc21   99-23550
                                                 CIP

© 2000 by Chapman & Hall/CRC

No claim to original U.S. Government works
International Standard Book Number 0-8493-1659-6
Library of Congress Card Number 99-23550
Printed in the United States of America   1 2 3 4 5 6 7 8 9 0
Printed on acid-free paper

Preface

This text grew out of a set of lecture notes for a course entitled "Introduction to Fuzzy Logic" taught several times at New Mexico State University and at the Tokyo Institute of Technology in the Spring of 1993. It is an introduction to the theory of fuzzy sets. Fuzzy sets are mathematical objects modeling the vagueness present in our natural language when we describe phenomena that do not have sharply defined boundaries. By using the concept of partial degrees of membership to give a mathematical definition of fuzzy sets, we increase the number of objects encountered in human reasoning which can be subjected to scientific investigation.

Fuzzy concepts need to be modeled mathematically for the purpose of automation such as in expert systems, computer vision, control engineering, and pattern recognition. Fuzzy set theory provides a machinery for carrying out approximate reasoning processes when available information is uncertain, incomplete, imprecise, or vague. With the emergence of new tools in the area of computational intelligence, such as nonstandard logics, neural networks, and symbolic reasoning, this new theory is a welcome addition to the repertoire of appropriate tools. This is especially true when the intrinsic property of fuzziness and the heterogeneity of objects under study have to be taken into account in order to do a better job of representing knowledge-fitting data. The success of this methodology has been demonstrated in a variety of fields, such as control of complex systems, where mathematical models are difficult to specify; in expert systems, where rules expressing knowledge and facts are linguistic in nature; and even in some areas of statistics, exemplified by categorical data analysis, where classes of objects are more fuzzy than crisp, and where the variability across objects needs to be modeled.

The material in this book has been chosen with a view to providing background for various areas of applications. The material in Chapters 1, 2, 3, 5, 6, 7, 8 and 13 is pertinent in engineering fuzzy logic, by which we mean the use of simple components of fuzzy theory, such as membership functions and fuzzy connectives, in the modeling of engineering knowledge and in the design of fuzzy control rules. Chapter 4 deals with that part of fuzzy logic that actually lies within the field of logic. Several propositional logics are discussed, including fuzzy propositional logic. This material should provide the reader with a clear way to think about this aspect of fuzzy theory, and should be of interest in theoretical computer science and artificial intelligence. The material in Chapters 9, 10, 11 and 12 is pertinent in decision making, in particular in fields such as expert systems and computer vision where the uncertainty involved can take on various facets, such as probability, possibility, belief functions, and more generally, fuzzy measures.

The material is drawn from many sources, including the novel series of papers of L. A. Zadeh, who is the founder of the theory of fuzzy sets [106]. For further reading, there is a bibliography. The text is designed for a one-semester course at the advanced undergraduate or beginning graduate level. The minimum prerequisite is some calculus, some set theory and Boolean logic, and some probability and statistics. However, we start from the ground up and background material will be reviewed at the appropriate places. The course is designed for students from fields such as artificial intelligence, computer science, engineering, cognitive science, mathematics, and probability and statistics, who seek a strong background for further study.

The exercises at the end of each chapter will deepen the students' understanding of the concepts and test their ability to make the necessary calculations. Exercises with an asterisk might not be suitable for homework, but should provide food for thought; they convey some advanced aspects of the topics treated. After completing the course, the students should be able to read more specialized and advanced books on the subject as well as articles in technical and professional journals.

This second edition is a fairly extensive revision and expansion of the first. Many exercises have been added. Chapter 5 of the first edition has been expanded into Chapters 5 and 6, with much new material on t-norms. Sections on rough sets and on conditional events have been added to Chapter 10. Material on distributions of random sets has been added to Chapter 11, and Chapter 12 contains a section on Radon-Nikodym derivatives of fuzzy measures.

We would like to express our thanks to Professor Lotfi A. Zadeh for his constant support of our research. Special thanks are due to Professor Mai Gehrke of the Department of Mathematical Sciences at New Mexico State University for her penetrating remarks and suggestions concerning the final version of this text. Professor N. Prasad of the Department of Electrical and Computer Engineering at New Mexico State University read the manuscript and made many valuable suggestions, for which we are grateful. We are also grateful to Professor Carol Walker of the Department of Mathematical Sciences at New Mexico State University who read the manuscript at various stages of its preparation and made many corrections and suggestions. The first named author would like to thank Professors T. Terano and M. Sugeno for providing him with excellent working conditions during his stay at Tokyo Institute of Technology as a visiting Professor holding the LIFE Chair of Fuzzy Theory.

We thank all those who used the first edition and gave us comments which led to substantial improvements. In particular, we are grateful to Professor S. Gudder of the Department of Mathematics and Computer Science at the University of Denver who did a thorough and thoughtful review of the first edition. Thanks are especially due to Professor Darel Hardy of the Department of Mathematics at Colorado State University, Professor Vladik Kreinovich of the Department of Computer Science at the University of Texas at El Paso, Professor N. Prasad of the Department of Electrical and Computer Engineering at New Mexico State University, and Professor Carol Walker of the Department of Mathematical Sciences at New Mexico State University all of whom read the second edition in its entirety and offered many penetrating comments and suggestions.

Hung T. Nguyen and Elbert A. Walker
Las Cruces, New Mexico
Spring 1999

Contents

CONTENTS

Chapter 1

THE CONCEPT OF FUZZINESS

In this opening chapter, we will discuss the intrinsic notion of fuzziness in natural language. Following Lotfi Zadeh, fuzzy concepts will be modeled as fuzzy sets, which are generalizations of ordinary (crisp) sets.

1.1 Examples

In using our everyday natural language to impart knowledge and information, there is a great deal of imprecision and vagueness, or fuzziness. Such statements as "John is tall" and "Fred is young" are simple examples. Our main concern is representing, manipulating, and drawing inferences from such imprecise statements.

We begin with some examples.

Example 1.1.1 The description of a human characteristic such as *healthy;*

Example 1.1.2 The classification of patients as *depressed;*

Example 1.1.3 The classification of certain objects as *large;*

Example 1.1.4 The classification of people by age such as *old;*

Example 1.1.5 A rule for driving such as "if an obstacle is *close*, then brake *immediately*".

In the examples above, terms such as *depressed* and *old* are fuzzy in the sense that they cannot be sharply defined. However, as humans, we do make sense out of this kind of information, and use it in decision

1

making. These "fuzzy notions" are in sharp contrast to such terms as *married, over 39 years old*, or *under 6 feet tall*. In ordinary mathematics, we are used to dealing with collections of objects, say certain subsets of a given set such as the subset of even integers in the set of all integers. But when we speak of the subset of *depressed* people in a given set of people, it may be impossible to decide whether a person is in that subset or not. Forcing a yes-or-no answer is possible and is usually done, but there may be information lost in doing so because no account is taken of the *degree* of depression. Although this situation has existed from time immemorial, the dominant context in which science is applied is that in which statements are precise (say either true or false)—no imprecision is present. But in this time of rapidly advancing technology, the dream of producing machines that mimic human reasoning, which is usually based on uncertain and imprecise information, has captured the attention of many scientists. The theory and application of fuzzy concepts are central in this endeavor but remain to a large extent in the domain of engineering and applied sciences.

With the success of automatic control and of expert systems, we are now witnessing an endorsement of fuzzy concepts in technology. The mathematical elements that form the basis of fuzzy concepts have existed for a long time, but the emergence of applications has provided a motivation for a new focus for the underlying mathematics. Until the emergence of fuzzy set theory as an important tool in practical applications, there was no compelling reason to study its mathematics. But because of the practical significance of these developments, it has become important to study the mathematical basis of this theory.

1.2   Mathematical modeling

The primitive notion of fuzziness as illustrated in the examples above needs to be represented in a mathematical way. This is a necessary step in getting to the heart of the notion, in manipulating fuzzy statements, and in applying them. This is a familiar situation in science. A good example is that of "chance". The outcome produced by many physical systems may be "random", and to deal with such phenomena, the theory of probability came into being and has been highly developed and widely used.

The mathematical modeling of fuzzy concepts was presented by Zadeh in 1965, and we will now describe his approach. His contention is that meaning in natural language is a matter of degree. If we have a proposition such as "John is young", then it is not always possible to assert that it is either true or false. When we know that John's age is x, then the "truth", or more correctly, the "compatibility" of x with "is young", is a matter of degree. It depends on our understanding of the concept "young". If the proposition is "John is under 22 years old" and we know John's age, then we can give a yes or no answer to whether the proposition is true or not. This can be formalized a bit by considering possible ages to be the interval $[0, \infty)$, letting A be the subset $\{x : x \in [0, \infty) : x < 22\}$, and then determining whether or not John's age is in A. But "young" cannot be defined as an ordinary subset of $[0, \infty)$. Zadeh was led to the notion of a fuzzy subset. Clearly, 18 and 20 year olds are young, but with different degrees: 18 is younger than 20. This suggests that membership in a fuzzy subset should not be on a 0 or 1 basis, but rather on a 0 to 1 scale, that is, the membership should be an element of the interval $[0, 1]$. This is handled as follows. An ordinary subset A of a set U is determined by its **indicator function**, or **characteristic function** χ_A defined by

$$\chi_A(x) = \begin{cases} 1 \text{ if } x \in A \\ 0 \text{ if } x \notin A \end{cases}$$

The indicator function of a subset A of a set U specifies whether or not an element is in A. It either is or is not. There are only two possible values the indicator function can take. This notion is generalized by allowing images of elements to be in the interval $[0, 1]$ rather than being restricted to the two element set $\{0, 1\}$.

Definition 1.2.1 *A **fuzzy subset** of a set U is a function $U \rightarrow [0, 1]$.*

Those functions whose images are contained in the two element set $\{0, 1\}$ correspond to ordinary, or **crisp** subsets of U, so ordinary subsets are special cases of fuzzy subsets. It is common to refer to a fuzzy subset simply as a **fuzzy set**, and we will do that.

It is customary in the fuzzy literature to have two notations for fuzzy sets, the letter A, say, and the notation μ_A. The first is called a "linguistic label". For example, one might say "Let A be the set of young people." A specific function $U \rightarrow [0, 1]$ representing this notion would be denoted μ_A. The notation A stands for the concept of "young", and μ_A spells out the degree of youngness that has been assigned to each member of U. We choose to make A stand for the actual fuzzy set which is always a function from a set U into $[0, 1]$, and thus we have no need for the notation μ_A. Fuzzy sets of course serve as models of concepts such as "young", but we have found no real need for special "linguistic labels" for these concepts. Such labels would not represent mathematical objects, so could not be manipulated as such. In any case, we will not use any special symbols for "linguistic labels", and by a fuzzy set we always mean a function from some set U into $[0, 1]$.

For a fuzzy set $A : U \rightarrow [0, 1]$, the function A is called the **membership function,** and the value $A(u)$ is called the **degree of membership** of u in the fuzzy set A. *It is not meant to convey the likelihood or probability that u has some particular attribute* such as "young".

Of course, for a fuzzy concept, different functions A can be considered. The choice of the function A is subjective and context dependent and can be a delicate one. But the flexibility in the choice of A is useful in applications, as in the case of fuzzy control, treated in Chapter 13.

Here are two examples of how one might model the fuzzy concept "young". Let the set of all possible ages of people be the positive real numbers. One such model, decided upon by a teenager might be

$$Y(x) = \begin{cases} 1 & \text{if} \quad x < 25 \\ \frac{40-x}{15} & \text{if} \quad 25 \leq x \leq 40 \\ 0 & \text{if} \quad 40 < x \end{cases}$$

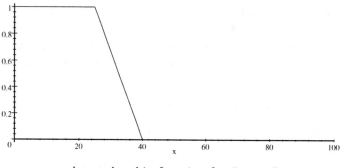

A membership function for "young"

An older person might model it differently, say with the function

$$Z(x) = \begin{cases} 1 & \text{if} \quad x < 40 \\ \frac{80-x}{40} & \text{if} \quad 40 \leq x \leq 60 \\ \frac{70-x}{20} & \text{if} \quad 60 < x \leq 70 \\ 0 & \text{if} \quad 70 < x \end{cases}$$

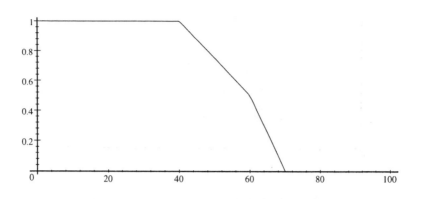

Another membership function for "young"

There are various ways to get reasonable membership functions. Here is an illustration of one way. Suppose we want to model the notion of "high income" with a fuzzy set. Again, let the set U be the positive real numbers, representing the totality of possible incomes. We survey a large number of people and find out that no one thought that an income under $20,000 was high, but that the proportion p of people that thought that an income x between $20,000 and $75,000 was high was approximately

$$p = \frac{x - 20}{55}$$

Of course, everyone thought that an income over $75,000 was high. Measuring in thousands of dollars, one reasonable model of the fuzzy set "high income" would be

$$H(x) = \begin{cases} 0 & \text{if} \quad x < 20 \\ \frac{x-20}{55} & \text{if} \quad 20 \leq x \leq 75 \\ 1 & \text{if} \quad 75 < x \end{cases}$$

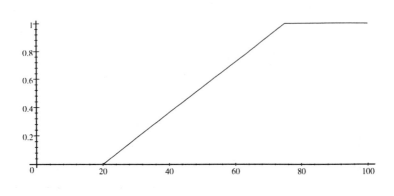

A membership function for "high income"

This last example bears a few comments. To get a reasonable model for the fuzzy set, a survey was made and some probabilities determined. The value $H(50) = 30/55$ *does not mean that the probability of an income of* $50,000$ *is high.* It is true that of the people surveyed, the proportion of them who would classify such an income as high was 30/55. We have simply taken that proportion to represent the *degree* to which such an income is considered high. It is not the probability of having a high income. There is no such probability around in this discussion. If an income is $50,000$, then the probability that a random person from those surveyed would classify that income as high is 30/55. But probabilities should not be confused with degrees of membership in a fuzzy set. *Membership functions are not probability distributions.*

We have modeled these membership functions with some very simple functions—piecewise linear ones. This is common practice.

1.3   Some operations on fuzzy sets

As we have noted, a subset A of a set U can be represented by a function $\chi_A : U \to \{0,1\}$, and a fuzzy subset of U has been defined to be a function $A : U \to [0,1]$. On the set $\mathcal{P}(U)$ of all subsets of U there are the familiar operations of union, intersection, and complement. These are given by the rules

$$
\begin{aligned}
A \cup B &= \{x : x \in A \text{ or } x \in B\} \\
A \cap B &= \{x : x \in A \text{ and } x \in B\} \\
A' &= \{x \in U : x \notin A\}
\end{aligned}
$$

Writing these in terms of indicator functions, we get

$$\begin{aligned}
\chi_{A \cup B}(x) &= \max\{\chi_A(x), \chi_B(x)\} = \chi_A(x) \vee \chi_B(x) \\
\chi_{A \cap B}(x) &= \min\{\chi_A(x), \chi_B(x)\} = \chi_A(x) \wedge \chi_B(x) \\
\chi_{A'}(x) &= 1 - \chi_A(x)
\end{aligned}$$

A natural way to extend these operations to the fuzzy subsets of U is by the membership functions

$$\begin{aligned}
(A \vee B)(x) &= \max\{A(x), B(x)\} = A(x) \vee B(x) \\
(A \wedge B)(x) &= \min\{A(x), B(x)\} = A(x) \wedge B(x) \\
A'(x) &= 1 - A(x)
\end{aligned}$$

There are many other generalizations of these operations, and some will be presented in Chapter 5. One remark about notation: we will use $\vee$ for max and for sup. Some authors denote the fuzzy set $A \vee B$ by $A \cup B$. This function is the smallest that is greater or equal to both A and B, that is, is the sup, or supremum, of the two functions. This notation conforms to lattice theoretic notation which we will have many occasions to use later. Similar remarks apply to using $\wedge$ for min and for inf (short for 'infimum').

Here are a couple of examples illustrating these operations between fuzzy sets. Consider the two fuzzy sets $A(x)$ and $B(x)$ of the nonnegative real numbers given by the formulas

$$A(x) = \begin{cases} 1 & \text{if} \quad x < 20 \\ \frac{40-x}{20} & \text{if} \quad 20 \leq x < 40 \\ 0 & \text{if} \quad 40 \leq x \end{cases} \tag{1.1}$$

and

$$B(x) = \begin{cases} 1 & \text{if} \quad x \leq 25 \\ \left(1 + \left(\frac{x-25}{5}\right)^2\right)^{-1} & \text{if} \quad 25 < x \end{cases} \tag{1.2}$$

Here are the plots of these two membership functions.

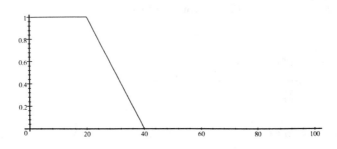

The membership function A

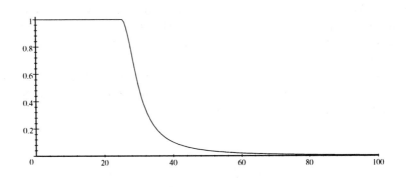

The membership function B

The plots for $A \vee B$, $A \wedge B$, and A' are the following. We leave as exercises the writing out of formulas for these membership functions.

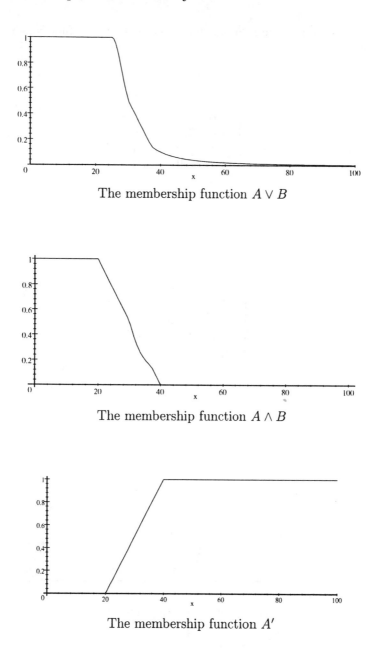

The membership function $A \vee B$

The membership function $A \wedge B$

The membership function A'

Here are two more membership functions for fuzzy subsets of the non-negative real numbers, their plots, union, intersection, and complements.

Again, writing down the formulas for these membership functions is left as an exercise.

$$C(x) = \begin{cases} 0 & \text{if} \quad 0 \leq x < 1 \\ x - 1 & \text{if} \quad 1 \leq x < 2 \\ 1 & \text{if} \quad 2 \leq x < 3 \\ 4 - x & \text{if} \quad 3 \leq x \leq 4 \\ 0 & \text{if} \quad 4 < x \end{cases} \tag{1.3}$$

$$D(x) = \begin{cases} e^{x-3} & \text{if} \quad 0 \leq x < 3 \\ 1 & \text{if} \quad 3 \leq x < 5 \\ 1 - \frac{x-5}{2} & \text{if} \quad 5 \leq x \leq 7 \\ 0 & \text{if} \quad 7 < x \end{cases} \tag{1.4}$$

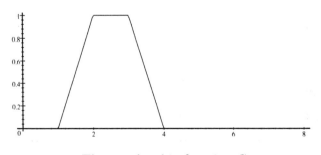

The membership function C

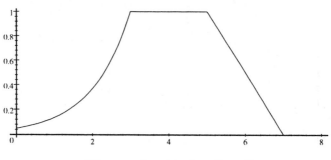

The membership function D

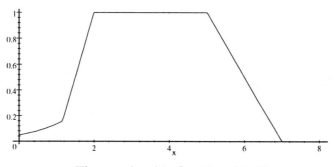

The membership function $C \vee D$

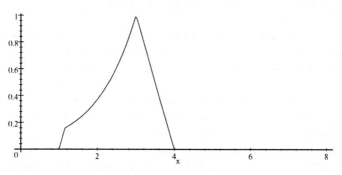

The membership function $C \wedge D$

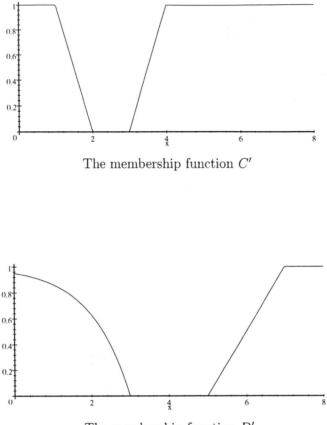

The membership function C'

The membership function D'

1.4   Fuzziness as uncertainty

There are many kinds of uncertainty arising in real-world problems and a variety of techniques are needed for modeling them. What are some of these techniques, and when does fuzzy set theory provide appropriate models?

Fuzzy sets deal with the type of uncertainty that arises when the boundaries of a class of objects are not sharply defined. We have seen several examples of such vagueness already: "young" and "high income", for instance. Membership in such classes is a matter of degree rather than certainty one way or another, and it is specified mathematically by fuzzy sets.

Ambiguity is another kind of uncertainty. This may come about in various ways. For example, if some parameter in a control system is only known to lie within a given interval, then there is uncertainty about any nominal value chosen from that interval for that parameter.

Another example is that of randomness, as exemplified by the uncertainty of the outcome of some experiment such as rolling a pair of dice, or of the observations made of some physical system. Randomness is typically modeled using probability theory. That is, outcomes are assumed to be observations of random variables and these random variables have distribution laws. These laws may not be known, of course, but each random variable has a unique one. This is in contrast to the fact that many different membership functions can be assigned to the same fuzzy concept. Again, *probability* and *degrees of membership* are distinct things.

As in the case of uncertainty modeled by probability theory, probability logic as discussed in Chapter 4 is used as a vehicle for making inferences from data. Probability logic is used in situations where events of interest are either true or false, but the information available is incomplete and prevents such a determination. Propositions will correspond to events, and the probability of an event is used as a measure of the truth of its corresponding proposition.

In confidence interval estimation in statistics, the situation is this. A model parameter, for example, the mean μ of a random variable X, is unknown. For each subset A of the range U of X, either $\mu \in A$ or $\mu \notin A$, but it is not known which. One would like to construct a small subset A such that μ is very likely to be in A. Since μ is fixed, a probabilistic statement to this effect is meaningful only when the set A is random. Thus the probability that $\mu \in A$ is the probability α that the random set contains μ. After constructing A from data, it is a nonrandom set, and we interpret the probability α as our degree of confidence that μ is in A.

Now consider a similar situation. Let U denote a set of all possible answers to a specific question, only one of which, say u_0 is correct. Which of the answers is the correct one is unknown. For each crisp subset A of U, we would like to ask an expert, or use evidence of some kind, to assign a value $Q(A) \in [0,1]$ which represents our degree of belief that A contains u_0. This type of assignment is a mathematical modeling of fuzzy concepts by fuzzy sets. Here the fuzzy sets are fuzzy subsets of the set of all subsets of U.

In complicated real-world cases, several types of uncertainty can coexist. For example, to each population of humans, chosen at random, one might be interested in its "morality", its "political spirit"; to each town chosen at random, one might be interested in its "shape", its "beauty", and so on. Viewing such attributes as fuzzy concepts, these are examples of fuzzy concepts that can be formulated rigorously as random fuzzy sets.

Each type of uncertainty has its mathematical representation or model, and associated calculus. Different mathematical theories are like tools in a toolbox. One may be more advantageous to use than another in a given situation. Sometimes several may apply, and one may even want to use several in conjunction. The practitioner has to be creative and use understanding in order to choose the right combination of mathematical theories to apply. We turn now to some typical useful aspects of fuzzy sets.

The modeling of fuzzy concepts by fuzzy sets leads to the possibility of giving mathematical meaning to natural language statements. For example, when modeling the concept "young" as a fuzzy subset of $[0, \infty)$ with a membership function $A : [0, \infty) \to [0, 1]$, we described the meaning of "young" in a mathematical way. It is a function, and can be manipulated mathematically and combined with other functions, for example. The fuzzy concept has been put into a useful form.

Even in areas where statistical techniques are dominant, such as in multivariate categorical data analysis, fuzzy sets are not only useful in various cases, but may be more efficient. They have the capacity, for example, to model variability across objects. We illustrate: if the categories are like "gender" or "marital status", there is no fuzzy concept involved. But if the categories are "depressed", "mentally unbalanced", and "stable", then instead of using an ordinary partition of a set of people into these categories, a fuzzy partition might be more realistic. This is a typical case where fuzziness is involved and the ordinary mathematical concept of partition needs to be generalized to its fuzzy counterpart. (See Section 7.4.) For example, fuzzy partitions are essential in the design of fuzzy controllers, which is the topic of Chapter 13.

There is a more formal relation between randomness and fuzziness. Let $A : U \to [0, 1]$ be a fuzzy set. For $\alpha \in [0, 1]$, let $A_\alpha = \{u \in U : A(u) \geq \alpha\}$. The set A_α is called the α-cut of A. Now let us view α as a random variable uniformly distributed on $[0, 1]$. That is, let $(\Omega, \mathcal{A}, P)$ be a probability space and $\alpha : \Omega \to \mathbb{R}$ a random variable with

$$P\{\omega : \alpha(\omega) \leq a\} = \begin{cases} 0 \text{ if } a < 0 \\ a \text{ if } 0 \leq a \leq 1 \\ 1 \text{ if } a > 1 \end{cases}$$

Then $A_{\alpha(\omega)}$ is a **random set**. (Random sets are discussed in Chapter 11.) The **covering function**, or **one-point covering function**, of the random set $A_{\alpha(\omega)}$ is defined to be

$$\pi : U \to [0, 1] : u \to P\{\omega : u \in A_{\alpha(\omega)}\}$$

which is $P\{\omega : \alpha(\omega) \leq A(u)\} = A(u)$. That is, $\pi(u) = A(u)$. This means that a fuzzy set can be written as the covering function of a random set. A random set is characterized by its distribution. But specifying a covering function of a random set is not sufficient to determine its distribution, much the same as the fact that moments of a random variable do not in general specify its distribution. In any case, a possible interpretation of the formal connection between fuzzy sets and covering functions of random sets is that fuzziness is a weakened form of randomness. It does not mean that probability theory subsumes fuzzy set theory.

The relation $A(u) = P\{\omega : u \in A_{\alpha(\omega)}\}$ is interesting in suggesting ways to obtain membership functions. First, A is a membership function, being a function $U \to [0,1]$. Of course this is true if A_α is replaced by any mapping $S : \Omega \to \mathcal{P}(U)$ and the probability function is replaced by any mapping $\mu : \mathcal{P}(\Omega) \to [0,1]$. Then $u \to \mu\{\omega : u \in S(\omega)\}$ is certainly a mapping $U \to [0,1]$ and hence is a fuzzy set.

We close this section with such an example. Suppose we are interested in describing the fuzzy concept of the seriousness of some illness. Suppose that the illness under consideration is manifested as subsets of the set $\Omega = \{\omega_1, \omega_2, ..., \omega_n\}$ of possible symptoms. Let U be a set of humans, and let $S : \Omega \to \mathcal{P}(U)$ be given by $S(\omega) = \{u \in U : u$ has symptom $\omega\}$. For $u \in U$, we are interested in some numerical measure of the set $\{\omega \in \Omega : u \in S(\omega)\}$. This is to be a measure of the seriousness of the illness of u. Medical experts often can provide assessments which can be described mathematically as a function $\mu : \mathcal{P}(\Omega) \to [0,1]$, where $\mu(B)$ is the degree of seriousness of the illness of a person having all the symptoms in B. So a membership function can be taken to be

$$A(u) = \mu\{\omega \in \Omega : u \in S(\omega)\}$$

Since μ is subjective, there is no compelling reason to assume that it is a measure, for example that it is additive. However, it is obvious that it should be monotone increasing, that is, $B \subseteq C$ should imply that $\mu(B) \leq \mu(C)$. Such functions as these are called *fuzzy measures*, which are a topic of discussion in Chapter 11.

1.5 Exercises

1. Give several statements in natural language which involve fuzzy concepts.

2. From your own experience, describe situations where fuzzy concepts are needed.

3. Using just common sense, give a reasonable membership function for the following fuzzy sets. (Of course, you must first specify exactly what the underlying set is.)

(a) n is a large integer.

(b) The mean of a random variable is around 5.

(c) x is much larger than y.

(d) These are very young people.

(e) x is between -3 and 2.

(f) x is approximately equal to y.

4. Let $U = \{0, 1, 2, 3, 4, 5, 6, 7, 8, 9\}$. Let two fuzzy subsets A and B of U be given by

u	0	1	2	3	4	5	6	7	8	9
$A(u)$	0	0	0.1	0.2	0.3	0.8	0.9	1	1	1

and

u	0	1	2	3	4	5	6	7	8	9
$B(u)$	1	1	0.9	0.8	0.7	0.5	0.4	0.2	0.2	0

Determine $A \vee B$, $A \wedge B$, and A'.

5. Let C and D be fuzzy subsets of the nonnegative real numbers. Draw the graph of $C \wedge D'$ when

$$A(x) = \begin{cases} 0 & \text{if } 0 < x < 1 \\ x - 1 & \text{if } 1 \leq x < 2 \\ 1 & \text{if } 2 \leq x < 3 \\ 4 - x & \text{if } 3 \leq x < 4 \\ 0 & \text{if } 4 \leq x \end{cases}$$

and

$$B(x) = \begin{cases} e^{x-3} & \text{if } 0 \leq x < 3 \\ 1 & \text{if } 3 \leq x < 5 \\ 1 - \frac{x-5}{2} & \text{if } 5 \leq x < 10 \\ 0 & \text{if } 10 \leq x \end{cases}$$

6. Write out the formulas for $A \vee B$, $A \wedge B$, A', and B', where A and B are given by formulas 1.1 and 1.2 in Section 1.3.

7. Write out the formulas for $C \vee D$, $C \wedge D$, C', and D', where C and D are given by formulas 1.3 and 1.4 in Section 1.3.

8. For fuzzy sets A, B, and C, prove that $(A \vee B) \vee C = A \vee (B \vee C)$, and that $(A \wedge B) \wedge C = A \wedge (B \wedge C)$.

9. For fuzzy sets A, B, and C, prove that $(A \vee B)' = A' \wedge B'$.

10. *Let $f : U \to [0,1]$ and let $\alpha : (\Omega, \mathcal{A}, P) \to [0.1]$ be a uniformly distributed random variable. Let $S_f : \Omega \to \mathcal{P}(U) : \omega \to \{u \in U : f(u) \geq \alpha(\omega)\}$.

   (a) Verify that the range $S_f(\Omega) = \{S_f(\omega) : \omega \in \Omega\}$ is totally ordered by set inclusion.

   (b) Show that for $A \in S_f(\Omega)$, $\{\omega \in \Omega : A \subseteq S_f(\omega)\} \in \mathcal{A}$.

   (c) Let $S : \Omega \to \mathcal{P}(U)$ be any random set in U. We say that S is **nested** if $S(\Omega)$ is totally ordered and for $A \in S(\Omega)$, $\{\omega : A \subseteq S(\omega)\} \in \mathcal{A}$. Show that if S is nested and its covering function coincides with f, then for $A \in S(\Omega)$, we have $P(\omega : A \subseteq S(\omega)) = P(\omega : A \subseteq S_f(\omega))$.

11. Let U be a finite set. The **cardinality**, or **sigma count**, of fuzzy subset A of U is $\#(A) = \sum_{u \in U} A(u)$. For fuzzy subsets A and B of U, the **degree of subsethood** of A in B is defined as $s(A|B) = \#(A \wedge B)/\#(A)$.

   (a) Verify that $s(A|B) = 1$ if and only if $A(u) \leq B(u)$ for every $u \in U$.

   (b) Compute the degree of subsethood of A in B in Exercise 4.

Chapter 2

SOME ALGEBRA OF FUZZY SETS

In Chapter 1, we have discussed modeling fuzzy concepts such as uncertainty with fuzzy sets. Applications demand combining these fuzzy sets in various ways. This means that we must understand the set $\mathcal{F}(U)$ of all fuzzy subsets of a set U as a mathematical object. The basic mathematical structure of $\mathcal{F}(U)$ comes from the fact that the unit interval $[0, 1]$ is *ordered*. This ordering on $[0, 1]$ induces a partial order on $\mathcal{F}(U)$ which in turn gives $\mathcal{F}(U)$ the algebraic structure of a *lattice*. So we need some background material about partially ordered sets, lattices, and related mathematical notions. These notions are fundamental, and are absolutely essential in understanding the mathematics of fuzzy sets.

2.1 Boolean algebras and lattices

We begin by discussing the familiar properties of the system of subsets of a set. Let $\mathcal{P}(U)$ be the set of all subsets of the set U. It is called the **power set** of U. Sometimes $\mathcal{P}(U)$ is identified with 2^U, which is the set of all mappings of U into $\{0, 1\}$. A subset $A \in \mathcal{P}(U)$ is identified with the function $U \to \{0, 1\}$ that maps each element of A to 1 and the other elements of U to 0. The set U may be finite or infinite. Perhaps the most basic thing about $\mathcal{P}(U)$ is that it is a *partially ordered* set. This simply comes from the familiar notion of set inclusion. If A and B are in $\mathcal{P}(U)$ then we write $A \subseteq B$ if A is a subset of B, that is, if every element of A is an element of B. To make the definition of partially ordered set formal, we use the notion of *relation*. The **Cartesian product** of a set S with a set T is the set $S \times T = \{(s, t) : s \in S, t \in T\}$.

Definition 2.1.1 *A **relation** on a set U is a subset R of the Cartesian product $U \times U$.*

The notion of relation is a very general one. For an element $(x, y) \in U \times U$, either $(x, y) \in R$ or it is not. Standard notation to denote that $(x, y) \in R$ is xRy. Taking this view, $\subseteq$ is a relation on $\mathcal{P}(U)$, and $A \subseteq B$ is the notation that we use to say that the pair $(A, B) \in \subseteq$.

The relation $\subseteq$ satisfies the following properties.

- $A \subseteq A$. (The relation is **reflexive.**)

- If $A \subseteq B$ and $B \subseteq C$, then $A \subseteq C$. (The relation is **transitive.**)

- If $A \subseteq B$ and $B \subseteq A$, then $A = B$. (The relation is **antisymmetric.**)

A **partial order on a set** is a relation on that set that is reflexive, transitive, and antisymmetric.

Definition 2.1.2 *A **partially ordered set** is a pair $(U, \leq)$ where U is a set and $\leq$ is a partial order on U.*

We often just say that U is a partially ordered set if it is clear what the relation $\leq$ is. Partially ordered sets abound. As already indicated, $(\mathcal{P}(U), \subseteq)$ is a partially ordered set. If $(U, \leq)$ is a partially ordered set, then so is $(A, \leq)$ where A is any subset of U and $\leq$ is the relation induced on A by the relation $\leq$ on U. If $(U, \leq)$ is a partially ordered set and for any two elements $x, y \in U$ either $x \leq y$ or $y \leq x$, then we say U is a **chain** or is **linearly ordered,** or **totally ordered**. Any subset U of the set of real numbers is a chain under the usual ordering. In particular, the interval $[0, 1]$ is a chain and will play a fundamental role throughout this book. Two elements x and y in a partially ordered set may have a sup. That is, there may be an element s such that $x \leq s$, $y \leq s$, and $s \leq t$ for all elements t such that $x \leq t$ and $y \leq t$. By antisymmetry, there is at most one such s, and it is the smallest element greater than both x and y. Such an element is denoted $\sup\{x, y\}$, or quite commonly $x \vee y$. Similarly, x and y may have an inf, denoted $x \wedge y$. In a chain, $x \vee y$ and $x \wedge y$ always exist. In this special case, the sup is one of the two elements x and y, and similarly for the inf.

Definition 2.1.3 *A **lattice** is a partially ordered set $(U, \leq)$ in which every pair of elements of U has a sup and an inf in U.*

Chains are always lattices, as noted above. The partially ordered set $(\mathcal{P}(U), \subseteq)$ is a lattice. The sup of two elements in $\mathcal{P}(U)$ is their union, and the inf is their intersection. The interval $[0, 1]$ is a lattice, being a chain.

A **binary operation** on a set U is a map $U \times U \to U$. For example, addition and multiplication are binary operations on the real numbers, that is, are functions from $\mathbb{R} \times \mathbb{R}$ to $\mathbb{R}$. Binary operations are often written between their arguments. That is, the image of $(2, 3)$ under the operation of addition is written $2 + 3$. We will adhere to this standard practice. Also, throughout this book, we will use $\mathbb{R}$ to denote the set of real numbers.

If $(U, \leq)$ is a lattice, then it comes equipped with the two binary operations $\vee$ and $\wedge$. For $x, y \in U$, $x \vee y$ is the sup and $x \wedge y$ is the inf of x and y. The operations $\vee$ and $\wedge$ are also called **join** and **meet**, respectively. From either of these binary operations one can reconstruct $\leq$. In fact, $a \leq b$ if and only if $a \wedge b = a$ if and only if $a \vee b = b$. These binary operations satisfy a number of properties.

Theorem 2.1.4 *If $(U, \leq)$ is a lattice, then for all a, b, $c \in U$,*

1. *$a \vee a = a$ and $a \wedge a = a$. ($\vee$ and $\wedge$ are **idempotent**.)*

2. *$a \vee b = b \vee a$ and $a \wedge b = b \wedge a$. ($\vee$ and $\wedge$ are **commutative**.)*

3. *$(a \vee b) \vee c = a \vee (b \vee c)$ and $(a \wedge b) \wedge c = a \wedge (b \wedge c)$. ($\vee$ and $\wedge$ are **associative**.)*

4. *$a \vee (a \wedge b) = a$ and $a \wedge (a \vee b) = a$. (These are the **absorption identities**.)*

The proof is quite easy and is left as an exercise. A pertinent fact is that two binary operations satisfying conditions 1-4 define a lattice.

Theorem 2.1.5 *If U is a set with binary operations $\vee$ and $\wedge$ which satisfy the properties of Theorem 2.1.4, then defining $a \leq b$ if $a \wedge b = a$ makes $(U, \leq)$ into a lattice whose sup and inf operations are $\vee$ and $\wedge$.*

Proof. We first show that $a \wedge b = a$ if and only if $a \vee b = b$. Thus defining $a \leq b$ if $a \wedge b = a$ is equivalent to defining $a \leq b$ if $a \vee b = b$. Indeed, if $a \wedge b = a$, then $a \vee b = (a \wedge b) \vee b = b$ by one of the absorption laws. Similarly, if $a \vee b = b$, then $a \wedge b = a$. We show the existence of sups, and claim that $\sup\{a, b\} = a \vee b$. Now $a \leq a \vee b$ since $a \wedge (a \vee b) = a$ by one of the absorption laws. Similarly $b \leq b \vee a = a \vee b$, so that $a \vee b$ is an upper bound of a and b. For any other upper bound x, $a = a \wedge x$ and $b = b \wedge x$, whence $x = a \vee x = b \vee x$. Therefore, $x = a \vee x \vee b \vee x = (a \vee b) \vee x$, and so $a \vee b \leq x$. Thus $a \vee b = \sup\{a, b\}$. The rest of the proof is left as an exercise. ∎

The upshot is that a lattice may be thought of as a partially ordered set in which every pair of elements has a sup and an inf, or as a set with two binary operations satisfying the conditions of Theorem 2.1.4. So we could say "$(U, \leq)$ is a lattice," or "$(U, \vee, \wedge)$ is a lattice." The latter would mean, of course, that the binary operations $\vee$ and $\wedge$ satisfy the conditions in Theorem 2.1.4.

Here are some pertinent additional properties that a lattice $(U, \leq)$ *may* have.

1. There is an element 0 in U such that $0 \vee a = a$ for all $a \in U$. There is an element $1 \in U$ such that $1 \wedge a = a$ for all $a \in U$. (0 and 1 are **identities** for $\vee$ and $\wedge$, respectively.)

2. U has identities, and for each element a in A, there is an element a' in U such that $a \wedge a' = 0$ and $a \vee a' = 1$. (Each element in U has a **complement**, or A is **complemented**.)

3. $a \vee (b \wedge c) = (a \vee b) \wedge (a \vee c)$ and $a \wedge (b \vee c) = (a \wedge b) \vee (a \wedge c)$. (The binary operations $\vee$ and $\wedge$ **distribute** over each other. If one of these distributive laws holds in a lattice, then so does the other.)

4. Every subset T of U has a sup. That is, there is an element $a \in U$ such that $t \leq a$ for all $t \in T$, and $a \leq x$ for any x such that $t \leq x$ for all $t \in T$. Similarly, every subset T of U may have an inf. That is, there is an element $b \in U$ such that $b \leq t$ for all $t \in T$, and $b \geq x$ for any x such that $x \leq t$ for all $t \in T$.

If a lattice has an identity for $\vee$ and an identity for $\wedge$, then it is a **bounded lattice**. From the equation $a \wedge 1 = a$, we get that 1 is the largest element in the lattice, and $0 \wedge a = 0 \wedge (0 \vee a) = 0$, so 0 is the smallest. This condition could have been stated just by saying that the lattice has a largest and a smallest element. A bounded lattice satisfying the second condition is a **complemented lattice**. A lattice satisfying both distributive laws is a **distributive lattice**. A bounded distributive lattice that is complemented is a **Boolean lattice**, or **Boolean algebra**. Chains are distributive lattices, and $(\mathcal{P}(U), \subseteq)$ is a Boolean lattice. If a lattice satisfies condition 4, that is, if every subset has both a sup and an inf, it is a **complete** lattice. For a subset T of a complete lattice, $\sup T$ is often written $\bigvee T$, or $\bigvee_{t \in T} t$, and similarly $\inf T = \bigwedge T$, or $\bigwedge_{t \in T} t$. If $\{t_i : i \in I\}$ is a family of elements of a complete lattice, then it should be clear that there is a unique smallest element x such that $t_i \leq x$ for all $i \in I$. (Some of the t_i may be equal.) We write $x = \bigvee_{i \in I} t_i$. Similar remarks apply to the inf of the family $\{t_i : i \in I\}$. The interval $[0, 1]$ is a complete lattice, and so is $(\mathcal{P}(U), \subseteq)$.

The lattice $([0,1], \leq)$ plays a fundamental role. It is a bounded distributive lattice. It is not complemented. For $x, y \in [0,1]$, $x \vee y = \sup\{x, y\} = \max\{x, y\}$, and similarly $x \wedge y = \inf\{x, y\}$. Distributivity is easy to check. This lattice has another important operation on it: $[0,1] \to [0,1] : x \to 1 - x$. We denote this operation by $'$ even though it is not a complement. The operation has the following properties.

- $(x')' = x$.

- $x \leq y$ implies that $y' \leq x'$.

Such an operation on a bounded lattice is called an **involution**, or a **duality**. It follows that $'$ is one-to-one and onto, and that $0' = 1$ and $1' = 0$. If $'$ is an involution, the equations

$$
\begin{aligned}
(x \vee y)' &= x' \wedge y' \\
(x \wedge y)' &= x' \vee y'
\end{aligned}
$$

are called the **De Morgan laws**. They may or may not hold. But $[0,1]$ is a bounded distributive lattice which has an involution, namely $x' = 1 - x$, satisfying the De Morgan laws. Such a system $(V, \vee, \wedge, ', 0, 1)$ is a **De Morgan algebra**. Every Boolean algebra is a De Morgan algebra, and in particular, the set of all subsets $\mathcal{P}(U)$ of a set U is a De Morgan algebra. A De Morgan algebra that satisfies $x \wedge x' \leq y \vee y'$ for all x and y is a **Kleene algebra**.

Theorem 2.1.6 *Let $(V, \vee, \wedge, ', 0, 1)$ be a De Morgan algebra and let U be any set. Let f and g be mappings from U into V. We define*

1. $(f \vee g)(x) = f(x) \vee g(x)$,

2. $(f \wedge g)(x) = f(x) \wedge g(x)$,

3. $f'(x) = (f(x))'$,

4. $0(x) = 0$,

5. $1(x) = 1$.

Let V^U be the set of all mappings from U into V. Then $(V^U, \vee, \wedge, ', 0, 1)$ is a De Morgan algebra. If V is a complete lattice, then so is V^U.

Proof. The proof is routine in all respects. For example, the fact that $\vee$ is an associative operation on V^U comes directly from the fact that $\vee$ is associative on V. (The two $\vee$s are different, of course.) Using the definition of $\vee$ on V^U and that $\vee$ is associative on V, we get

$$
\begin{aligned}
(f \vee (g \vee h))(x) &= f(x) \vee (g \vee h)(x) \\
&= f(x) \vee (g(x) \vee h(x)) \\
&= (f(x) \vee g(x)) \vee h(x) \\
&= (f \vee g)(x) \vee h(x) \\
&= ((f \vee g) \vee h)(x)
\end{aligned}
$$

whence $f \vee (g \vee h) = (f \vee g) \vee h$, and so $\vee$ is associative on V^U. The rest of the proof is left as an exercise. ∎

The set V^U of all mappings from U into V is really the same as the Cartesian product $\prod_{u \in U} V$ of $|U|$ copies of V, where $|U|$ denotes the number of elements of U. Viewed in this way, the operations $\vee$ and $\wedge$ are coordinatewise, and it is easy to see that $\prod_{u \in U} V$ is a De Morgan algebra simply because V is.

Let U be any set and let $\mathcal{F}(U)$ be the set of all fuzzy subsets of U. We have defined operations on $\mathcal{F}(U)$ in Chapter 1. Those operations come from operations on $[0,1]$ just as the ones in Theorem 2.1.6 come from operations on V. We have the following.

Corollary 2.1.7 $(\mathcal{F}(U), \vee, \wedge, ', 0, 1)$ *is a complete De Morgan algebra.*

When the operations are clear, it is customary to write simply V for a De Morgan algebra $(V, \vee, \wedge, ', 0, 1)$. Thus we would write $\mathcal{F}(U)$ for the De Morgan algebra in the corollary.

There are many ways to construct new lattices from old. One of the most fundamental is this. Let X and Y be partially ordered sets. In the Cartesian product $X \times Y = \{(x,y) : x \in X, y \in Y\}$, let $(a,b) \leq (c,d)$ if $a \leq c$ and $b \leq d$. Then $X \times Y$ becomes a partially ordered set, and if X and Y are lattices, then $X \times Y$ is a lattice and this lattice is respectively, bounded, distributive, complemented, Boolean, or De Morgan if and only if X and Y are. (See Exercise 10.) The set $X \times Y$ with this componentwise ordering is the **product** of the lattices X and Y. This notion can be extended to the product of any family of lattices and gives a way to make new lattices from old. If $(X, \leq)$ is a lattice and Y is a subset of X, then the partial order on X induces one on Y. If this induced partial order on Y makes it into a lattice, and if the sup and inf of two elements of Y taken in X are the same as the sup and inf taken in Y, then Y is called a **sublattice** of X.

Let X be a lattice and let $X^{[2]} = \{(x,y) \in X \times X : x \leq y\}$. Then $X^{[2]}$ is a sublattice of $X \times X$. (See Exercise 11.) If B and C are Boolean algebras, then $B \times C$ is a Boolean algebra. But $B^{[2]}$ is not a Boolean algebra, since it does not have complements. (See Exercise 12.) However, $B^{[2]}$ does have pseudocomplements.

Definition 2.1.8 *Let X be a bounded lattice, and let $x \in X$. Then an element x^* is a **pseudocomplement** of x if $x \wedge x^* = 0$, and $y \leq x^*$ whenever $x \wedge y = 0$. That is, for each $x \in X$, there is a largest element whose meet with x is 0.*

An element in a bounded lattice has at most one pseudocomplement since two pseudocomplements must each be less or equal to the other, and hence equal. If every element has a pseudocomplement, then the bounded lattice is **pseudocomplemented,** and the unary operation * is called a **pseudocomplement.** Clearly, every finite lattice is pseudocomplemented. The equation $x^* \vee x^{**} = 1$ is called **Stone's identity,** and a **Stone algebra** is a pseudocomplemented *distributive* lattice satisfying this identity. If $(S, \vee, \wedge, ^*, 0, 1)$ is a Stone algebra, then for $S^* = \{s^* \in S : s \in S\}$, $(S^*, \vee, \wedge, ^*, 0, 1)$ is a Boolean algebra. That is, * is a complement on S^*. The sublattice S^* consists precisely of the complemented elements of S, and is sometimes called the **center** of S. If B is a Boolean algebra, then $B^{[2]}$ is a Stone algebra (Exercise 12). Stone algebras have a fairly extensive theory [44] and only a few facts are cited here and in the exercises. The connection with fuzzy sets follows.

The bounded distributive lattice $(\mathcal{F}(U), \vee, \wedge, 0, 1)$ of all fuzzy subsets of a set U is pseudocomplemented. If $A \in \mathcal{F}(U)$, then

$$A^*(u) = \begin{cases} 0 & \text{if } A(u) \neq 0 \\ 1 & \text{if } A(u) = 0 \end{cases}$$

is the pseudocomplement of A. It is totally straightforward to check that this is indeed the case. What is the center of $\mathcal{F}(U)$?

Theorem 2.1.9 $(\mathcal{F}(U), \vee, \wedge, ^*, 0, 1)$ *is a Stone algebra whose center consists of the crisp (ordinary) subsets of U.*

2.2 Equivalence relations and partitions

There are many instances in which we would like to consider certain elements of a set to be the same. For example, in the set of integers there are occasions where all we care about an integer is whether it is even or odd. Thus we may as well consider all even integers to be the same, and likewise all odd integers to be the same. This considering of certain subsets of a set as one element is one of the most fundamental notions in mathematics. It generalizes the notion of equality. The appropriate embodiment of this notion is a special kind of relation on a set, an *equivalence relation.* A standard notation for such a relation is some symbol such as $\sim$, or $\equiv$, and it is customary to write $a \sim b$ for $(a, b) \in \sim$.

Definition 2.2.1 *A relation $\sim$ on a set U is an* **equivalence relation** *if for all a, b, and c in U,*

- $a \sim a$,

- $a \sim b$ implies $b \sim a$, and

- $a \sim b$, $b \sim c$ imply that $a \sim c$.

The first and third conditions we recognize as reflexivity and transitivity. The second is that of **symmetry**. Thus an equivalence relation is a relation that is reflexive, symmetric, and transitive. Before giving some examples, there are two more pertinent definitions.

Definition 2.2.2 *Let $\sim$ be an equivalence relation on a set U and let $a \in U$. The* **equivalence class** *of an element a is the set $[a] = \{u \in U : u \sim a\}$.*

We defined a finite partition in Chapter 1. Here is the definition in general.

Definition 2.2.3 *Let U be a nonempty set. A* **partition** *of U is a set of nonempty pairwise disjoint subsets of U whose union is U.*

There is an intimate connection between equivalence relations and partitions. Here are some examples illustrating these notions and this connection.

Example 2.2.4 Let U be a set, and define $x \sim y$ if $x = y$. This example is just meant to point out that equality is an equivalence relation. For any $x \in U$, $[x] = \{x\}$. That is, the equivalence classes are just singletons. But do notice that the equivalence classes form a partition of U.

Example 2.2.5 Let U be a set and define $x \sim y$ for any two elements of U. That is, any two elements are equivalent. For any $x \in U$, $[x] = U$. Again the equivalence classes form a partition, but the partition has only one member, namely U itself.

Example 2.2.6 Let $\mathbb{Z}$ be the set of all integers. Let $m \sim n$ if $m - n$ is even, that is, is divisible by 2 in $\mathbb{Z}$. This is an equivalence relation because $m - m = 0$ is even, if $m - n$ is even then so is $-(m - n) = n - m$ and if $m - n$ and $n - p$ are even then so is $m - n + n - p = m - p$. So $\sim$ is reflexive, symmetric, and transitive. For $m \in \mathbb{Z}$,

$$\begin{aligned} [m] &= \{n \in \mathbb{Z} : n \sim m\} \\ &= \{n \in \mathbb{Z} : n - m \text{ is even}\} \\ &= \{m + 2k : k \in \mathbb{Z}\} \end{aligned}$$

So, for example,

$$[5] = \{..., 5 - 6, 5 - 4, 5 - 2, 5 - 0, 5 + 2, 5 + 4, 5 + 6, ...\}$$
$$= \{..., -1, 1, 3, 5, 7, 9, 11, 13, ...\}$$

which is just the set of odd integers. The equivalence class $[m]$ of any odd integer m will be this same set. Similarly, the equivalence class of any even integer will be the set of all even integers. So there are just two equivalence classes, the set of odd integers and the set of even integers. Again, the set of equivalence classes forms a partition.

Example 2.2.7 Let $f : U \to V$ be any function from the set U to the set V. On U, let $x \sim y$ if $f(x) = f(y)$. Then $\sim$ is an equivalence relation and the equivalence class $[x]$ consists of all those elements of U that have the same image as x. It should be clear that these classes form a partition of U. So any function induces an equivalence relation on its domain, two elements being equivalent if they have the same image.

Here is the formal connection between equivalence relations and partitions.

Theorem 2.2.8 *Let $\sim$ be an equivalence relation on the set U. Then the set of equivalence classes of $\sim$ is a partition of U. This association of an equivalence relation $\sim$ with the partition consisting of the equivalence classes of $\sim$ is a one-to-one correspondence between the set of equivalence relations on U and the set of partitions of U.*

Proof. The union of the equivalence classes $[u]$ is U since $u \in [u]$. We need only that two equivalence classes be equal or disjoint. If $x \in [u] \cap [v]$, then $x \sim u$, $x \sim v$, and so $u \sim x$ and $x \sim v$. By transitivity, $u \sim v$. If $y \in [u]$, then $y \sim u$, and since $u \sim v$, it follows from transitivity that $y \sim v$. Thus $y \in [v]$. This means that $[u] \subseteq [v]$. Similarly, $[v] \subseteq [u]$ and hence $[u] = [v]$. So if two equivalence classes are not disjoint, they are equal. Therefore the equivalence classes form a partition. Notice that two elements are equivalent if and only if they are in the same member of the partition, that is, in the same equivalence class. So this map from equivalence relations to partitions is one-to-one.

Given a partition, declaring two elements equivalent if they are in the same member of the partition is an equivalence relation whose associated equivalence classes are the members of the partition. So our map from equivalence relations to partitions is onto. ∎

Definition 2.2.9 *Let $\sim$ be an equivalence relation on the set U. The set of equivalence classes of $\sim$ is denoted $U/\sim$ and is called the **quotient space of** $\sim$. The map $U \to U/\sim : u \to [u]$ is the **natural map** of U onto $U/\sim$.*

The natural map simply associates an element with the equivalence class it is in. There is another map that is fundamental in algebra. In its various settings, it is called the **first isomorphism theorem**. The setting here is simply for sets. We state it, leaving its proof as an exercise.

Theorem 2.2.10 *Let f be a mapping from U onto V. On U let $x \sim y$ if $f(x) = f(y)$. Then $\sim$ is an equivalence relation, and*

$$U/\sim \to V : [x] \to f(x)$$

is a one-to-one map from $U/\sim$ onto V.

An equivalence relation on U is a subset of $U \times U$, so the set $\mathcal{E}(U)$ of all equivalence relations on U comes equipped with a partial order, namely set inclusion in $U \times U$.

Theorem 2.2.11 *Let $\mathcal{E}(U)$ be the set of all equivalence relations on the set U. Then $(\mathcal{E}(U), \subseteq)$ is a complete lattice.*

Proof. There is a biggest and smallest element of $\mathcal{E}(U)$, namely $U \times U$ and $\{(u, u) : u \in U\}$, respectively. We need to show that any nonempty family $\{E_i : i \in I\}$ of elements of $\mathcal{E}(U)$ has a sup and an inf. Now certainly $\bigwedge\{E_i : i \in I\} = \bigcap_{i \in I} E_i$ if $\bigcap_{i \in I} E_i$ is an equivalence relation. Let (u, v) and $(v, w) \in \bigcap_{i \in I} E_i$. Then (u, v) and (v, w) belong to each E_i and hence (u, w) belongs to each E_i. Therefore, $(v, w) \in \bigcap_{i \in I} E_i$. Thus $\bigcap_{i \in I} E_i$ is a transitive relation on U. That $\bigcap_{i \in I} E_i$ is reflexive and symmetric is similar. What we have shown is that the intersection of any family of equivalence relations on a set is an equivalence relation on that set. This is clearly the inf of that family. Now $\bigvee\{E_i : i \in I\}$ of a family of equivalence relations on U is

$$\bigcap\{E \in \mathcal{E}(U) : E \supseteq E_i \text{ for all } i \in I\}$$

Note that $U \times U$ is an equivalence containing all the E_i. This intersection is an equivalence relation on U by what we just proved, and it is clearly the least equivalence relation containing all the E_i. Therefore it is the desired sup. ∎

What we have just seen is an instance of a lattice L, namely $\mathcal{E}(U)$, contained in a lattice M, namely $\mathcal{P}(U \times U)$, where the lattice L gets its order from M, the meet in L agrees with the meet in M, and the join in L is different from the join in M. The join of elements E and F of $\mathcal{E}(U)$ is $E \cup F$ considered as elements of $\mathcal{P}(U \times U)$, but (unless one is contained in the other) is not $E \cup F$ considered in the ordered set $\mathcal{E}(U)$. Thus $\mathcal{E}(U)$ is not a sublattice of $\mathcal{P}(U \times U)$.

To see just what the elements are in the sup of two equivalence relations E and F on a set U is a good exercise. It is not $E \cup F$ unless one contains the other. The elements can be described and the sup is the so-called **transitive closure**. (See Exercise 24 at the end of this chapter.)

The lattice $\mathcal{E}(U)$ is not distributive if U has at least three elements. By Theorem 2.2.8, the set $\mathcal{E}(U)$ is in one-to-one correspondence with the set $\mathbb{P}(U)$ of partitions of U. Thus the lattice order $\subseteq$ on $\mathcal{E}(U)$ induces a lattice order on $\mathbb{P}(U)$, making it into a complete lattice. (See Exercise 25.)

2.3 Composing mappings

In dealing with fuzzy sets, it will be necessary to combine mappings, or functions, in various ways. This section is a collection of a few things about mappings, and mappings induced by mappings.

Perhaps the most basic thing about mappings is that sometimes they can be composed. Let $f : U \to V$, and $g : V \to W$. Then $g \circ f$, or more simply gf, is the mapping $U \to W$ defined by $(gf)(u) = g(f(u))$. This is called the **composition** of the mappings f and g. Any two functions of a set into itself can be composed. The notation gf will be given preference. The function $f : U \to U$ such that $f(u) = u$ for all u is denoted by 1_U and is called the **identity function** on U. The set of all functions from U to V is denoted $Map(U, V)$, or by V^U.

We have denoted the set of all subsets, or the **power set**, of U by $\mathcal{P}(U)$, or by 2^U. Both are standard notations, with 2^U reminding us that the set of subsets of U may be identified with the set of mappings from U into $\{0, 1\}$. Let $f : U \to V$. The mapping f induces a mapping $\mathcal{P}(U) \to \mathcal{P}(V)$, also denoted by f, given by $f(X) = \{f(x) : x \in X\}$. In addition, the mapping f induces a mapping $f^{-1} : \mathcal{P}(V) \to \mathcal{P}(U)$ defined by

$$f^{-1}(Y) = \{u \in U : f(u) \in Y\}$$

It should be noted that $f^{-1}(Y)$ might be empty. We also use f^{-1} to denote the restriction of f^{-1} to the one element subsets of V, and for an element $v \in V$, we write $f^{-1}(v)$ instead of $f^{-1}(\{v\})$ and view this restriction of f^{-1} as a mapping from V into $\mathcal{P}(U)$. The context will make it clear what is meant, and using f^{-1} in these various ways cuts down on the proliferation of notation.

Now consider mappings from U to L, where U is a set and L is a complete lattice. If L is $[0, 1]$ with the usual order on it, then mappings we are considering are fuzzy subsets of U. Sometimes functions from U

to L are called **L-fuzzy sets**. In any case, in what follows, L will be a complete lattice.

A mapping $A : U \to L$ induces a mapping $A : \mathcal{P}(U) \to \mathcal{P}(L)$. So with a subset X of U, $A(X)$ is a subset of L. But since L is a complete lattice, we may take the sup of $A(X)$. This sup is denoted $\bigvee (A(X))$. One should view $\bigvee$ as a mapping $\mathcal{P}(L) \to L$. The composition $\bigvee A$ is a mapping $\mathcal{P}(U) \to L$, namely the mapping given by

$$\mathcal{P}(U) \xrightarrow{A} \mathcal{P}(L) \xrightarrow{\vee} L$$

In particular, *a fuzzy subset of U yields a fuzzy subset of $\mathcal{P}(U)$.*

For sets U and V, a subset of $U \times V$ is called **a relation in $U \times V$**. Now, a relation R in $U \times V$ induces a mapping $R^{-1} : V \to \mathcal{P}(U)$ given by

$$R^{-1}(v) = \{u : (u, v) \in R\}$$

Thus with $A : U \to L$ we have the mapping

$$V \xrightarrow{R^{-1}} \mathcal{P}(U) \xrightarrow{A} \mathcal{P}(L) \xrightarrow{\vee} L$$

Thus a relation R in $U \times V$ associates with a mapping $A : U \to L$ a mapping $\vee A R^{-1} : V \to L$. This latter mapping is sometimes denoted $R(A)$. When $L = [0, 1]$, we then have a mapping $\mathcal{F}(U) \to \mathcal{F}(V)$ sending A to $R(A) = \vee A R^{-1}$. If R is actually a function from U to V, then R has been *extended* to a function $\mathcal{F}(U) \to \mathcal{F}(V)$ sending A to $\vee A R^{-1}$. In fuzzy set theory, this is called the **extension principle**. One should note that if $R^{-1}(v) = \varnothing$, then $\vee A R^{-1}(v) = 0$.

At this point we need some notation. Suppose that $f_1 : X_1 \to Y_1$ and $f_2 : X_2 \to Y_2$. Then $f_1 \times f_2$ is standard notation for the mapping

$$X_1 \times X_2 \to Y_1 \times Y_2 : (x_1, x_2) \to (f_1(x_1), f_2(x_2))$$

Now if A and B are fuzzy subsets of U and V, respectively, then $A \times B$ maps $U \times V$ into $[0, 1] \times [0, 1]$, and the image $(A(u), B(v))$ of an element of $U \times V$ is a pair of elements of $[0, 1]$ and hence has a min. Thus the composition $\wedge(A \times B)$ is a fuzzy subset of $U \times V$. Sometimes in fuzzy set theory, the mapping $\wedge(A \times B)$ is denoted simply $A \times B$, but there are other binary operations besides $\wedge$ that we will have occasion to follow $A \times B$ with. If A is a fuzzy subset of U, and V is any set, then letting B be the constant map $V \to [0, 1] : v \to 1$ yields a fuzzy subset of $U \times V$ called the **cylindrical extension** of A to $U \times V$. In any case, given fuzzy subsets of U and V, we get a fuzzy subset of $U \times V$, and thus a mapping $\mathcal{F}(U) \times \mathcal{F}(V) \to \mathcal{F}(U \times V)$ given by $(A, B) \to \wedge(A \times B)$.

We now look at some special cases.

- A function $f : U \to V$ is the relation $\{(u, v) : f(u) = v\}$ in $U \times V$, so it induces the mapping $\mathcal{F}(U) \to \mathcal{F}(V)$ that sends A to $\vee A f^{-1}$, sometimes denoted $f(A)$, that is, to the composition

$$V \xrightarrow{f^{-1}} 2^U \xrightarrow{A} 2^{[0,1]} \xrightarrow{\vee} [0, 1]$$

- A relation in $(U \times V) \times W$ induces a mapping $\mathcal{F}(U \times V) \to \mathcal{F}(W)$. But we have the mapping $\mathcal{F}(U) \times \mathcal{F}(V) \to \mathcal{F}(U \times V)$ just described. Thus a relation R in $(U \times V) \times W$ induces the mapping

$$\mathcal{F}(U) \times \mathcal{F}(V) \to \mathcal{F}(W)$$

which sends (A, B) to $\vee (\wedge (A \times B)) R^{-1}$. When $U = V = W$, then the relation R in $(U \times U) \times U$ induces a binary operation on $\mathcal{F}(U)$. In the case $U = \mathbb{R}$, there are the familiar arithmetic binary operations such as addition and multiplication on $\mathbb{R}$, and each induces a binary operation on the fuzzy subsets $\mathcal{F}(\mathbb{R})$. These particular operations will be taken up in Chapter 3.

2.4   Isomorphisms and homomorphisms

We introduce here a concept that is basic in algebra and one which we will meet in a nontrivial way in the next section. Also it will be of particular significance for us in Chapter 5. We begin with a couple of examples. Suppose $(U, \leq)$ and $(V, \leq)$ are two partially ordered sets. (We are using the same symbol, namely $\leq$, for the partial orders in both the sets U and V.) When are these partially ordered sets "just alike"? For example, it is intuitively clear that as partially ordered sets, there is no difference between $([0, 1], \leq)$ and $([1, 2], \leq)$. For them to be alike, there must be a one-to-one mapping from U onto V that respects, in some sense, the ordering of the two sets. Precisely, there must be a one-to-one onto mapping $f : U \to V$ such that $x \leq y$ if and only if $f(x) \leq f(y)$. Such a mapping f is an (order) **isomorphism,** and if there is such a mapping from U to V, the partially ordered sets $(U, \leq)$ and $(V, \leq)$ are **isomorphic**. The mapping $f(x) = x + 1$ is an order isomorphism from $[0, 1]$ to $[1, 2]$. A mapping $g : U \to V$ such that $g(x) \leq g(y)$ whenever $x \leq y$ is called a **homomorphism**, or an **order homomorphism**, emphasizing that the order relation is being respected. The condition on g that if $x \leq y$ then $g(x) \leq g(y)$ is expressed by saying that g preserves order, or is **order preserving**.

Suppose that $(U, \vee, \wedge)$ and $(V, \vee, \wedge)$ are lattices. Here, instead of having sets with one relation, as in the case of a partially ordered set, we have sets each with two binary operations on them. (Again, we are using the same symbols for different binary operations.) A mapping $f : U \to V$ is an **isomorphism** of these two lattices if f is one-to-one and onto, $f(x \vee y) = f(x) \vee f(y)$, and $f(x \wedge y) = f(x) \wedge f(y)$. That is, f must be one-to-one and onto and preserve both lattice operations. If the one-to-one and onto conditions are dropped, then f is a **lattice homomorphism**. If U and V are complete lattices, then an isomorphism $f : U \to V$ is a **complete lattice homomorphism** if and only if $f(\bigvee S) = \bigvee \{f(s) : s \in S\}$ and $f(\bigwedge S) = \bigwedge \{f(s) : s \in S\}$ for every subset S of U. An isomorphism of a lattice (or any algebraic structure) with itself is called an **automorphism**.

Compositions of homomorphisms are homomorphisms and compositions of isomorphisms are isomorphisms. For example, if $f : U \to V$ and $g : V \to W$ are lattice homomorphisms, then $g \circ f : V \to W$ is a lattice homomorphism. These facts are left as exercises. In later chapters, especially in Chapter 5, we will see many examples.

The general theme then is this. Suppose that U is a set on which we have various structures — relations, binary operations, unary operations such as complements in Boolean algebras, and so on. If V is another set with corresponding structures, then the system U with its operations is isomorphic to the system V with its operations if there is a one-to-one mapping from U onto V preserving these structures. A homomorphism just preserves the structure; it is not required to be one-to-one or onto. This all can be made more precise, but would lead us too far afield at the moment.

Example 2.4.1 Consider the lattice $([0,1], \vee, \wedge, ')$ with involution, where $\vee$ is sup, $\wedge$ is inf, and $x' = 1 - x$, and the lattice $\{0, \frac{1}{2}, 1\}$ with the same operations. Then the mapping $f : [0,1] \to \{0, \frac{1}{2}, 1\}$ that sends endpoints to endpoints and the interior points of $[0,1]$ to $\frac{1}{2}$ is a homomorphism. Note that one requirement is that $f(x') = f(x)'$, and that this does hold.

Suppose that $f : U \to V$ is a homomorphism from a lattice $(U, \vee, \wedge)$ to a lattice $(V, \vee, \wedge)$. Then the relation $\sim$ on U defined by $a \sim b$ if $f(a) = f(b)$ is an equivalence relation. But also, if $a \sim b$ and $c \sim d$, then $f(a \vee c) = f(a) \vee f(c) = f(b) \vee f(d) = f(b \vee d)$, so $a \vee c \sim b \vee d$. Similarly $a \wedge c \sim b \wedge d$. So this equivalence relation has these two additional properties: if $a \sim b$ and $c \sim d$ then $a \vee c \sim b \vee d$ and $a \wedge c \sim b \wedge d$. Such an equivalence relation on a lattice is called a **congruence.** And congruences on lattices give rise to homomorphisms.

Theorem 2.4.2 *If $\sim$ is a congruence on the lattice U, then the set of equivalence classes $U/\sim$ forms a lattice under the operations $[a] \vee [b] = [a \vee b]$ and $[a] \wedge [b] = [a \wedge b]$. The mapping $U \to U/\sim : a \to [a]$ is a lattice homomorphism.*

The proof is left as an exercise. The lattice $U/\sim$ is the **quotient lattice** of U relative to the congruence $\sim$. Congruences are defined analogously on other algebraic systems, such as Boolean algebras, De Morgan algebras, Stone algebras, and so on. And in each instance, these congruences give rise in an analogous way to quotient structures. We will see instances of these concepts in later chapters.

There is another algebraic concept that will be important for us, especially in Chapters 5 and 6, and that is the concept of a **group**. Before giving the definition, we illustrate with an example that will be pertinent. Consider the partially ordered set $\mathbb{I} =. ([0,1], \leq)$, and let $Aut(\mathbb{I})$ be the set of all automorphisms of $\mathbb{I}$ with itself. That is, $Aut(\mathbb{I})$ is the set of all functions f from $[0,1]$ to $[0,1]$ that are one-to-one and onto, and such that $f(x) \leq f(y)$ if and only if $x \leq y$. We remarked above that compositions of homomorphisms are homomorphisms, and certainly compositions of one-to-one and onto functions are one-to-one and onto. Thus the composition of two elements of $Aut(\mathbb{I})$ is an element of $Aut(\mathbb{I})$, and thus composition is a binary operation on $Aut(\mathbb{I})$. The composition of f and g will be written simply as fg, meaning the function given by $(fg)(x) = f(g(x))$. This composition has some special properties. It is associative, has an identity, and every element has an inverse. This means that

- $f(gh) = (fg)h$. (Composition is **associative**.)

- There is an element 1 in $Aut(\mathbb{I})$ such that $1f = f1 = f$ for all f. (The function 1 is the function given by $1(x) = x$ for all x. It is called the **identity** of the group.)

- For each $f \in Aut(\mathbb{I})$, there is an element $f^{-1} \in Aut(\mathbb{I})$ such that $ff^{-1} = f^{-1}f = 1$. (Each element of $Aut(\mathbb{I})$ has an **inverse**. The element f^{-1} is simply the inverse of f as a function on $[0,1]$.)

Now, any set G with a binary operation that is associative, has an identity, and for which every element has an inverse, is a **group.** $Aut(\mathbb{I})$ is a set, namely the set of all automorphisms of $\mathbb{I}$, and composition of its elements as functions is a binary operation on it satisfying the requisites to be a group. This group is called the **group of automorphisms** of $\mathbb{I}$. Groups abound in mathematics, and have a huge theory. Our main interest will be in groups that are automorphism groups of other algebraic structures, such as the lattice $\mathbb{I}$. We will elaborate on various aspects of groups, such as subgroups, congruences, quotient structures, and isomorphisms between groups themselves as the need arises.

For us, one important aspect is this: mathematical models of real-world situations are often algebraic structures, that is, sets on which there are defined various operations or relations. Homomorphisms, automorphisms, and quotient structures are basic notions in mathematics. This is especially true in fuzzy set theory, as we will see. If two models are isomorphic, then abstractly, one is as good as the other. But one may be simpler to implement, for computational reasons, for example. In any case, it pays to know whether or not two such models are isomorphic, or equivalent.

2.5   Alpha-cuts

This notion plays a fairly big role in fuzzy set theory. Let A be a fuzzy subset of U, and let $\alpha \in [0,1]$. The α-cut of A is simply the set of those $u \in U$ such that $A(u) \geq \alpha$. All that is needed to make this definition is that the image of the mapping A be in a partially ordered set. In the fuzzy case, it is in the complete lattice $[0,1]$. The fact that this image is a complete lattice is important in the theory because we will need to take the sup of infinite subsets of $[0,1]$. We will present the basic facts about α-cuts in a more general context than usual, requiring at first just a partially ordered set C instead of the complete lattice $[0,1]$.

There is a bit of notation that is standard and convenient. If $\alpha \in C$, then $\uparrow\alpha = \{c \in C : c \geq \alpha\}$. Thus $\uparrow$ is a mapping from C into $\mathcal{P}(C)$ and $\uparrow\alpha$ is called the **up set** of α. We write $\uparrow\alpha$ rather than $\uparrow(\alpha)$. If $C = [0,1]$, then for $\alpha \in [0,1]$, $\uparrow\alpha = [\alpha,1]$.

Definition 2.5.1 *Let U be a set, let C be a partially ordered set and let $A : U \to C$. For $\alpha \in C$, the α-cut of A, or the α-level set of A, is $A^{-1}(\uparrow\alpha) = \{u \in U : A(u) \geq \alpha\}$. This subset of U will be denoted by A_α.*

Thus the α-cut of a function $A : U \to C$ is the subset $A_\alpha = A^{-1}(\uparrow\alpha)$ of U, and we have one such subset for each $\alpha \in C$. A fundamental fact about the α-cuts A_α is that they determine A and this is easy to see. It follows immediately from the equation

$$A^{-1}(\alpha) = A_\alpha \bigcap (\bigcup_{\beta > \alpha} A_\beta)'$$

Here, $'$ means set complement in the set U. This equation just says that the left side, $\{u : A(u) = \alpha\}$, namely the set of those elements that A takes to α, is the intersection of $\{u : A(u) \geq \alpha\}$ with the set $\{u : A(u) \not\geq \alpha\}$. But these two sets are given strictly in terms of α-cuts. So knowing all the α-cuts of A is the same as knowing A itself. We can state this as follows.

Theorem 2.5.2 *Let A and B be mappings from a set U into a partially ordered set C. If $A_\alpha = B_\alpha$ for all $\alpha \in C$, then $A = B$.*

Thus we have the following situation. A function $A : U \to C$ gets a function $A^{-1}\uparrow : C \to \mathcal{P}(U)$ which is just the composition

$$C \xrightarrow{\uparrow} \mathcal{P}(C) \xrightarrow{A^{-1}} \mathcal{P}(U)$$

We already know from the theorem just above that associating A with the function $A^{-1}\uparrow$ is one-to-one. What we want first is a simple description of the functions $C \to \mathcal{P}(U)$ that can come from functions $A : U \to C$ in this way. For example, in the fuzzy case, this would allow us to identify the set of fuzzy subsets of U with a set of maps from $[0,1]$ into $\mathcal{P}(U)$ with certain properties.

In Exercise 37 we note that if C is a complete lattice, then $\bigcap_{\alpha \in D} A_\alpha = A_{\vee D}$ for any subset D of C. In different notation, this says that the mapping $A^{-1}\uparrow$ has the property that

$$A^{-1}\uparrow(\vee D) = \bigcap_{d \in D} A^{-1}\uparrow d$$

Letting $g = A^{-1}\uparrow$, this says that the following diagram of mappings commutes.

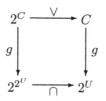

This is the condition that functions $g : C \to \mathcal{P}(U)$ must satisfy to be $A^{-1}\uparrow$ for some A.

Theorem 2.5.3 *Let C be a complete lattice and U a set. Let $\mathcal{F}(U)$ be the set of all mappings from U into C, and $\mathcal{L}(U)$ be the set of all mappings $g : C \to \mathcal{P}(U)$ such that the diagram above commutes, or equivalently such that for all subsets D of C,*

$$g(\vee D) = \bigcap_{d \in D} g(d)$$

Then the mapping $\Phi : \mathcal{F}(U) \to \mathcal{L}(U)$ given by $\Phi(A) = A^{-1}\uparrow$ is one-to-one and onto.

Proof. We have already observed that Φ maps $\mathcal{F}(U)$ into $\mathcal{L}(U)$ and that this mapping is one-to-one. Let $g \in \mathcal{L}(U)$. We must show that $g = A^{-1}\uparrow$ for some $A \in \mathcal{F}(U)$. For $u \in U$, define

$$h(u) = \{d \in C : g(d) \supseteq \bigcap_{u \in g(x)} g(x)\} = \{d \in C : u \in g(d)\}$$

Let $A = \bigvee \circ h$. Then

$$A^{-1}{\uparrow}(c) = \{u \in U : A(u) \geq c\}$$

Now if $u \in g(c)$, then $g(c) \supseteq \bigcap_{u \in g(x)} g(x)$ which implies that $c \in h(u)$ and thus that $u \in A^{-1}{\uparrow}(c)$. Thus $g(c) \subseteq A^{-1}{\uparrow}(c)$. Now suppose that $u \in A^{-1}{\uparrow}(c)$, so that $A(u) \geq c$. Then $u \in \bigcap_{u \in g(x)} g(x) \subseteq g(d)$ for all $d \in h(u)$. Thus

$$u \in \bigcap_{d \in A(u)} g(d) = g(A(u)) \subseteq g(c)$$

It follows that $g(c) = A^{-1}{\uparrow}(c)$, whence $g = A^{-1}{\uparrow} = \Phi(A)$. ■

The system $\mathcal{F}(U)$ is a complete lattice, and this one-to-one correspondence with $\mathcal{L}(U)$ suggests that it too is a complete lattice in a natural way. This is indeed the case.

First, $\mathcal{F}(U)$ is a complete lattice via the ordering $A \leq B$ if $A(u) \leq B(u)$ for all $u \in U$. Also $\mathcal{L}(U)$ is partially ordered by $f \leq g$ if $f(x) \subseteq g(x)$ for all $x \in C$. This is just the pointwise ordering of the maps in $\mathcal{L}(U)$. For A and B in $\mathcal{F}(U)$, $A(u) \leq B(u)$ for all $u \in U$ if and only if

$$\begin{aligned} A^{-1}{\uparrow}(x) &= \{u : A(u) \geq x\} \\ &\subseteq \{u : B(u) \geq x\} \end{aligned}$$

if and only if $A^{-1}{\uparrow} \leq B^{-1}{\uparrow}$ if and only if $\Phi(A) \leq \Phi(B)$. Since Φ is one-to-one and onto, Φ is an order isomorphism and thus a lattice isomorphism where the lattice operations come from the orders. The orders are complete, so Φ is a complete order isomorphism and thus a complete lattice isomorphism. In particular, $\mathcal{L}(U)$ is a complete lattice.

Corollary 2.5.4 *The complete lattices $\mathcal{F}(U)$ and $\mathcal{L}(U)$ are isomorphic.*

The upshot of all this is that studying the complete lattice $\mathcal{F}(U)$ is the same as studying the complete lattice $\mathcal{L}(U)$. In particular, the lattice of fuzzy sets may be viewed as a special set of functions from $[0, 1]$ to $\mathcal{P}(U)$ with appropriate operations on them. However, it should be realized that $\mathcal{L}(U)$ is not necessarily a sublattice of $\mathcal{P}(U)^C$. This happens if and only if C is a chain. If C is not a chain, the sup of two elements of $\mathcal{L}(U)$ taken in $\mathcal{L}(U)$ may not be the sup taken in $\mathcal{P}(U)^C$.

2.6   Images of alpha-level sets

Let $f : U \to V$ and let A be a fuzzy subset of U. Then $\bigvee A f^{-1}$ is a fuzzy subset of V by the extension principle. It is the mapping that is the composition

$$V \xrightarrow{f^{-1}} \mathcal{P}(U) \xrightarrow{A} \mathcal{P}([0,1]) \xrightarrow{\vee} [0,1]$$

There is a connection between the α-level sets, or the α-cuts, of these two fuzzy sets. This relation is important in the calculus of fuzzy quantities. It holds in more generality than for fuzzy sets.

Theorem 2.6.1 *Let C be a complete lattice, U and V be sets, $A : U \to C$, and $f : U \to V$. Then*

1. *$f(A_\alpha) \subseteq (\bigvee Af^{-1})_\alpha$ for all $\alpha \in C$.*

2. *$f(A_\alpha) = (\bigvee Af^{-1})_\alpha$ for $\alpha > 0$ if and only if for each member P of the partition induced by f, $\bigvee A(P) \geq \alpha$ implies $A(u) \geq \alpha$ for some $u \in P$.*

3. *$f(A_\alpha) = (\bigvee Af^{-1})_\alpha$ for all $\alpha > 0$ if and only if for each member P of the partition induced by f, $\bigvee A(P) = A(u)$ for some $u \in P$.*

Proof. The theorem follows immediately from the equalities below.

$$
\begin{aligned}
f(A_\alpha) &= \{f(u) : A(u) \geq \alpha\} \\
&= \{v \in V : A(u) \geq \alpha, \ f(u) = v\} \\
(\vee Af^{-1})_\alpha &= \{v \in V : \bigvee Af^{-1}(v) \geq \alpha\} \\
&= \{v \in V : \bigvee\{A(u) : f(u) = v\} \geq \alpha\}
\end{aligned}
$$

One should notice that for some α, it may not be true that $\bigvee A(P) = \alpha$ for any P. $\blacksquare$

When $U = V \times V$, f is a binary operation on V. If A and B are fuzzy subsets of V and the binary operation f is denoted $\circ$ and written in the usual way, then the theorem specifies exactly when $A_\alpha \circ B_\alpha = (A \circ B)_\alpha$, namely when certain sups are realized. These α-level images for fuzzy subsets of $\mathbb{R}$ will be discussed further in Chapter 3, where also some conditions will be given which are sufficient for the realization of these sups.

The function $\bigvee Af^{-1}$ is sometimes written $f(A)$, and in this notation, the theorem relates $f(A_\alpha)$ and $f(A)_\alpha$. Of special interest is when $U = U_1 \times U_2 \times \cdots \times U_n$. In that case, let $A^{(i)}$ be a fuzzy subset of U_i. Then $A^{(1)} \times ... \times A^{(n)}$ is a fuzzy subset of U, and trivially $\left(A^{(1)} \times ... \times A^{(n)}\right)_\alpha = A_\alpha^{(1)} \times ... \times A_\alpha^{(n)}$. The fuzzy subset $\bigvee(A^{(1)} \times ... \times A^{(n)})f^{-1}$ is sometimes written $f(A^{(1)}, ..., A^{(n)})$. In this notation, the third part of the theorem may be stated as

- $f(A_\alpha^{(1)}, ..., A_\alpha^{(n)}) = f(A^{(1)}, ..., A^{(n)})_\alpha$ for all $\alpha > 0$ if and only if for each member P of of the partition induced by f,

$$\bigvee (A^{(1)} \times ... \times A^{(n)})(P) = (A^{(1)} \times ... \times A^{(n)})(u)$$

for some $u \in P$. That is, for any $v \in V$,

$$\bigvee \left\{ \bigwedge_{i=1}^{n} A^{(i)}(u_i) : (u_1, u_2, ..., u_n) \in f^{-1}(v) \right\}$$

is attained.

This result is often referred to as **Nguyen's Theorem** in the literature. It is valid for L-fuzzy sets. Note that the extension of f : $U^{(1)} \times ... \times U^{(n)} \to V$ to fuzzy sets, namely

$$f(A^{(1)}, ..., A^{(n)})(v) = \bigvee \left\{ \bigwedge_{i=1}^{n} A^{(i)}(u_i) : (u_1, u_2, ..., u_n) \in f^{-1}(v) \right\}$$

is defined in terms of $\wedge$, a binary operation on $[0, 1]$. The operation $\wedge$ may be replaced by a t-norm T (see Chapter 5). This leads to the sup-T **convolution** for defining the fuzzy set $f(A^{(1)}, ..., A^{(n)})$. Nguyen's theorem holds in this more general setting, and says that for all $v \in V$,

$$\bigvee \left\{ T(A^{(1)}(u_1), ..., A^{(n)}(u_n)) : (u_1, u_2, ..., u_n) \in f^{-1}(v) \right\}$$

is attained if and only if for all $\alpha \in L$,

$$[f(A^{(1)}, ..., A^{(n)})]_\alpha = \bigcup_{T(t_1, ..., t_n) \geq \alpha} f(A_{t_1}^{(1)}, ..., A_{t_n}^{(n)})$$

See exercise 45.

2.7 Exercises

1. Let U be a set and let $\mathcal{P}(U)$ be the set of all subsets of U. Verify in detail that $(\mathcal{P}(U), \subseteq)$ is a Boolean algebra. Show that it is complete.

2. Show that a chain with more than two elements is not complemented.

3. A relation on U is a **preorder** if it is reflexive and transitive. Let $U = [0, \infty) \times [0, \infty)$, the product of the set of non-negative reals with itself. Let $\leq$ be the usual order relation in $[0, \infty)$.

(a) On U, define $(x, y) \preceq (u, v)$ if $xy \leq uv$. Show that $\preceq$ is a preorder. Show that it is linear, that is, that for a and b in U, either $a \preceq b$ or $b \preceq a$.

(b) On U, define $(x, y) \preceq (u, v)$ if either $xy = 0$, or $\frac{u}{x} = \frac{v}{y} \geq 1$. Show that $\preceq$ is a preorder, and is not linear.

(c) Let Γ denote the set of all preorders $\preceq$ on U such that

    i. If $(x, y) \preceq (u, v)$, then $(ax, by) \preceq (au, bv)$ for $a, b \in [0, \infty)$.

    ii. If $(x, y) \preceq (u, v)$, then $(y, x) \preceq (v, u)$.

    iii. $(x, x) \preceq (u, u)$ if and only if $x \leq u$.

Show that the preorders in (a) and (b) are in Γ, and that the preorder in (a) is the only linear one in Γ.

4. Prove that in a lattice, if $\vee$ distributes over $\wedge$, then $\wedge$ distributes over $\vee$, and conversely.

5. Prove Theorem 2.1.4.

6. Complete the proof of Theorem 2.1.5.

7. Let $\mathbb{N}$ be the set of positive integers and let R be the relation mRn if m divides n. Show that this makes $\mathbb{N}$ into a distributive lattice.

8. Complete the proof of Theorem 2.1.6

9. Show that the De Morgan algebra $(\mathcal{F}(U), \vee, \wedge, ', 0, 1)$ satisfies $A \wedge A' \leq B \vee B'$ for all $A, B \in \mathcal{F}(U)$, that is, is a Kleene algebra. Show that $[0, 1]$ is a Kleene algebra. Show that $[0, 1]^{[2]}$ is not a Kleene algebra.

10. Show that the product $X \times Y$ of lattices X and Y is a lattice. Show that $X \times Y$ is respectively, bounded, complete, distributive, complemented, Boolean, or De Morgan if and only if X and Y are. Show that if S and T are sublattices of X and Y, respectively, then $S \times T$ is a sublattice of $X \times Y$, but that not every sublattice of $X \times Y$ need be of this form.

11. Let X be a lattice. Show that $X^{[2]}$ is a lattice. If $'$ is an involution of X, show that $(x, y)' = (y', x')$ is an involution of $X^{[2]}$ and that $X^{[2]}$ with this involution is a De Morgan algebra if and only if X with the involution $'$ is a De Morgan algebra.

12. Let B be a Boolean algebra. Show that $B^{[2]}$ is a Stone algebra but not a Boolean algebra.

13. Show that if S is a Stone algebra, then so is $S^{[2]}$.

14. Show that every bounded chain is a Stone algebra.

15. Show that if S is a Stone algebra with pseudocomplement *, then $S^* = \{s^* : s \in S\}$ is a Boolean algebra.

16. Prove Theorem 2.1.9 in detail.

17. Show that an involution of a bounded lattice is one-to-one and onto.

18. Let U be an infinite set and let $\mathcal{F}$ be the set of all subsets of U that are either finite or have finite complement. Show that $(\mathcal{F}, \subseteq)$ is a Boolean algebra, but is not complete.

19. Show that the lattices pictured are not distributive.

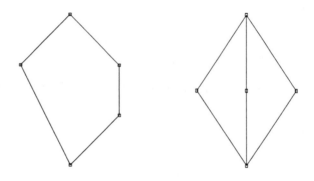

20. Show that for relations, reflexive, symmetric, and transitive are independent. That is, show that for any subset of these three, there is a relation satisfying those conditions and not satisfying the others. For example, there is a relation which is symmetric but not reflexive and not transitive.

21. Let $\mathbb{N}$ be the set of positive integers. On $\mathbb{N} \times \mathbb{N}$ let $(a, b) \sim (c, d)$ if $a + d = b + c$. Show that $\sim$ is an equivalence relation. What is the quotient space $(\mathbb{N} \times \mathbb{N})/ \sim$?

22. Let $U = \mathbb{Z} \times \mathbb{Z}^*$, where $\mathbb{Z}$ is the set of integers and $\mathbb{Z}^*$ is the set of nonzero integers. On U, let $(m, n) \sim (p, q)$ if $mq = np$. Verify that $\sim$ is an equivalence relation. What is the quotient space $U/ \sim$?

23. Prove Theorem 2.2.10 in detail.

24. Let $\mathcal{E}(U)$ be the set of all equivalence relations on the set U. Under $\subseteq$, $\mathcal{E}(U)$ is a complete lattice by Theorem 2.2.11. Describe the sup of two equivalence relations directly in terms of the elements of those two sets. Describe the sup of a set of equivalence relations directly in terms of the elements of those sets.

25. Let $\mathbb{P}(U)$ be the set of all partitions of the set U. For P and Q in $\mathbb{P}(U)$, put $P \le Q$ if every member of P is a subset of a member of Q.

    (a) Let $\Phi : \mathcal{E}(U) \to \mathbb{P}(U)$ be the mapping that takes an equivalence relation to its set of equivalence classes. This is the map in Theorem 2.2.8. Show that for A and B in $\mathcal{E}(U)$, $A \subseteq B$ if and only if $\Phi(A) \le \Phi(B)$.

    (b) Show that $(\mathbb{P}(U), \le)$ is a complete lattice.

    (c) Show that Φ is a lattice isomorphism.

26. Prove Theorem 2.4.2 in detail.

27. Suppose that U has three elements. Show that the lattice $\mathbb{P}(U)$ is the lattice pictured on the right in Exercise 19. Show that $\mathbb{P}(U)$ is not distributive if U has at least three elements.

28. Let R, S, and T be relations in $U \times V$, $V \times W$, and $W \times X$, respectively. Show that $R(ST) = (RS)T$. That is, show that composition of relations is associative.

29. Let $f : U \to V$, $g : V \to W$, and $h : W \to X$. Show that $h(gf) = (hg)f$. That is, show that composition of functions is associative.

30. For any sets X and Y, let $Map(X, Y)$ be the set of all mappings from X to Y. Show that

$$\Phi : Map(W, Map(X, Y)) \to Map(W \times X, Y) : \Phi(f)(w, x) = f(w)(x)$$

    is a one-to-one correspondence.

31. Let $f : U \to V$.

    (a) Show that $ff^{-1} = 1_{\mathcal{P}(V)}$ if f is onto.
    (b) Show that $f^{-1}f = 1_{\mathcal{P}(U)}$ if f is one-to-one.

32. Let $f : U \to V$. Show that $\{f^{-1}f(x) : x \in U\}$ is a partition of U. Why is $\{f^{-1}(v) : v \in V\}$ not a partition of U?

33. Let A and B be fuzzy subsets of U and $\alpha \in [0,1]$.

    (a) Show that

    $$\begin{aligned}
    (A \vee B)_\alpha &= A_\alpha \cup B_\alpha \\
    (A \wedge B)_\alpha &= A_\alpha \cap B_\alpha
    \end{aligned}$$

    (b) Is $(A')_\alpha = (A_\alpha)'$?

    (c) Let $\mathcal{S}$ be a set of fuzzy subsets of U. Show that $(\bigwedge_{A \in \mathcal{S}} A)_\alpha = \bigcap_{A \in \mathcal{S}} A_\alpha$.

    (d) Is $(\bigvee_{A \in \mathcal{S}} A)_\alpha = \bigcup_{A \in \mathcal{S}} A_\alpha$?

34. Let A be a fuzzy subset of a set U. For $\alpha \in [0,1]$, the **strong α-cut** of A is

    $$A_\alpha^* = \{u : u \in U, A(u) > \alpha\}$$

    Show that for $u \in U$,

    $$A(u) = \sup_\alpha \alpha \chi_{A_\alpha^*}(u)$$

35. Let A be a fuzzy subset of the set U and let A_α be the α-cut of A.

    (a) Verify that for all $x \in U$,

    $$A(x) = \bigvee_{\alpha \in [0,1]} \alpha \chi_{A_\alpha}(x)$$

    This is called the **resolution of the identity**.

    (b) Verify that for all $x \in U$,

    $$A(x) = \int_0^1 \chi_{A_\alpha}(x) d\alpha$$

    (c) Suppose that $\{B_\alpha : \alpha \in [0,1]\}$ is a family of subsets of U such that for all $x \in U$,

    $$A(x) = \bigvee_{\alpha \in [0,1]} \alpha \chi_{B_\alpha}(x)$$

    Show that for all $\alpha \in [0,1]$, $B_\alpha \subseteq A_\alpha$.

    (d) The support of a fuzzy set A is given by

    $$\text{supp}(A) = \{x \in U : A(x) > 0\}$$

    With B as in part (b), verify that

    $$\bigcup_{\alpha \in (0,1]} A_\alpha = \bigcup_{\alpha \in (0,1]} B_\alpha = \text{supp}(A)$$

36. Show that $\mathcal{L}(U)$ in Theorem 2.5.3 is a complete lattice via the ordering $f \leq g$ if $f(x) \subseteq g(x)$ for all $x \in C$.

37. Let C be a complete lattice and $\mathcal{P}(U)$ the power set of U. There are two complete lattices around: the complete lattice $\mathcal{P}(U)$ of all subsets of U, and of course C. We use $\cup$ and $\cap$ for the operations in $\mathcal{P}(U)$ and $\vee$ and $\wedge$ for the operations in C. Let D be a subset of C and let $A : U \to C$. Prove the following:

    (a) For α and β in C, and $\alpha \leq \beta$ then $A_\alpha \supseteq A_\beta$.

    (b) $\bigcup_{\alpha \in D} A_\alpha \subseteq A_{\wedge D}$.

    (c) $\bigcap_{\alpha \in D} A_\alpha = A_{\vee D}$.

    (d) The converse of (a) does not always hold.

    (e) The opposite inclusion in (b) does not always hold.

38. Show that $\mathcal{L}(U)$ is not a sublattice of $\mathcal{P}(U)^C$. See Corollary 2.5.4.

39. In the context of Theorem 2.6.1, show that $f(A_\alpha) = (\vee A f^{-1})_\alpha$ if either

    (a) f is one-to-one, or

    (b) $C = V$ and $f = A$.

40. In the notation following Theorem 2.6.1, show that

$$f(A^{(1)}, A^{(2)}, ..., A^{(n)})(y) = \bigvee_{\alpha \in [0,1]} \alpha f(A_\alpha^{(1)}, A_\alpha^{(2)}, ..., A_\alpha^{(n)}) (y)$$

41. Let A be the fuzzy set and $f : \mathbb{R} \to \mathbb{R}$ defined by the equations

$$A(x) = \chi_{\{0\}}(x) + e^{-\frac{1}{x}} \chi_{(0,\infty)}(x)$$
$$f(x) = x \chi_{(0,1)}(x) + \chi_{[0,\infty)}(x)$$

Recall that the fuzzy subset $f(A)$ is defined by

$$f(A)(y) = (\vee A f^{-1})(y) = \sup\{A(x) : x \in f^{-1}(y)\}$$

where $f^{-1}(y) = \{x : y = f(x)\}$.

    (a) Show that for each $y \in [0,1)$, $\sup\{A(x) : x \in f^{-1}(y)\}$ is attained.

    (b) Show that $\sup\{A(x) : x \in f^{-1}(1)\}$ is not attained.

    (c) Show that $f(A)_1 \neq f(A_1)$.

42. Let A and B be fuzzy sets, and $f : \mathbb{R} \times \mathbb{R} \to \mathbb{R}$ defined by the equations

$$\begin{aligned} A(x) &= \chi_{[3,4]}(x) + e^{-\frac{1}{x}}\chi_{(0,3)\cup(4,\infty)}(x) \\ B(x) &= \chi_{[-2,-1]}(x) + e^{-\frac{1}{x}}\chi_{(-\infty,-2)\cup(-1,0)}(x) \\ f(x) &= x + y \end{aligned}$$

Recall that $f(A, B)$ is the fuzzy subset of $\mathbb{R}$ defined by $f(A, B)(z) = \sup\{A(x) \wedge B(y) : x + y = z\}$.

  (a) Show that $\sup\{A(x) \wedge B(y) : x + y = 0\}$ is not attained.

  (b) Show that $[f(A, B)]_1 \neq f(A_1, B_1)$.

43. Let A and B be fuzzy subsets of $\mathbb{R}$, and $f : \mathbb{R} \times \mathbb{R} \to \mathbb{R}$ defined by the equations

$$\begin{aligned} A(x) &= \chi_{[0,1]}(x) + \tfrac{1}{2}\chi_{(5,\infty)}(x) \\ B(x) &= \chi_{\{0\}}(x) + \tfrac{1}{4}e^{-\frac{1}{x}}\chi_{(0,\infty)}(x) \\ f(x) &= x \wedge y \end{aligned}$$

Show that $A_{\frac{1}{4}} \wedge B_{\frac{1}{4}} \neq (A \wedge B)_{\frac{1}{4}}$.

44. Let A and B be fuzzy subsets of $\mathbb{R}$, and $f : \mathbb{R} \times \mathbb{R} \to \mathbb{R}$ be defined by

$$\begin{aligned} A(x) &= B(-x) = \begin{cases} 0 & \text{if } x \leq 0 \\ e^{-\frac{1}{x}} & \text{if } x > 0 \end{cases} \\ f(x, y) &= x + y \end{aligned}$$

Show that $f(A_1, B_1)$ is strictly contained in $[f(A, B)]_1$.

45. *Let A and B be fuzzy subsets of U and V, respectively, and let $f : U \times V \to W$. Let $T : [0, 1] \times [0, 1] \to [0, 1]$ be symmetric, associative, non-decreasing in each variable, and $T(x, 1) = x$ for all $x \in [0, 1]$. Define $f_T(A, B) : W \to [0, 1]$ by the formula

$$f_T(A, B)(w) = \sup\{T(A(u), B(v)) : (u, v) \in f^{-1}(w)\}$$

This is a sup-T convolution.

  (a) Show that the following are equivalent.

    i. For each $w \in W$,

$$\sup\{T(A(u), B(v)) : (u, v) \in f^{-1}(w)\}$$

    is attained.

    ii. For each $\alpha \in (, 1]$, $[f_T(A, B)]_\alpha = \bigcup_{T(t,s) \geq \alpha} f(A_t, B_s)$.

(b) If $T(x, y) = x \wedge y$, verify that $[f_T(A, B)]_\alpha = f(A_\alpha, B_\alpha)$.

(c) If $T(x, y) = xy$, verify that $[f_T(A, B)]_\alpha = \bigcup_{t \in [\alpha, 1]} f(A_t, B_{\alpha/t})$.

(d) If $T(x, y) = x$ if $y = 1$, y if $x = 1$, and 0 otherwise, verify that

$$[f_T(A, B)]_\alpha = f(A_1, B_\alpha) \cup f(A_\alpha, B_1)$$

(e) If $T(x, y) = \max\{0, x + y - 1\}$, verify that

$$[f_T(A, B)]_\alpha = \bigcup_{t \in [\alpha, 1]} f(A_t, B_{\alpha + 1 - t})$$

46. *Let $f : \mathbb{R} \times \mathbb{R} \to \mathbb{R}$ be continuous and let T be as in the previous exercise and upper semicontinuous. Upper semicontinuous means that for each $\alpha \in [0, 1]$, $\{(s, t) \in [0, 1] \times [0, 1] : T(s, t) \geq \alpha\}$ is closed. Let A and B be fuzzy subsets of $\mathbb{R}$ which are upper semicontinuous and have compact support. This means that the closure $S(A) = \{x \in \mathbb{R} : A(x) > 0\}$ of A is compact and similarly for B. Show that for $\alpha \in (0, 1]$, $[f_T(A, B)]_\alpha = \bigcup_{T(s,t) \geq \alpha} f(A_s, B_t)$.

47. *Let $f : \mathbb{R}^+ \times \mathbb{R}^+ \to \mathbb{R}^+$ be continuous, and A and B be continuous fuzzy subsets of $\mathbb{R}^+$ such that $\lim_{x \to \infty} A(x) = \lim_{x \to \infty} B(x) = 0$. Show that $[f(A, B)]_\alpha = f(A_\alpha, B_\alpha)$ for $\alpha \in (0, 1]$.

Chapter 3

FUZZY QUANTITIES

This chapter is devoted to the study of a concrete class of fuzzy sets, namely those of the real line $\mathbb{R}$. Fuzzy quantities are fuzzy subsets of $\mathbb{R}$, generalizing ordinary subsets of $\mathbb{R}$. In order to define operations among fuzzy quantities, we will evoke the. *extension principle*, which was discussed in Chapter 2. This principle provides a means for extending operations on $\mathbb{R}$ to those of $\mathcal{F}(\mathbb{R})$. In later sections, we will look at special fuzzy quantities, in particular, fuzzy numbers and fuzzy intervals.

3.1   Fuzzy quantities

Let $\mathbb{R}$ denote the set of real numbers. The elements of $\mathcal{F}(\mathbb{R})$, that is, the fuzzy subsets of $\mathbb{R}$, are **fuzzy quantities**. A relation R in $U \times V$, which is simply a subset R of $U \times V$, induces the mapping $R : \mathcal{F}(U) \to \mathcal{F}(V)$ defined by $R(A) = \vee AR^{-1}$. This is the mapping given by

$$R(A)(v) = \bigvee \{A(\{u : (u,v) \in R\})\}$$

as expressed by the extension principle at work. In particular, a mapping $f : \mathbb{R} \to \mathbb{R}$ induces a mapping $f : \mathcal{F}(\mathbb{R}) \to \mathcal{F}(\mathbb{R})$. A binary operation $\circ : \mathbb{R} \times \mathbb{R} \to \mathbb{R}$ on $\mathbb{R}$ gives a mapping $\mathcal{F}(\mathbb{R} \times \mathbb{R}) \to \mathcal{F}(\mathbb{R})$, and we have the mapping $\mathcal{F}(\mathbb{R}) \times \mathcal{F}(\mathbb{R}) \to \mathcal{F}(\mathbb{R} \times \mathbb{R})$ sending (A, B) to $\wedge(A \times B)$. Remember that $\wedge(A \times B)(r, s) = A(r) \wedge B(s)$. The composition

$$\mathcal{F}(\mathbb{R}) \times \mathcal{F}(\mathbb{R}) \to \mathcal{F}(\mathbb{R} \times \mathbb{R}) \to \mathcal{F}(\mathbb{R})$$

of these two is the mapping that sends (A, B) to $\vee(\wedge(A \times B))\circ^{-1}$, where $\circ^{-1}(x) = \{(a, b) : a \circ b = x\}$. We denote this binary operation by $A \circ B$.

This means that

$$(A \circ B)(x) \quad = \quad \bigvee \wedge (A \times B) \circ^{-1} (x)$$

$$= \quad \bigvee_{a \circ b = x} \wedge (A \times B)(a, b)$$

$$= \quad \bigvee_{a \circ b = x} \{A(a) \wedge B(b)\}$$

For example, for the ordinary arithmetic binary operations of addition and multiplication on $\mathbb{R}$, we then have corresponding operations $A + B = \vee \wedge (A \times B) +^{-1}$ and $A \cdot B = \vee \wedge (A \times B) \cdot^{-1}$ on $\mathcal{F}(\mathbb{R})$. Thus

$$(A + B)(z) = \bigvee_{x + y = z} \{A(x) \wedge B(y)\}$$

$$(A \cdot B)(z) = \bigvee_{x \cdot y = z} \{A(x) \wedge B(y)\}$$

The mapping $\mathbb{R} \to \mathbb{R} : r \to -r$ induces a mapping $\mathcal{F}(\mathbb{R}) \to \mathcal{F}(\mathbb{R})$ and the image of A is denoted $-A$. For $x \in \mathbb{R}$,

$$(-A)(x) = \bigvee_{x = -y} \{A(y)\} = A(-x)$$

If we view $-$ as a binary operation on $\mathbb{R}$, we get

$$(A - B)(z) = \bigvee_{x - y = z} \{A(x) \wedge B(y)\}$$

It turns out that $A + (-B) = A - B$, as is the case for $\mathbb{R}$ itself.

Division deserves some special attention. It is not a binary operation on $\mathbb{R}$ since it is not defined for pairs $(x, 0)$, but it is the relation

$$\{((r, s), t) \in (\mathbb{R} \times \mathbb{R}) \times \mathbb{R} : r = st\}$$

By the extension principle, this relation induces the binary operation on $\mathcal{F}(\mathbb{R})$ given by the formula

$$\frac{A}{B}(x) = \bigvee_{y = zx} (A(y) \wedge B(z))$$

So division of any fuzzy quantity by any other fuzzy quantity is possible. In particular, a real number may be divided by 0 *in* $\mathcal{F}(\mathbb{R})$. Recall that $\mathbb{R}$ is viewed inside $\mathcal{F}(\mathbb{R})$ as the characteristic functions $\chi_{\{r\}}$ for elements r of $\mathbb{R}$. We note the following easy proposition.

Proposition 3.1.1 *For any fuzzy set A, $A/\chi_{\{0\}}$ is the constant function whose value is $A(0)$.*

Proof. The function $A/\chi_{\{0\}}$ is given by the formula

$$\left(A/\chi_{\{0\}}\right)(u) = \bigvee_{s=t\cdot u}\left(A(s)\wedge\chi_{\{0\}}(t)\right)$$
$$= \bigvee_{s=0\cdot u}\left(A(s)\wedge\chi_{\{0\}}(0)\right)$$
$$= A(0)$$

∎

Thus $\chi_{\{r\}}/\chi_{\{0\}}$ is the constant function 0 if $r \neq 0$ and 1 if $r = 0$. Neither of these fuzzy quantities are real numbers, that is, neither is a characteristic function of a real number.

We note that performing operations on $\mathbb{R}$ is the same as performing the corresponding operations on $\mathbb{R}$ viewed as a subset of $\mathcal{F}(\mathbb{R})$. For binary operations $\circ$, this means that $\chi_{\{r\}} \circ \chi_{\{s\}} = \chi_{\{r \circ s\}}$. This last equation just expresses the fact that the mapping $\mathbb{R} \to \mathcal{F}(\mathbb{R}) : r \to \chi_{\{r\}}$ is a homomorphism. More generally, the following holds.

Theorem 3.1.2 *Let $\circ$ be any binary operation on a set U, and let S and T be subsets of U. Then*

$$\chi_S \circ \chi_T = \chi_{\{s \circ t : s \in S, t \in T\}}$$

Proof. For $u \in U$,

$$(\chi_S \circ \chi_T)(u) = \bigvee_{s \circ t = u}(\chi_S(s) \wedge \chi_T(t))$$

The sup is either 0 or 1 and is 1 exactly when there is an $s \in S$ and a $t \in T$ with $s \circ t = u$. The result follows. ∎

Thus if U is a set with a binary operation $\circ$, then $\mathcal{F}(U)$ contains a copy of U with this binary operation. In particular, if $U = \mathbb{R}$, then $\mathbb{R}$ with its various binary operations is contained in $\mathcal{F}(\mathbb{R})$. We identify $r \in \mathbb{R}$ with its corresponding element $\chi_{\{r\}}$.

The characteristic function $\chi_\varnothing$ has some special properties, where $\varnothing$ denotes the empty set. From the theorem, $\chi_\varnothing \circ \chi_T = \chi_\varnothing$, but in fact, $\chi_\varnothing \circ A = \chi_\varnothing$ for any fuzzy set A. It is simply the function that is 0 everywhere.

Binary operations on a set induce binary operations on its set of subsets. For example, if S and T are subsets of $\mathbb{R}$, then

$$S + T = \{s + t : s \in S, t \in T\}$$

These operations on subsets S of $\mathbb{R}$ carry over exactly to operations on
the corresponding characteristic sets χ_S in $\mathcal{F}(\mathbb{R})$. This is the content of
the previous theorem. Subsets of particular interest are intervals of $\mathbb{R}$
such as $[a, b]$, (a, b), and so on.

Now we go to the algebraic properties of fuzzy quantities with respect
to the operations induced from $\mathbb{R}$. Not all properties of binary operations
on $\mathbb{R}$ carry over to the ones induced on $\mathcal{F}(\mathbb{R})$, but many do. We identify
a real number r with its characteristic function $\chi_{\{r\}}$.

Theorem 3.1.3 *Let A, B, and C be fuzzy quantities. The following
hold.*

1. $0 + A = A$ 2. $0 \cdot A = 0$

3. $1 \cdot A = A$ 4. $A + B = B + A$

5. $A + (B + C) = (A + B) + C$ 6. $AB = BA$

7. $(AB)C = A(BC)$ 8. $r(A + B) = rA + rB$

9. $A(B + C) \leq AB + AC$ 10. $(-r)A = -(rA)$

11. $-(-A) = A$ 12. $(-A)B = -(AB) = A(-B)$

13. $\dfrac{A}{1} = A$ 14. $\dfrac{A}{r} = \dfrac{1}{r}A$

15. $\dfrac{A}{B} = A\dfrac{1}{B}$ 16. $A + (-B) = A - B$

Proof. We prove some of these, leaving the others as exercises. The
equations

$$\begin{aligned}
1 \cdot A(x) &= \bigvee\nolimits_{yz=x} \chi_{\{1\}}(y) \wedge A(z) \\
&= \bigvee\nolimits_{1x=x} \chi_{\{1\}}(1) \wedge A(x) \\
&= A(x)
\end{aligned}$$

show that $1 \cdot A = A$. If $(A(B + C))(x) > (AB + AC)(x)$, then there exist
u, v, y with $y(u + v) = x$ and such that

$$A(y) \wedge B(u) \wedge C(v) > A(p) \wedge B(q) \wedge A(h) \wedge C(k)$$

for all p, q, h, k with $pq + hk = x$. But this is not so for $p = h = y$,
$q = u$, and $v = k$. Thus $(A(B + C))(x) \leq (AB + AC)(x)$ for all x, whence
$A(B + C) \leq AB + AC$.

However, $r(A + B) = rA + rB$ since

$$
\begin{aligned}
\left(\chi_{\{r\}}(A + B)\right)(x) &= \bigvee_{uv=x}(\chi_{\{r\}}(u) \wedge (A + B)(v)) \\
&= \bigvee_{rv=x}(\chi_{\{r\}}(r) \wedge (A + B)(v)) \\
&= \bigvee_{\substack{s+t=v \\ rv=x}}(A(s) \wedge B(t)) \\
&= \bigvee_{\substack{s+t=v \\ rv=x}}(\chi_{\{r\}}(r)A(s) \wedge \chi_{\{r\}}(r)B(t)) \\
&= (rA + rB)(x)
\end{aligned}
$$

∎

There are a number of special properties that fuzzy quantities may have, and we need a few of them in preparation for dealing with fuzzy numbers and intervals. A subset A of the plane, that is, of $\mathbb{R}^2 = \mathbb{R} \times \mathbb{R}$, is **convex** if it contains the straight line connecting any two of its points. This can be expressed by saying that for $t \in [0, 1]$, $tx + (1 - t)y$ is in A whenever x and y are in A. This definition applies to $\mathbb{R}^n$ actually. For $\mathbb{R}$, convex subsets are just intervals, which may be infinite, and α-cuts of indicator functions of convex sets are convex.

Definition 3.1.4 *A fuzzy quantity A is* **convex** *if its α-cuts are convex, that is, if its α-cuts are intervals.*

Theorem 3.1.5 *A fuzzy quantity A is convex if and only if $A(y) \geq A(x) \wedge A(z)$ whenever $x \leq y \leq z$.*

Proof. Let A be convex, $x \leq y \leq z$, and $\alpha = A(x) \wedge A(z)$. Then x and z are in A_α, and since A_α is an interval, y is in A_α. Therefore $A(y) \geq A(x) \wedge A(z)$.

Suppose that $A(y) \geq A(x) \wedge A(z)$ whenever $x \leq y \leq z$. Let $x < y < z$ with $x, z \in A_\alpha$. Then $A(y) \geq A(x) \wedge A(z) \geq \alpha$, whence $y \in A_\alpha$ and A_α is convex. ∎

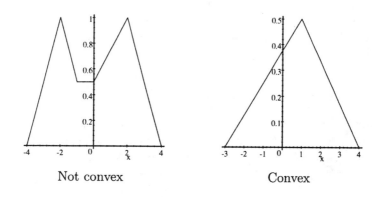

Not convex · Convex

Theorem 3.1.6 *If A and B are convex, then so are $A + B$ and $-A$.*

Proof. We show that $A + B$ is convex, leaving the other parts as exercises. Let $x < y < z$. We need that $(A + B)(y) \geq (A + B)(x) \wedge (A + B)(z)$. Let $\varepsilon > 0$. There are numbers x_1, x_2, z_1 and z_2 with $x_1 + x_2 = x$ and $z_1 + z_2 = z$ and satisfying

$$A(x_1) \wedge B(x_2) \geq (A + B)(x) - \varepsilon$$
$$A(z_1) \wedge B(z_2) \geq (A + B)(z) - \varepsilon$$

Now $y = \alpha x + (1 - \alpha)z$ for some $\alpha \in [0, 1]$. Let $x' = \alpha x_1 + (1 - \alpha)z_1$ and $z' = \alpha x_2 + (1 - \alpha)z_2$. Then $x' + z' = y$, x' lies between x_1 and z_1, and z' lies between x_2 and z_2. Thus we have

$$
\begin{aligned}
(A + B)(y) &\geq A(x') \wedge B(z') \\
&\geq A(x_1) \wedge A(z_1) \wedge B(x_2) \wedge B(z_2) \\
&\geq [(A + B)(x) - \varepsilon] \wedge [(A + B)(z) - \varepsilon] \\
&\geq [(A + B)(x) \wedge (A + B)(z)] - \varepsilon
\end{aligned}
$$

It follows that $A + B$ is convex. ■

A function $f : \mathbb{R} \to \mathbb{R}$ is **upper semicontinuous** if $\{x : f(x) \geq \alpha\}$ is closed. The following definition is consistent with this terminology.

Definition 3.1.7 *A fuzzy quantity is **upper semicontinuous** if its α-cuts are closed.*

Theorem 3.1.8 *A fuzzy quantity A is upper semicontinuous if and only if whenever $x \in \mathbb{R}$ and $\epsilon > 0$ there is a $\delta > 0$ such that $|x - y| < \delta$ implies that $A(y) < A(x) + \epsilon$.*

Proof. Suppose that A_α is closed for all α. Let $x \in \mathbb{R}$ and $\epsilon > 0$. If $A(x) + \epsilon > 1$, then $A(y) < A(x) + \epsilon$ for any y. If $A(x) + \epsilon \leq 1$ then for $\alpha = A(x) + \epsilon$, $x \notin A_\alpha$ and so there is $\delta > 0$ such that $(x - \delta, x + \delta) \cap A_\alpha = \varnothing$. Thus $A(y) < \alpha = A(x) + \epsilon$ for all y with $|x - y| < \delta$.

Conversely, take $\alpha \in [0, 1]$, $x \notin A_\alpha$, and $\epsilon = \frac{\alpha - A(x)}{2}$. There is $\delta > 0$ such that $|x - y| < \delta$ implies that $A(y) < A(x) + \frac{\alpha - A(x)}{2} < \alpha$, and so $(x - \delta, x + \delta) \cap A_\alpha = \varnothing$. Thus A_α is closed. ■

The following theorem is the crucial fact that enables us to use α-cuts in computing with fuzzy quantities.

Theorem 3.1.9 *Let $\circ : \mathbb{R} \times \mathbb{R} \to \mathbb{R}$ be a continuous binary operation on $\mathbb{R}$ and let A and B be fuzzy quantities with closed α-cuts and bounded supports. Then for each $u \in \mathbb{R}$, $(A \circ B)(u) = A(x) \wedge B(y)$ for some x and y with $u = x \circ y$.*

Proof. By definition,

$$(A \circ B)(u) = \bigvee_{x \circ y = u} (A(x) \wedge B(y))$$

The equality certainly holds if $(A \circ B)(u) = 0$. Suppose $\alpha = (A \circ B)(u) > 0$, and $A(x) \wedge B(y) < \alpha$ for all x and y such that $x \circ y = u$. Then there is a sequence $\{A(x_i) \wedge B(y_i)\}_{i=1}^{\infty}$ in the set $\{A(x) \wedge B(y) : x \circ y = u\}$ having the following properties.

1. $\{A(x_i) \wedge B(y_i)\}$ converges to α.

2. Either $\{A(x_i)\}$ or $\{B(y_i)\}$ converges to α.

3. Each x_i is in the support of A and each y_i is in the support of B.

Suppose that it is $\{A(x_i)\}$ that converges to α. Since the support of A is bounded, the set $\{x_i\}$ has a limit point x and hence a subsequence converging to x. Since the support of B is bounded, the corresponding subsequence of y_i has a limit point y and hence a subsequence converging to y. The corresponding subsequence of x_i converges to x. Thus we have a sequence $\{\{A(x_i) \wedge B(y_i)\}_{i=1}^{\infty}$ satisfying the three properties above and with $\{x_i\}$ converging to x and $\{y_i\}$ converging to y. If $A(x) = \gamma < \alpha$, then for $\delta = \frac{\alpha+\gamma}{2}$ and for sufficiently large i, $x_i \in A_\delta$, x is a limit point of those x_i, and since all cuts are closed, $x \in A_\delta$. But it is not, so $A(x) = \alpha$. In a similar vein, $B(y) \geq \alpha$, and we have $(A \circ B)(u) = A(x) \wedge B(y)$. Finally, $u = x \circ y$ since $u = x_i \circ y_i$ for all i, and $\circ$ is continuous. ∎

Corollary 3.1.10 *If A and B are fuzzy quantities with bounded support, all α-cuts are closed, and $\circ$ is a continuous binary operation on $\mathbb{R}$, then $(A \circ B)_\alpha = A_\alpha \circ B_\alpha$.*

Proof. Applying the theorem, for $u \in (A \circ B)_\alpha$, $(A \circ B)(u) = A(x) \wedge B(y)$ for some x and y with $u = x \circ y$. Thus $x \in A_\alpha$ and $y \in B_\alpha$, and therefore $(A \circ B)_\alpha \subseteq A_\alpha \circ B_\alpha$. The other inclusion is easy. ∎

Corollary 3.1.11 *If A and B are fuzzy quantities with bounded support and all α-cuts are closed, then*

1. $(A + B)_\alpha = A_\alpha + B_\alpha$,

2. $(A \cdot B)_\alpha = A_\alpha \cdot B_\alpha$,

3. $(A - B)_\alpha = A_\alpha - B_\alpha$.

We end this section with a note about division of sets of real numbers. We have no obvious way to divide a set S by a set T. We cannot take $S/T = \{s/t : s \in S \text{ and } t \in T\}$ since t may be 0. But we can perform the operation χ_S/χ_T. Therefore fuzzy arithmetic gives a natural way to divide sets of real numbers one by the other, and in particular to divide intervals. (A similar definition is actually used in computer arithmetic.) We also note that if S and T are closed and bounded, then $(\chi_S/\chi_T)(u) = \chi_S(ux) \wedge \chi_T(x)$ for a suitable x. The proof of this is left as an exercise, but mimics the proof of the theorem.

3.2   Fuzzy numbers

We are going to specify a couple of special classes of fuzzy quantities, the first being the class of **fuzzy numbers**. A fuzzy number is a fuzzy quantity A that represents a generalization of a real number r. Intuitively, $A(x)$ should be a measure of how well $A(x)$ "approximates" r, and certainly one reasonable requirement is that $A(r) = 1$ and that this holds only for r. A fuzzy number should have a picture somewhat like the one following.

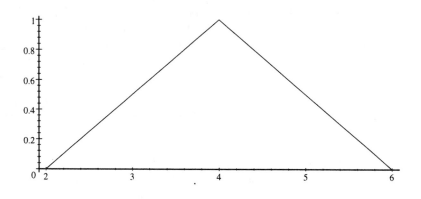

A fuzzy number 4

Of course we do not want to be so restrictive as to require that fuzzy numbers all look like triangles, even with different slopes. A number of definitions have been proffered, and one of the simplest and most reasonable seems to be the following.

Definition 3.2.1 *A **fuzzy number** is a fuzzy quantity A that satisfies the following conditions.*

1. $A(x) = 1$ *for exactly one* x.

2. *The support* $\{x : A(x) > 0\}$ *of A is bounded.*

3. *The α-cuts of A are closed intervals.*

There are many ways this definition could be stated, and the third condition may seem arbitrary. However, it is easily stated and visualized, and has some nice consequences. This definition is not universally used. Sometimes the second condition is not required.

The following proposition is an easy consequence of the definition.

Proposition 3.2.2 *The following hold:*

1. *Real numbers are fuzzy numbers.*

2. *A fuzzy number is a convex fuzzy quantity.*

3. *A fuzzy number is upper semicontinuous.*

4. *If A is a fuzzy number with $A(r) = 1$, then A is non-decreasing on $(-\infty, r]$ and non-increasing on $[r, \infty)$.*

Proof. It should be clear that real numbers are fuzzy numbers. A fuzzy number is convex since its α-cuts are intervals, and is upper semicontinuous since its α-cuts are closed. If A is a fuzzy number with $A(r) = 1$ and $x < y < r$, then since A is convex and $A(y) < A(r)$, we have $A(x) \leq A(y)$, so A is monotone increasing on $(-\infty, r]$. Similarly, A is monotone decreasing on $[r, \infty)$. ∎

Theorem 3.2.3 *If A and B are fuzzy numbers then so are $A + B$, $A \cdot B$, and $-A$.*

Proof. That these fuzzy quantities have bounded support and assume the value 1 in exactly one place is easy to show. The α-cuts of $A + B$ and $A \cdot B$ are closed intervals by Corollary 3.1.11. Since $-A = (-1) \cdot A$, the rest follows. ∎

Carrying out computations with fuzzy quantities, and in particular with fuzzy numbers, can be complicated. There are some special classes of fuzzy numbers for which computations of their sum, for example, is easy. One such class is that of triangular fuzzy numbers. They are the ones with pictures like the one above. It is clear that such a fuzzy number is uniquely determined by a triple (a, b, c) of numbers with $a \leq b \leq c$. So computing with triangular fuzzy numbers should be reduced to operating on such triples.

Definition 3.2.4 *A **triangular fuzzy number** is a fuzzy quantity A whose values are given by the formula*

$$A(x) = \begin{cases} 0 & if \quad x < a \\ \dfrac{x-a}{b-a} & if \quad a \le x \le b \\ \dfrac{x-c}{b-c} & if \quad b < x \le c \\ 0 & if \quad c < x \end{cases}$$

for some $a \le b \le c$.

For example, if $a = -1$, $b = 2$, and $c = 3$, the picture is

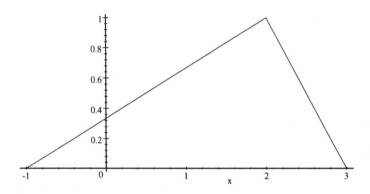

The formula is just a precise way to say what we want triangular fuzzy numbers to be. It is clear that there is a one-to-one correspondence between triangular numbers and triples (a, b, c) with $a \le b \le c$, and we identify a triangular number with its triple. Then (a, b, c) is the fuzzy quantity given by the formula in the definition above. If $a = b$, then $A(b)$ in the formula above is $\frac{b-a}{b-a}$ with numerator and denominator both 0. We take this quotient to be 1. Similar remarks apply to triangular numbers (a, c, c). Thus triangular numbers (r, r, r) are those functions which are 1 at r and 0 elsewhere, the image of the real number r in $\mathcal{F}(\mathbb{R})$. For $a < c$, the support of (a, b, c) is the open interval (a, c), and the value 1 is assumed at b and only at b.

Theorem 3.2.5 *For triangular numbers,*

$$(a, b, c) + (d, e, f) = (a + d, b + e, c + f)$$

Proof. Using $((a, b, c) + (d, e, f))_\alpha = (a, b, c)_\alpha + (d, e, f)_\alpha$, it follows that the support of the sum is the interval $(a + d, c + f)$ and that 1 is assumed exactly at $b + e$. Suppose that $\alpha > 0$, the left endpoint of the α-cut of (a, b, c) is u and that of (d, e, f) is v. Then $a \leq u \leq b$, $d \leq v \leq e$, and

$$\alpha = \frac{u - a}{b - a} = \frac{v - d}{e - d}$$

An easy calculation shows that

$$\alpha = \frac{u + v - (a + d)}{b + e - (a + d)}$$

which shows that $u+v$ is the left endpoint of the α-cut of $(a+d, b+e, c+f)$. But we know that the left endpoint of the α-cut of $(a, b, c) + (d, e, f)$ is $u + v$. Similarly for right endpoints of cuts, and hence $(a, b, c) + (d, e, f)$ and $(a + d, b + e, c + f)$ have the same cuts and so are equal. ■

Thus adding two triangular numbers is the same as adding two triples of real numbers coordinatewise, a particularly simple thing to do.

Products of triangular numbers are not necessarily triangular. However, we do have $(AB)_\alpha = A_\alpha B_\alpha$. So we can compute the cuts of a product of triangular numbers easily. The problem is that the product is not piecewise linear. For example, the square of the triangular number $(-1, 0, 1)$ looks like the following figure.

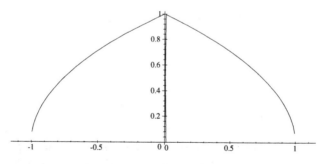

The square of the triangular number $(-1, 0, 1)$

3.3 Fuzzy intervals

A subset S of $\mathbb{R}$ is identified with χ_S, and in particular, intervals $[a, b]$ are identified with their characteristic functions, namely the fuzzy quantities $\chi_{[a,b]}$. The use of intervals with their arithmetic is appropriate in some situations involving impreciseness. When the intervals themselves are not sharply defined, we are driven to the concept of fuzzy interval. Thus we want to generalize intervals to fuzzy intervals, and certainly a fuzzy quantity generalizing the interval $[a, b]$ should have value 1 on $[a, b]$. A fuzzy quantity that attains the value 1 is called **normal**. The other defining properties of fuzzy intervals should be like those of fuzzy numbers. Thus a fuzzy interval should look something like the following picture.

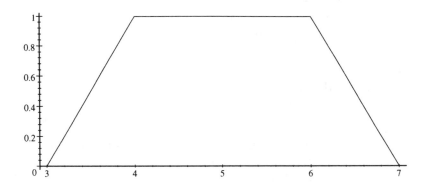

A membership function for the fuzzy interval $[4, 6]$

This fuzzy interval has a trapezoidal form, representing "approximately between 4 and 6". Our definition is this:

Definition 3.3.1 *A **fuzzy interval** is a fuzzy quantity A satisfying the following:*

1. *A is normal.*

2. *The support $\{x : A(x) > 0\}$ of A is bounded.*

3. *The α-cuts of A are closed intervals.*

Theorem 3.2.3 holds for fuzzy intervals, with only minor changes in the proofs. In fact, fuzzy numbers are fuzzy intervals. The only difference is that a fuzzy number can attain the value 1 at only one place, while a fuzzy interval can have an interval of such places.

3.4   Exercises

1. Show that $\chi_S / \chi_{\{0\}}$ is the constant function 1 or 0 according to whether or not $0 \in S$.

2. Complete the proof of Theorem 3.1.3.

3. Show that for relation on a set U, the resulting binary operation $\circ$ on $\mathcal{F}(U)$ has the property that $A \circ \chi_\varnothing = \chi_\varnothing$ for all $A \in \mathcal{F}(U)$.

4. Show that there are fuzzy quantities A and B, such that

   (a) $A - A \neq 0$
   (b) $(A + B) - B \neq A$
   (c) $\frac{A}{A} \neq 1$
   (d) $\frac{A}{B} B \neq A$

5. Show that for fuzzy quantities, multiplication does not distribute over addition. That is, $A(B + C) \neq AB + AC$.

6. Let S and T be closed and bounded subsets of $\mathbb{R}$. Show that $(\chi_S/\chi_T)(u) = \chi_S(ux) \wedge \chi_T(x)$ for some x.

7. Compute the α-cuts of the sum of two triangular numbers.

8. For $f : \mathbb{R} \to \mathbb{R}$ and $A \in \mathcal{F}(\mathbb{R})$, write down the membership function of $f(A)$ when

$$f(x) = -x, \qquad f(x) = x^2$$
$$f(x) = x^5, \qquad f(x) = |x|$$

9. Let $\varphi_L(x) = 0 \vee (1 - x)$ and $\varphi_R(x) = e^{-x}$, and

$$A(x) = \begin{cases} \varphi_L(\frac{a-x}{\alpha}) & \text{if } x \leq a \\ 1 & \text{if } a < x < b \\ \varphi_R(\frac{x-b}{\beta}) & \text{if } x \geq b \end{cases}$$

$$B(x) = \begin{cases} \varphi_L(\frac{c-x}{\alpha'}) & \text{if } x \leq c \\ 1 & \text{if } c < x < d \\ \varphi_R(\frac{x-d}{\beta'}) & \text{if } x \geq d \end{cases}$$

Compute $-A$, $A + B$ and A/B. In general, when a membership function $A(x)$ is obtained from two functions $\varphi_L(x)$ and $\varphi_R(x)$ as above, these functions $\varphi_L(x)$ and $\varphi_R(x)$ are referred to as **shape functions**.

10. Show that if A is a convex fuzzy quantity, then so is $-A$.

11. Is it true that if A and B are convex fuzzy quantities, then so is AB?

12. Let $\mathcal{C}$ be the set of all convex fuzzy quantities. Order these functions by the usual pointwise order. That is, $A \le B$ if $A(u) \le B(u)$ for all $u \in \mathbb{R}$. Show that $(\mathcal{C}, \le)$ is a lattice. Show that if A and B are in $\mathcal{C}$, then $A \vee B$ taken in $\mathcal{C}$ may not be $A \vee B$ taken in $\mathcal{F}(\mathbb{R})$.

13. Define the fuzzy quantities A and B by

$$A(x) = \frac{1}{2(1 + e^{-\frac{x}{2}})}$$
$$B(x) = 1$$

Show that A and B are convex, $A + B$ is convex, but $(A + B)_{\frac{3}{4}} \ne A_{\frac{3}{4}} + B_{\frac{3}{4}}$.

14. Let $A \in \mathcal{F}(\mathbb{R})$ be continuous. Show that A is convex if and only if the A_α are closed intervals.

15. The condition that $A(y) \ge A(x) \wedge A(z)$ whenever $x < y < z$ is sometimes called **quasiconcavity**. Show that $A \in \mathcal{F}(\mathbb{R})$ is convex if and only if for $x, y \in \mathbb{R}$, the function $\lambda \to A(\lambda x + (1 - \lambda)y)$ is quasiconcave.

16. A function $f : \mathbb{R} \to \mathbb{R}$ is **pseudoconcave** if for $x, y \in U$ with $f(x) \ne f(y)$ and for $z = \lambda x + (1 - \lambda)y$ with $\lambda \in (0, 1)$, we have $f(z) > f(x) \wedge f(y)$. A fuzzy set $A \in \mathcal{F}(U)$ is **strongly convex** if A is convex and pseudoconcave. Show that a convex fuzzy subset A of U is strongly convex if its membership function A is one-to-one on $\{x \in U : A(x) < 1\}$.

Chapter 4

LOGICAL ASPECTS OF FUZZY SETS

The phrase "fuzzy logic" has two meanings. On the one hand it refers to the use of fuzzy sets in the representation and manipulation of vague information for the purpose of making decisions or taking actions. It is the theory of fuzzy sets, the study of the system of all mappings of a set U into the unit interval. This involves not only the usual connectives of max and min on fuzzy sets, but the theory of t-norms, t-conorms, negations, and many other related concepts. Also there are generalizations of ordinary set theoretical concepts to the fuzzy setting, such as that of equivalence relation. Many topics can be "fuzzified". Some of these appear throughout this book.

On the other hand, fuzzy logic means the extension of ordinary logic with truth values in the two-element Boolean algebra $(\{0, 1\}, \vee, \wedge,', 0, 1)$ to the case where they are in the Kleene algebra $([0, 1], \vee, \wedge,', 0, 1)$. There are of course many extensions of two-valued logic to multivalued ones, generally with the truth values being finite in number. A standard reference for this is [86].

This chapter focuses on fuzzy logic in this second sense. First, we present the basics of the two-valued propositional calculus, next the corresponding material for a well-known three-valued propositional calculus due to Lukasiewicz, and then for the propositional calculus in which the truth values consist of the Kleene algebra above, that is, for fuzzy propositional calculus. The fact that these last two propositional calculi are equal does not seem to be widely known. But it is a useful fact. It enables one to determine in finitely many steps, following a specific algorithm, whether two expressions in fuzzy sets are equal, where the expressions involve just the fuzzy sets and the connectives max, min, and the com-

61

plement $x \to 1 - x$. An analogous fact holds for fuzzy sets whose values are themselves intervals in $[0, 1]$.

4.1 Classical two-valued logic

First, we will look at the simplest formal logic, the classical two-valued propositional logic, or calculus. In it, propositions can take on only two truth values. This will not be the case in the other logics we consider. Here are the basics of **propositional calculus.**

The building blocks consist of a set of formal entities $V = \{a, b, c, ...\}$ often called the **variables** of the logic. The elements of V are to be thought of as primitive propositions, or simple statements, or variables. We specify three basic logical connectives: $\vee$ (*or*), $\wedge$ (*and*), and $'$ (*not*). From the elements of V and these connectives, we build up expressions such as $a \vee b$, $a \wedge b'$, $a' \wedge (b \vee c)$, and so on. Such expressions are **formulas.** Formulas are defined inductively as follows:

- If a is a variable, then a is a formula.

- If u and v are formulas, then $u \vee v$, $u \wedge v$ and u' are formulas.

The set of all formulas is denoted $\mathbb{F}$. Thus if we have only three variables, a, b, and c, at the first step in this recursive process, we get the formulas a, b, c. Then at the next step, we get $a', b', c', a \vee a, a \wedge a, a \vee b, a \vee c, b \vee a$, and several more. At the third step, we get such new formulas as $a \vee b'$, $a \vee (b \wedge c)$, $(a \vee c)'$, and $(b')' \wedge c'$. Of course there are many many more. The set $\mathbb{F}$ is the set of all formulas defined. If $\mathbb{F}_n$ is the set of formulas gotten at the n-th stage, then $\cup_n \mathbb{F}_n = \mathbb{F}$. It is an infinite set, even if V has only one element. For example, if a is a variable, then $a, a \wedge a, (a \wedge a) \wedge a, a \wedge ((a \wedge a) \wedge a), a \wedge (a \wedge ((a \wedge a) \wedge a)), \ldots$ are all distinct formulas, as are $a, a', (a')', ((a')')'$, At this point, elements of $\mathbb{F}$ are just certain strings of symbols. No meaning has been attached to anything.

For any function $t : V \to \{0, 1\}$ we get a function $\tilde{t} : \mathbb{F} \to \{0, 1\}$ as follows: for each variable a appearing in a formula, substitute $t(a)$ for it. Then we have an expression in the symbols $0, 1, \vee, \wedge$, and $'$, together with balanced sets of parentheses. The tables below define the operations of $\vee, \wedge$, and $'$ on the truth values $\{0, 1\}$.

$\vee$	0	1
0	0	1
1	1	1

$\wedge$	0	1
0	0	0
1	0	1

	$'$
0	1
1	0

Using these tables, which describe the two element Boolean algebra, we get an extension to $\mathbb{F}$. For example, if $t(a) = 0$ and $t(b) = t(c) = 1$, then

$$
\begin{aligned}
\tilde{t}\left(((a \vee b) \wedge c) \wedge (b' \vee c)\right) &= (((t(a) \vee t(b)) \wedge t(c)) \wedge (t(b)' \vee t(c))) \\
&= ((0 \vee 1) \wedge 1) \wedge (1' \vee 1) \\
&= (1 \wedge 1) \wedge (0 \vee 1) \\
&= 1 \wedge 1 \\
&= 1
\end{aligned}
$$

Such a mapping $\mathbb{F} \rightarrow \{0, 1\}$ is called a **truth evaluation**. We have exactly one for each mapping $V \rightarrow \{0, 1\}$. Expressions that are assigned the value 1 by every t are called **tautologies**, such as $a \vee a'$ and $b \vee b'$.

There are two other common logical connectives, $\Rightarrow$ (implies) and $\Leftrightarrow$ (implies and is implied by, or if and only if), and we could write down the usual truth tables for them. However, in classical two-valued logic, $a \Rightarrow b$ is taken to mean $a' \vee b$, and $a \Leftrightarrow b$ to mean $(a \Rightarrow b) \wedge (b \Rightarrow a)$. Thus they can be defined in terms of the three connectives we used. The formula $a \Rightarrow b$ is called **material implication**.

Two formulas a and b are called **(logically) equivalent** if they have the same image for every truth evaluation. This is the same thing as $a \Leftrightarrow b$ being a tautology, for which we write $a \equiv b$. The expressions a, $a \vee a$, and $(a \vee a) \vee a$ are equivalent, always having truth evaluation the same as a. Likewise, the expressions $a \wedge (b \vee c)$ and $(a \wedge b) \vee (a \wedge c)$ are equivalent. Both have truth evaluation 1 when a and either b or c have truth evaluation 1, and have truth evaluation 0 otherwise. We want to consider two formulas equal if they are logically equivalent, and indeed, being logically equivalent is an equivalence relation, and so partitions $\mathbb{F}$. Two elements of $\mathbb{F}$ are in the same member of this partition if they are equivalent, that is, in the same equivalence class.

Consider now the set $\mathbb{F}/\equiv$ ($\mathbb{F}$ "modulo" $\equiv$) of all equivalence classes of this equivalence relation. Let $[a]$ denote the equivalence class containing the formula a. Then setting

$$
\begin{aligned}
{[a] \vee [b]} &= [a \vee b] \\
{[a] \wedge [b]} &= [a \wedge b] \\
{[a]'} &= [a']
\end{aligned}
$$

makes $\mathbb{F}/\equiv$ into a Boolean algebra. That these operations are well defined, and actually do what is claimed takes some checking and we will not give the details. *This Boolean algebra is the classical propositional calculus.* If the set V of variables, or atomic formulas, is finite, then $\mathbb{F}/\equiv$ is finite, even though $\mathbb{F}$ is infinite. It is a fact that if V has n elements, then $\mathbb{F}/\equiv$ has 2^{2^n} elements. If $\{v_1, v_2, \ldots, v_n\}$ is the set of variables, then the elements of the form

$$w_1 \wedge w_2 \wedge \ldots \wedge w_n$$

where w_i is either v_i or v_i' are called **monomials**, and every element of $\mathbb{F}$ is logically equivalent to the join of a unique set of these monomials. (The element $[0]$ is the join of the empty set of monomials.) Elements written in this fashion are said to be in **disjunctive normal form**. For example, if there are just two variables a and b, then $a \vee b$ is logically equivalent to $(a \wedge b) \vee (a \wedge b') \vee (a' \wedge b)$. This is easy to check: just check the truth evaluations for all four possible truth evaluations of the pair a and b.

The classical propositional calculus is associated with a Boolean algebra of sets in the following way. Suppose we associate with each equivalence class $[u] \in \mathbb{F}/\equiv$ the set

$$\{t : V \to \{0,1\} : \tilde{t}(u) = 1\}$$

This defines a function by the very definition of logically equivalent. Thus we have a mapping

$$T : \mathbb{F}/\equiv \;\to\; \mathcal{P}(\{0,1\}^V)$$

where $\{0,1\}^V$ denotes the set of all functions from V to $\{0,1\}$ and $\mathcal{P}(\{0,1\}^V)$ denotes the set of all subsets of $\{0,1\}^V$. The mapping T is one-to-one, again by the definition of being logically equivalent. The set of all subsets of $\{0,1\}^V$ is a Boolean algebra under ordinary set union, intersection, and complementation. The image of $\mathbb{F}/\equiv$ is a sub-Boolean algebra and T is a homomorphism, that is,

$$
\begin{aligned}
T(a \vee b) &= T(a) \cup T(b) \\
T(a \wedge b) &= T(a) \cap T(b) \\
T(a') &= T(a)' \\
T([0]) &= \varnothing \\
T([1]) &= \{0,1\}^V
\end{aligned}
$$

If V is finite, the mapping T is onto and we have a one-to-one correspondence between the equivalence classes of propositions and subsets of $\{0,1\}^V$. This is the connection between propositional calculus and set theory. It is isomorphic to a Boolean algebra of sets.

We illustrate the construction of $\mathbb{F}/\equiv$ above in the case $V = \{a, b\}$. We need only concern ourselves with writing down one formula for each equivalence class. First, we have the formulas a, b, a', b', and in the next iteration the formulas, $a \vee a, a \vee b, b \vee a, a \vee a', a \vee b', a''$, and so on. We will not write them all down. There can be at most 16 nonequivalent ones, that being the number of subsets of $\{0,1\}^{\{a,b\}}$. If we denote the equivalence class containing $a \vee a'$ by 1 and the one containing $a \wedge a'$ by 0, then these 16 are

0	1
a	a'
b	b'
$a \vee b$	$a' \wedge b'$
$a \vee b'$	$a' \wedge b$
$a' \vee b$	$a \wedge b'$
$a' \vee b'$	$a \wedge b$
$(a \wedge b') \vee (a' \wedge b)$	$(a' \vee b) \wedge (a \vee b')$

We have *not* written them all in disjunctive normal form. For any two of these formulas, it is easy to find a truth evaluation for which they take different values. For example, consider the formulas $a' \vee b$ and $(a \wedge b') \vee (a' \wedge b)$. If $t(a) = t(b) = 1$ then $t(a' \vee b) = 1$ and $t((a \wedge b') \vee (a' \wedge b)) = 0$. So this propositional calculus has only 16 formulas, and is isomorphic to the Boolean algebra with 16 elements.

In summary, propositional calculus is a logic of atomic propositions. These atomic propositions cannot be broken down. The validity of arguments does not depend on the meaning of these atomic propositions, but rather on the form of the argument. For example, the deduction rule known as **modus ponens** states that from $b \Rightarrow a$ and b one deduces a logically. That is, $((b \Rightarrow a) \wedge b) \Rightarrow a$ is a tautology. It is easily checked that this is indeed the case.

If we consider propositions of the form "all a's are b", which involves the quantifier "all" and the predicate b, then the validity of an argument should depend on the relationship between parts of the statement as well as the form of the statement. For example, "all men are mortal", "Napoleon is a man", and therefore "Napoleon is mortal". In order to reason with this type of proposition, propositional calculus is extended to predicate calculus, specifically to *first-order predicate calculus*. Its alphabet includes the quantifiers "for all" ($\forall$) and "there exists" ($\exists$), as well as predicates, or relation symbols. A predicate on a set S is a relation on S, for example, "x is a positive integer". This is a unary predicate. It can be identified with the subset $\{s \in S : P(s)$ is a positive integer$\}$, that is, a unary relation on S. More generally, a n-ary predicate on S can be identified with the subset $\{(s_1, s_2, \cdots, s_n) \in S_n : P(s_1, s_2, \cdots, s_n)$ is true$\}$, that is, with a n-ary relation on S. The formalism of such logic is

more elaborate than that of propositional calculus and we will not discuss
it further.

4.2   A three-valued logic

The construction carried out in the previous section can be generalized
in many ways. Perhaps the simplest is to let the set $\{0, 1\}$ of truth values
be larger. Thinking of 0 as representing **false** and 1 as representing **true**,
we add a third truth value u representing **undecided**. It is common to
use $\frac{1}{2}$ instead of u, but a truth value should not be confused with a
number, so we prefer u. Now proceed as before. Starting with a set
of variables, or primitive propositions V, build up formulas using this
set and some logical connectives. Such logics are called **three-valued**,
for obvious reasons. The set $\mathbb{F}$ of formulas is the same as in classical
two-valued logic. However, the truth evaluations t will be different, thus
leading to a different equivalence relation $\equiv$ on $\mathbb{F}$. There are a multitude
of three-valued logics, and their differences arise in the specification of
truth tables and implication.

In extending a mapping $V \to \{0, u, 1\}$ to a mapping $\mathbb{F} \to \{0, u, 1\}$, we
need to specify how the connectives operate on the truth values. Here
is that specification for a particularly famous three-valued logic, that of
Lukasiewicz.

$\vee$	0	u	1
0	0	u	1
u	u	u	1
1	1	1	1

$\wedge$	0	u	1
0	0	0	0
u	0	u	u
1	0	u	1

	$'$
0	1
u	u
1	0

Again, we have chosen the basic connectives to be $\vee, \wedge$, and $'$. These
operations $\vee$ and $\wedge$ come simply from viewing $\{0, u, 1\}$ as the three-
element chain with the implied lattice operations. The operation $'$ is the
duality of this lattice. The connectives $\Rightarrow$ and $\Leftrightarrow$ are defined as follows.

$\Rightarrow$	0	u	1
0	1	1	1
u	u	1	1
1	0	u	1

$\Leftrightarrow$	0	u	1
0	1	u	0
u	u	1	u
1	0	u	1

For this logical system, we still have that a and b are logically equivalent,
that is, $\tilde{t}(a) = \tilde{t}(b)$ for all truth valuations $t : V \to \{0, u, 1\}$ if and only if
$a \Leftrightarrow b$ is a three-valued tautology.

It is clear that a three-valued tautology is a two-valued tautology. That is, if $\widetilde{t}(a) = \widetilde{t}(b)$ for all truth valuations $t : V \to \{0, u, 1\}$, then $\widetilde{t}(a) = \widetilde{t}(b)$ for all truth valuations $t : V \to \{0, 1\}$. But the converse is not true. For example, $a \vee a'$ is a two-valued tautology, while it is not a three-valued one. Just consider $t : V \to \{0, u, 1\}$ with $t(a) = u$.

The set $\mathbb{F}/\equiv$ of equivalence classes of formulas under the operations induced by $\vee$, $\wedge$, and $'$ do not form a Boolean algebra since, in particular, for a variable a, it is not true that $a \vee a'$ always has truth value 1. If a has truth value u, then $a \vee a'$ has truth value u. The law of the **excluded middle** does not hold in this logic. These equivalence classes of logically equivalent formulas do form a Kleene algebra, however, and this Kleene algebra is the Lukasiewicz three-valued propositional calculus.

With each formula f, associate the mapping

$$\{0, u, 1\}^V \to \{0, u, 1\}$$

given by $t \to t(f)$. This induces a one-to-one mapping from $\mathbb{F}/\equiv$ into the set of all mappings from $\{0, u, 1\}^V$ into $\{0, u, 1\}$. This set of mappings is itself a Kleene algebra, and the one-to-one mapping from $\mathbb{F}/\equiv$ into it is a homomorphism. In the two-valued case, we had a one-to-one homomorphism from the Boolean algebra $\mathbb{F}/\equiv$ into the Boolean algebra of all mappings from $\{0, 1\}^V$ into $\{0, 1\}$, or equivalently into $\mathcal{P}(\{0, 1\}^V)$.

4.3 Fuzzy logic

Fuzzy propositional calculus generalizes classical propositional calculus by using the truth set $[0, 1]$ instead of $\{0, 1\}$. The construction parallels those in the last two sections. The set of building blocks in both cases is a set V of symbols representing atomic or elementary propositions. The set of formulas $\mathbb{F}$ is built up from V using the logical connectives $\wedge$, $\vee$, $'$ (*and, or,* and *not,* respectively) in the usual way. As in the two-valued and three-valued propositional calculi, a truth evaluation is gotten by taking any function $t : V \to [0, 1]$ and extending it to a function $\widetilde{t} : \mathbb{F} \to [0, 1]$ by replacing each element $a \in V$ which appears in the formula by its value $t(a)$, which is an element in $[0, 1]$. This gives an expression in elements of $[0, 1]$ and the connectives $\vee$, $\wedge$, $'$. This expression is evaluated by letting

$$
\begin{aligned}
x \vee y &= \max\{x, y\} \\
x \wedge y &= \min\{x, y\} \\
x' &= 1 - x
\end{aligned}
$$

for elements x and y in $[0, 1]$. We get an equivalence relation on $\mathbb{F}$ by letting two formulas be equivalent if they have the same truth evaluation for all $\tilde{t}$. A formula is a **tautology** if it always has truth value 1. Two formulas u and v are **logically equivalent** when $\tilde{t}(u) = \tilde{t}(v)$ for all truth valuations t. As in Lukasiewicz's three-valued logic, the law of the excluded middle fails. For an element $a \in V$ and a t with $t(a) = 0.3$, $t(a \vee a') = 0.3 \vee 0.7 = 0.7 \neq 1$. The set of equivalence classes of logically equivalent formulas forms a Kleene algebra, just as in the previous case.

The association of formulas with fuzzy sets is this. With each formula u, associate the fuzzy subset $[0,1]^V \rightarrow [0,1]$ of $[0,1]^V$ given by $t \rightarrow t(u)$. Thus we have a map from $\mathbb{F}$ to $\mathcal{F}\left([0,1]^V\right)$. This induces a one-to-one mapping from $\mathbb{F}/\equiv$ into the set of mappings from $[0,1]^V$ into $[0,1]$, that is into the set of fuzzy subsets of $[0,1]^V$. This one-to-one mapping associates fuzzy logical equivalence with equality of fuzzy sets.

4.4   Fuzzy and Lukasiewicz logics

The construction of $\mathbb{F}/\equiv$ for the three-valued Lukasiewicz propositional calculus and the construction of the corresponding $\mathbb{F}/\equiv$ for fuzzy propositional calculus were the same except for the truth values used. In the first case the set of truth values was $\{0, u, 1\}$ with the tables given, and in the second, the set of truth values was the interval $[0,1]$ with

$$
\begin{aligned}
x \vee y &= \max\{x, y\} \\
x \wedge y &= \min\{x, y\} \\
x' &= 1 - x
\end{aligned}
$$

We remarked that in each case, the resulting equivalence classes of formulas formed Kleene algebras. Now the remarkable fact is that the resulting propositional calculi are the same. They yield the same Kleene algebra. In fact, they yield the same equivalence relation. The set $\mathbb{F}$ of formulas is clearly the same in both cases, and in fact is the same as the set of formulas in the classical two-valued case. Any differences in the propositional calculi must come from the truth values and the operations of $\vee$, $\wedge$, and $'$ on them. Different truth values may or may not induce different equivalence relations on $\mathbb{F}$. It turns out that using truth values $\{0, u, 1\}$ with the operations discussed, and using $[0,1]$ with the operations displayed above *give the same equivalence relation on* $\mathbb{F}$, and thus the same resulting propositional calculi $\mathbb{F}/\equiv$.

Theorem 4.4.1 *The propositional calculus for three-valued Lukasiewicz logic and the propositional calculus for fuzzy logic are the same.* [34]

Proof. We outline a proof. Truth evaluations are mappings f from $\mathbb{F}$ into the set of truth values satisfying

$$\begin{aligned}
f(v \vee w) &= f(v) \vee f(w) \\
f(v \wedge w) &= f(v) \wedge f(w) \\
f(v') &= f(v)'
\end{aligned}$$

for all formulas v and w in $\mathbb{F}$. Two formulas in $\mathbb{F}$ are equivalent if and only if they have the same values for all truth valuations. So we need that two formulas have the same value for all truth valuations into $[0, 1]$ if and only if they have the same values for all truth valuations into $\{0, u, 1\}$. First, let $\prod$ be the Cartesian product $\prod_{x \in (0,1)} \{0, u, 1\}$ with $\vee$, $\wedge$ and $'$ defined componentwise. If two truth valuations from $\mathbb{F}$ into $\prod$ differ on an element, then these functions followed by the projection of $\prod$ into one of the copies of $\{0, u, 1\}$ differ on that element. If two truth valuations from $\mathbb{F}$ into $\{0, u, 1\}$ differ on an element, then these two functions followed by any lattice embedding of $\{0, u, 1\}$ into $[0, 1]$ differ on that element. There is a lattice embedding $[0, 1] \to \prod$ given by $y \to \{y_x\}_x$, where y_x is 0, u, or 1 depending on whether y is less than x, equal to x or greater than x. If two truth valuations from $\mathbb{F}$ into $[0, 1]$ differ on an element, then these two functions followed by this embedding of $[0, 1]$ into $\prod$ will differ on that element. The upshot of all this is that taking the truth values to be the lattices $\{0, u, 1\}$, $[0, 1]$, and $\prod$ all induce the same equivalence relation on $\mathbb{F}$, and hence yield the same propositional calculus. ∎

Now what import does this have for fuzzy set theory? One consequence is this. Suppose we wish to check the equality of two expressions involving fuzzy sets connected with $\wedge$, $\vee$, and $'$, with $'$ denoting the usual negation. For example, does the equality

$$A \wedge ((A' \wedge B) \vee (A' \wedge B') \vee (A' \wedge C)) = A \wedge A'$$

hold for fuzzy sets? That is, is it true that this equality holds no matter what the fuzzy sets A, B, and C are? Equivalently, for every $x \in U$ is it true that

$$[A \wedge ((A' \wedge B) \vee (A' \wedge B') \vee (A' \wedge C))](x) = [A \wedge A'](x)$$

which is asking whether the equality

$$\begin{aligned}
&A(x) \wedge ((A'(x) \wedge B(x)) \vee (A'(x) \wedge B'(x)) \vee (A'(x) \wedge C(x))) \\
&= A(x) \wedge A'(x)
\end{aligned} \tag{4.1}$$

holds for all $x \in U$. The theorem implies that it is enough to check the equality

$$A \wedge ((A' \wedge B) \vee (A' \wedge B') \vee (A' \wedge C)) = A \wedge A'$$

for all possible combinations of 0, $\frac{1}{2}$, and 1 substituted in for A, B, and C, so in this case there are 27 checks that need to be made. One such is for the case $A = \frac{1}{2}$, $B = 0$, and $C = \frac{1}{2}$. In that case, the right-hand side has value $\frac{1}{2}$ clearly, and the left side is

$$A \wedge ((A' \wedge B) \vee (A' \wedge B') \vee (A' \wedge C))$$
$$= \frac{1}{2} \wedge \left(\left(\frac{1}{2} \wedge 0\right) \vee \left(\frac{1}{2} \wedge 1\right) \vee \left(\frac{1}{2} \wedge \frac{1}{2}\right)\right)$$
$$= \frac{1}{2} \wedge \left(0 \vee \frac{1}{2} \vee \frac{1}{2}\right)$$
$$= \frac{1}{2}$$

The other 26 cases also yield equalities, so the expression is an identity for fuzzy sets. No matter what the fuzzy sets A, B, and C are, for every $x \in [0,1]$, the equality holds. It should be noted that checking equality for combinations of the values $\{0, \frac{1}{2}, 1\}$ is the same as checking equality when the values $\{0, u, 1\}$ are substituted into the expressions involved.

4.5   Interval valued fuzzy logic

A fuzzy subset of a set S is a mapping $A : U \rightarrow [0,1]$. The value $A(u)$ for a particular u is typically associated with a degree of belief of some expert. An increasingly prevalent view is that this method of encoding information is inadequate. Assigning an exact number to an expert's opinion is too restrictive. Assigning an interval of values is more realistic. This means replacing the interval $[0,1]$ of fuzzy values by the set $\{(a,b) : a, b \in [0,1], a \leq b\}$. A standard notation for this set is $[0,1]^{[2]}$. An expert's degree of belief for a particular element $u \in U$ will be associated with a pair $(a,b) \in [0,1]^{[2]}$. Now we can construct the propositional calculus whose truth values are the elements of $[0,1]^{[2]}$. But first we need the appropriate algebra of these truth values. It is given by the formulas

$$\begin{aligned}
(a,b) \vee (c,d) &= (a \vee c, b \vee d) \\
(a,b) \wedge (c,d) &= (a \wedge c, b \wedge d) \\
(a,b)' &= (b', a')
\end{aligned}$$

where the operations $\vee, \wedge$, and $'$ on elements of $[0,1]$ are the usual ones, commonly referred to in logic as the disjunction ($\vee$), conjunction ($\wedge$), and negation.

The resulting propositional calculus $\mathbb{F}/\equiv$ is a De Morgan algebra, and is not a Kleene algebra. The algebras $([0,1], \vee, \wedge, ')$ and $([0,1]^{[2]}, \vee, \wedge, ')$

are fundamentally different. In the first, the inequality $x \wedge x' \leq y \vee y'$ holds, but not in the second. That is, the first is a Kleene algebra and the second is not. However, it turns out that the propositional calculus is the same as the propositional calculus $\mathbb{F}/\equiv$ whose truth values are the elements of the lattice $\{0, u, v, 1\}$ with the order indicated in the diagram below, and with the negation which fixes u and v and interchanges 0 and 1. The details may be found in [34].

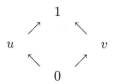

Again, as in the ordinary fuzzy case where values are in the unit interval, this implies that there is an algorithm for checking the equality of two expressions. It just comes down to checking the equality of the two expressions for all possible combinations of the values $\{0, u, v, 1\}$ substituted in for the interval-valued fuzzy sets involved. So there would be 4^n checks to be made, where n is the number of fuzzy sets involved.

In this particular case, there has been a canonical form developed. That is, any expression in these interval-valued fuzzy sets, using the connectives we have indicated, can be put into a form such that two expressions represent the same interval-valued fuzzy set if and only if their forms are equal. (See [75] for details.)

The result in Theorem 4.4.1 that fuzzy propositional calculus is the same as that of Lukasiewicz three-valued logic may seem paradoxical, as may also the fact that interval-valued fuzzy propositional calculus is the same as the propositional calculus with truth values pictured above. But these do not follow from the structure of fuzzy sets themselves, or interval-valued fuzzy sets, but from the algebraic structure on the set of truth values involved. The results have to do with the logical equivalence of formulas in the set $\mathbb{F}$. All this is determined by the algebra of truth values. The set of variables, which in the case of fuzzy propositional calculus we imagine to be fuzzy subsets of some set, is totally irrelevant.

One point should be totally clear. The fact that fuzzy propositional calculus is the same as Lukasiewicz's three-valued propositional calculus says nothing about the efficacy of using fuzzy sets to model physical phenomena, and of combining these fuzzy sets in the usual ways and to reach decisions based on the results of such combinations. It does **not** say that one may as well replace fuzzy sets by mappings into the three-element chain $\{0 < u < 1\}$. Similar remarks apply to interval-valued fuzzy sets.

Knowing fuzzy propositional calculus is the same as Lukasiewicz's

does give a way to test in finitely many steps the logical equivalence of two formulas in fuzzy sets. In a finite-valued logic, logical equivalence can be checked either by testing with truth values, or by comparing canonical forms, if such exist. We discuss canonical forms for these propositional calculi in the next section.

4.6 Canonical forms

As noted in Section 4.1, in classical two-valued propositional calculus, every formula, that is, every Boolean expression such as $a \wedge (b \vee c) \wedge d'$ has a canonical form, the well-known **disjunctive normal form**. For example, the disjunctive normal form for $(a \vee b) \wedge c'$ in the logic on the variables $\{a, b, c\}$ is

$$(a \wedge b \wedge c') \vee (a \wedge b' \wedge c') \vee (a' \wedge b \wedge c')$$

and that of $(a \wedge c') \vee (b \wedge c')$ is the same form exactly. Of course, we could have just used the distributive law and noted equality, but that is not the point here. In this classical case, two formulas can be checked for logical equivalence by putting them in their canonical forms and noting whether or not the two forms are identical. Alternately, one can check logical equivalence by checking equality for all truth evaluations of the two expressions. Since the set $\{0, 1\}$ of truth values is finite, this is a finite procedure.

Now for Lukasiewicz's three-valued logic, which is equal to fuzzy propositional calculus, two formulas may be similarly tested for logical equivalence, that is, by checking equality of all truth evaluations. Two formulas in fuzzy propositional calculus are logically equivalent if and only if they are logically equivalent in Lukasiewicz's three-valued propositional calculus. That is the implication of Theorem 4.4.1. Similar remarks hold for interval-valued fuzzy logic and the four-valued one discussed in Section 4.5. In this latter case, a normal form is well-known and is described in detail in the paper [73]. There is also a normal form for Lukasiewicz's three-valued logic [34]. We now discuss these three normal forms, that is, for the propositional calculi $\mathbb{F}/ \equiv$ for the two-valued case, the three-valued Lukasiewicz case, and for the four-valued case above. We refer to these three cases as $\mathbb{F}_1, \mathbb{F}_2$, and $\mathbb{F}_3$. The cases differ in that the first is a Boolean algebra, the second is a Kleene algebra, and the third is a De Morgan algebra. In each case we have variables, which we denote $x_1, x_2, x_3, ...$, binary operations $\vee$ and $\wedge$, a unary operation $'$, and the constants 0 and 1.

Variables $x_1, x_2, x_3, ...$ and their negations $x'_1, x'_2, x'_3, ...$ are called **literals**. An element a in a lattice is **join irreducible** if it cannot be written as the join $b \vee c$ with $b < a$ and $c < a$. For example, in the lattice $[0,1]$, every element is join irreducible, while in $[0,1] \times [0,1]$ there are lots of join reducible elements: $(a,a) = (a,0) \vee (0,a)$. The fact that is of fundamental use in finding all of these normal forms is that in each of the cases $\mathbb{F}_1, \mathbb{F}_2$, and $\mathbb{F}_3$ the algebra generated by a finite set $x_1, x_2, \ldots, x_n$ of variables is a bounded distributive lattice generated by the literals x_1, $x_2, \ldots, x_n, x_1', x_2', \ldots, x_n'$. It is well known that each element of a finite distributive lattice is the join of all the join irreducibles below it. Getting a normal form boils down to determining the join irreducibles and the ordering between them. The normal form for De Morgan algebras stems from realizing that all conjunctions of literals as well as 1, are join irreducible. The normal form for Boolean algebras stems from realizing that the only join irreducible elements in the Boolean case are the complete conjunctions of literals. Here a **complete conjunction of literals** is a conjunction of literals in which each variable occurs exactly once. For example, if the variables are x_1, x_2, x_3, then $x_1 \wedge x_2 \wedge x_3$ and $x_1 \wedge x_2' \wedge x_3'$ are complete disjunctions, while $x_1 \wedge x_2'$ and $x_2 \wedge x_2'$ are not. The empty disjunction is 0, and the disjunction of all the complete conjunctions is 1.

The join irreducibles in the Kleene case are a little more subtle. If the variables are $x_1, x_2, \ldots, x_n$, then a conjunction of literals is join irreducible if and only if it is 1, or it contains at most one of the literals for each variable, or it contains at least one of the literals for each variable. Suppose $n = 3$. Here are some examples.

1. $x_1 \wedge x_2 \wedge x_3$ is join irreducible. It contains at least one of the literals for each variable. (It also contains at most one of the literals for each variable, so qualifies on two counts.)

2. $x_1 \wedge x_2 \wedge x_3'$ is join irreducible for the same reasons as above.

3. $x_1 \wedge x_2 \wedge x_2'$ is not join irreducible. It does not contain at least one of the literals for each variable, and it contains two literals for the variable x_2.

4. $x_1 \wedge x_1' \wedge x_2 \wedge x_3$ is join irreducible. It contains at least one of the literals for each variable.

5. $x_1 \wedge x_1' \wedge x_2 \wedge x_2'$ is not join irreducible. It does not contain at least one of the literals for each variable, and it contains two literals for two variables.

6. $x_1 \wedge x_2$ is join irreducible. It contains at most one of the literals for each variable.

7. x_3 is join irreducible. It contains at most one of the literals for each variable.

In each of the three cases, Boolean, Kleene, and De Morgan, we need to know the ordering between the join irreducibles. In the Boolean case, no two join irreducibles are comparable. In the other two cases, join irreducibles x and y satisfy $x \leq y$ if $y = 1$ or if every literal in y appears in x. For example, $x_1 \wedge x_2 \wedge x_3' \leq x_1 \wedge x_2$.

The normal form for the Boolean algebra case, that is, for $\mathbb{F}_1$, is of course well-known: every element is uniquely a disjunction of complete conjunctions of literals. Instead of getting into this, we will describe the procedure for putting an arbitrary formula in Kleene normal form. In the examples illustrating the steps, we assume that there are three variables, x_1, x_2, x_3.

1. Given a formula w, first use De Morgan's laws to move all the negations in, so that the formula is rewritten as a formula w_1 which is just meets and joins of the literals, 0, and 1. For example, $x_1 \wedge (x_2' \wedge x_3)'$ would be replaced by $x_1 \wedge (x_2 \vee x_3')$.

2. Next use the distributive law to obtain a new formula w_2 from w_1 which is a disjunction of conjunctions involving the literals, 0, and 1. For example, replace $x_1 \wedge (x_2 \vee x_3')$ by $(x_1 \wedge x_2) \vee (x_1 \wedge x_3')$. At this point, discard any conjunction in which 0 or $1'$ appears as one of the conjuncts. Also discard any repetition of literals from any conjunction, as well as 1 and $0'$ from any conjunction in which they do not appear alone (if a conjunction consists entirely of 1's and $0'$'s, then replace the whole thing by 1). This yields a formula w_3.

3. Now discard all non-maximal conjunctions among the conjunctions that w_3 is a disjunction of. The type of conjunctions we now are dealing with are either conjunctions of literals or 1 by itself. Of course 1 is above all the others and one conjunction of literals is below another if and only if the former contains all the literals contained in the latter. This process yields a formula w_4.

4. At this point, replace any conjunction of literals, c, which contains both literals for at least one variable by the disjunction of all the conjunctions of literals below c that contain exactly one of the literals for each variable not occurring in c. For example, if one of the conjunctions is $x_1 \wedge x_1' \wedge x_2$, replace it by the disjunction $(x_1 \wedge x_1' \wedge x_2 \wedge x_3) \vee (x_1 \wedge x_1' \wedge x_2 \wedge x_3')$. ($x_3$ is the only variable not occurring in $x_1 \wedge x_1' \wedge x_2$.)

5. Finally, again discard all non-maximal conjunctions among the conjunctions that are left, and if no conjunctions are left, then replace the formula by 0. The formula thus obtained is now in the normal form described above.

We illustrate the Kleene normal form with the two equivalent expressions

$$w = A \wedge ((A' \wedge B) \vee (A' \wedge B') \vee (A' \wedge C))$$
$$w' = A \wedge A'$$

in the variables, A, B, and C.

1. There is nothing to do in this step.

2. Applications of the distributive law lead to disjunctions of conjunctions involving the literals:

$$w_2 = (A \wedge A' \wedge B) \vee (A \wedge A' \wedge B') \vee (A \wedge A' \wedge C)$$
$$w'_2 = A \wedge A'$$

3. Neither of the expressions in #2 contains any non-maximal conjunctions, so $w_3 = w_2$ and $w'_3 = w'_2$.

4. Replace

$$A \wedge A' \wedge B \quad \text{by} \quad (A \wedge A' \wedge B \wedge C) \vee (A \wedge A' \wedge B \wedge C')$$
$$A \wedge A' \wedge B' \quad \text{by} \quad (A \wedge A' \wedge B' \wedge C) \vee (A \wedge A' \wedge B' \wedge C')$$
$$A \wedge A' \wedge C \quad \text{by} \quad (A \wedge A' \wedge C \wedge B) \vee (A \wedge A' \wedge C \wedge B')$$

and

$$A \wedge A' \quad \text{by} \quad (A \wedge A' \wedge B \wedge C) \vee (A \wedge A' \wedge B' \wedge C)$$
$$\vee (A \wedge A' \wedge B \wedge C') \vee (A \wedge A' \wedge B' \wedge C')$$

to get

$$w_4 = (A \wedge A' \wedge B \wedge C) \vee (A \wedge A' \wedge B \wedge C')$$
$$\vee (A \wedge A' \wedge B' \wedge C) \vee (A \wedge A' \wedge B' \wedge C')$$
$$\vee (A \wedge A' \wedge C \wedge B) \vee (A \wedge A' \wedge C \wedge B')$$

$$w'_4 = (A \wedge A' \wedge B \wedge C) \vee (A \wedge A' \wedge B' \wedge C)$$
$$\vee (A \wedge A' \wedge B \wedge C') \vee (A \wedge A' \wedge B' \wedge C')$$

5. Discarding all non-maximal conjunctions among the conjunctions that are left means in this case, simply discarding repetitions, leading to the normal forms

$$w_5 = (A \wedge \neg A \wedge B \wedge C) \vee (A \wedge \neg A \wedge B \wedge C')$$
$$\vee (A \wedge \neg A \wedge B' \wedge C) \vee (A \wedge \neg A \wedge B' \wedge C')$$

$$w'_5 = (A \wedge \neg A \wedge B \wedge C) \vee (A \wedge \neg A \wedge B' \wedge C)$$
$$\vee (A \wedge \neg A \wedge B \wedge C') \vee (A \wedge \neg A \wedge B' \wedge C')$$

The procedure for getting the normal form in the De Morgan case is quite similar. The only real difference is that in the De Morgan case all conjunctions of literals are join irreducible, while this is not the case for Kleene.

4.7   Notes on probabilistic logic

In this section, we mention briefly probabilistic logic, pointing out similarities and differences with fuzzy logic. We wish to represent and reason with information of the following sort.

1. It is likely that John's height is over $5'10''$.

2. John's height is probably between $5'10''$ and $6'$.

3. Most birds fly.

4. Usually mathematicians know all fields of mathematics.

5. Few men are heroes.

6. If the patient has symptom a, then the patient likely has disease b.

The information presented is uncertain. The statements contain quantifiers, and these quantifiers are not the usual ones in first order logic such as $\exists$ and $\forall$. It contains other quantifiers such as "few", "most", and "usually". So extending propositional logic to first order logic will not suffice. But knowledge of the type above is very common and must be modeled if we wish to make machines mimic human behavior and reasoning. So we need a method to allow us to manipulate and to make inferences from such knowledge. The essential added tool used here is probability theory.

A typical example where propositions arise such as those above is in medical diagnosis. For example, a fact such as "this patient has disease b" might be uncertain for the physician since what he can observe is not the disease itself but symptoms. The relation between symptoms and diseases, usually expressed as conditional propositions of the form "if symptom a then disease b" is also uncertain. However, a physician or other "expert" can assign his "degree of belief" in the truth of such a proposition. The subjectivity of this assignment is inevitable.

Probabilistic logic is an approach to reasoning with this type of uncertain information. Viewing degrees of belief in the truth of propositions as subjective probabilities, one can think about using the standard calculus of probabilities to implement a "logic" of uncertain information.

The possibility of assigning probabilities to propositions is due to the fact that the set of all equivalence classes of propositions, which we have denoted $\mathbb{F}/\equiv$ does form a Boolean algebra, so that it is a suitable domain for probability measures. So on a suitable Boolean algebra of propositions, such as one containing those above with the quantifiers removed, we postulate a probability measure. For example, it makes some sense to assign the proposition "birds fly" the probability 0.90, and the proposition "men are heroes" the probability 0.01. The assignment of probabilities is of course subjective and can come from many sources: experts' opinions, common sense, and so on. (In this regard, see [68].) So the quantifiers are manifested in probability assignments.

We will not go into the technical details on implementation. Our purpose is to spell out some relevant points which are somewhat similar to fuzzy logic. This new mathematical tool has emerged from the requirements of practical applications, for example in decision support systems in which causal relationships are uncertain. Given a collection of facts and rules forming a knowledge domain for a problem, that is, propositions of the form

$$a: \quad \text{the patient has property } a$$
$$a \Rightarrow b: \quad \text{if } a \text{ then } b$$

one quantifies the degrees of belief by probabilities $P(a)$ and $P(a \Rightarrow b)$. The implication $a \Rightarrow b$ should not be interpreted as classical two-valued material implication. In assigning the value $P(a \Rightarrow b)$ to the rule "if a then b", the expert will assess the chance for b to occur under the condition a. Thus $P(a \Rightarrow b) = P(a|b)$, the conditional probability of b under a. But $a \Rightarrow b$ is not $a' \vee b$ since $P(a' \vee b) \neq P(a|b) = P(ab)/P(a)$ in general. (For the modeling of the implication operator in probabilistic inference, see Chapter 10 and the exercises at the end of that chapter.)

Essentially the strategy of probabilistic reasoning is this. Given a knowledge base consisting of rules and facts together with their probabilities, one proceeds to construct a joint probability measure on a suitable set of relevant events (a Boolean algebra of propositions) which will allow the computations of probabilities of events of interest. Since probabilities take values in the unit interval $[0, 1]$, probabilistic logic is multivalued. Its syntax is the same as for classical two-valued logic. For each proposition a, there are two sets of possible worlds, those in which a is true and those in which a is false. Not knowing the actual world, one has to consider the probability of a being true as a truth value of a. This is obviously a generalization of classical logic. In view of the axioms of probability measures, probabilities of combined or compound propositions cannot always be determined from those of component propositions, as opposed to the case of fuzzy logic. This is expressed by saying that this logical system is *non-truth functional*. For more detail on probability logic, especially on basic concepts such as probabilistic entailment, see [79].

4.8   Exercises

1. Write down the tables for $\Rightarrow$ and $\Leftrightarrow$ for classical two-valued propositional logic.

2. In two-valued propositional calculus, verify that two propositions a and b are logically equivalent if and only if $a \Leftrightarrow b$ is a tautology.

3. We write $a = b$ for $a \Leftrightarrow b$. Verify the following for two-valued propositional calculus.

   (a) $a'' = a$

   (b) $a \vee a' = 1$

   (c) $a \wedge a' = 0$

   (d) $a = a \vee a$

   (e) $a \vee b = b \vee a$

   (f) $a \wedge b = b \wedge a$

   (g) $a \vee (b \vee c) = (a \vee b) \vee c$

   (h) $a \wedge (b \wedge c) = (a \wedge b) \wedge c$

   (i) $a \wedge (b \vee c) = (a \wedge b) \vee (a \wedge c)$

   (j) $a \vee (b \wedge c) = (a \vee b) \wedge (a \vee c)$

   (k) $(a \vee b)' = a' \wedge b'$

   (l) $(a \wedge b)' = a' \vee b'$

4. In two-valued propositional calculus, show that $a \Rightarrow b$ is a tautology if and only if for every truth valuation t, $\widetilde{t}(a) = 1$ implies that $\widetilde{t}(b) = 1$.

5. In Lukasiewicz's three-valued logic, verify that $\wedge$ and $\vee$ agree with $\wedge$ and $\vee$ on $\{0, u, 1\}$ considered as a chain with $0 < u < 1$.

6. In Lukasiewicz's three-valued logic, show that a and b are logically equivalent if and only if $a \Leftrightarrow b$ is a tautology.

7. Show that the set $\mathbb{F}/\equiv$ of equivalence classes of formulas in the three-valued Lukasiewicz logic forms a Kleene algebra under the operations

$$
\begin{aligned}
[a] \vee [b] &= [a \vee b] \\
[a] \wedge [b] &= [a \wedge b] \\
[a]' &= [a']
\end{aligned}
$$

8. Show that the set $\mathbb{F}/\equiv$ of equivalence classes of formulas in the interval-valued fuzzy logic forms a De Morgan algebra under the operations

$$
\begin{aligned}
[a] \vee [b] &= [a \vee b] \\
[a] \wedge [b] &= [a \wedge b] \\
[a]' &= [a']
\end{aligned}
$$

Show that this De Morgan algebra does not satisfy the inequality $x \wedge x' \leq y \vee y'$.

9. In Bochvar's three-valued logic, $\Leftrightarrow$ is defined by

$\Leftrightarrow$	0	u	1
0	1	u	0
u	u	u	u
1	0	u	1

Verify that a and b being logically equivalent does not imply that $a \Leftrightarrow b$ is a three-valued tautology.

10. Show if $u \vee u = u$ is changed to $u \vee u = 1$ in the table for $\vee$ in Lukasiewicz's three-valued logic, then the law of the excluded middle holds.

11. Let a be a formula in fuzzy logic. Show that if $t(a \vee a') = 1$, then necessarily $t(a) \in \{0, 1\}$.

12. Show that $\{0, u, 1\}$ with $0 < u < 1$ is a Kleene algebra. For any set S, show that $\{0, u, 1\}^S$ is a Kleene algebra.

13. Show that in the algebra $([0, 1], \vee, \wedge,', 0, 1)$ the inequality $x \wedge x' \leq y \vee y'$ holds for all x and y in $[0, 1]$. Show that this inequality does not hold in $([0, 1]^{[2]}, \vee, \wedge,', 0, 1)$.

14. Show that

$$
A \wedge ((A' \wedge B) \vee (A' \wedge B') \vee (A' \wedge C)) = A \wedge A'
$$

is false for fuzzy sets taking values in $[0, 1]^{[2]}$.

15. In the three variables A, B, C, find the disjunctive normal form, the Kleene normal form, and the De Morgan normal form for

(a) $A \vee (A' \wedge B \wedge B')$

(b) $A \wedge (B \vee C)'$

(c) $A \wedge A'$

16. *An elementary polynomial in n variables $x_1, x_2, \cdots, x_n$ is an expression of the form $y_1 \wedge y_2 \wedge \cdots \wedge y_n$, where $y_i = x_i$ or the symbol x_i'. A Boolean polynomial in the n variables $x_1, x_2, \cdots, x_n$ is an expression of the form $E_1 \vee E_2 \cdots \vee E_m$ where the E_i's are distinct elementary Boolean polynomials.

   (a) Show that there are 2^{2^n} Boolean polynomials in n variables. (Assume that $E_1 \vee E_2 = E_2 \vee E_1$, etc.)

   (b) If f is a Boolean polynomial in n variables, then f induces a map $\{0,1\}^n \to \{0,1\}$ which we also call f. Spell out exactly what this map is. Show that every map $\{0,1\}^n \to \{0,1\}$ is induced by a Boolean polynomial. Show that two Boolean polynomials are equal if and only if they induce the same map $\{0,1\}^n \to \{0,1\}$. (Such a map is a "truth function".)

   (c) Let $\mathcal{A}$ be a Boolean algebra. A Boolean polynomial in n variables induces a map $\mathcal{A}^n \to \mathcal{A}$. Such a map is called a **Boolean function**. Spell out exactly what this map is. Show that two Boolean polynomials are equal if and only if they induce the same map $\mathcal{A}^n \to \mathcal{A}$. Show that there is a one-to-one correspondence between truth functions $\{0,1\}^n \to \{0,1\}$ and Boolean functions $\mathcal{A}^n \to \mathcal{A}$.

17. *Let $(\Omega, \mathcal{A}, P)$ be a probability space. For $a, b \in \mathcal{A}$, define a function $(a|b) : \Omega \to \{0, u, 1\}$ by

$$(a|b)(\omega) = \begin{cases} 0 & \text{if} \quad \omega \in a' \cap b \\ u & \text{if} \quad \omega \in b' \\ 1 & \text{if} \quad \omega \in a \cap b \end{cases}$$

Define the operations $\vee, \wedge$, and $'$ on $\{0, u, 1\}$ as in the truth tables for Lukasiewicz's three-valued logic. Define

$$\begin{array}{rcl} ((a|b) \vee (c|d))(\omega) & = & (a|b)(\omega) \vee (c|d)(\omega) \\ ((a|b) \wedge (c|d))(\omega) & = & (a|b)(\omega) \wedge (c|d)(\omega) \\ (a|b)'(\omega) & = & (a'|b)(\omega) \end{array}$$

   (a) Show that $(a|b) \to [a \cap b, b' \cup a]$ is a one-to-one correspondence between the set $\{(a|b) : a, b \in \mathcal{A}\}$ of functions and the set $\mathcal{I}$ $\{[a \cap b, b' \cup a] : a, b \in \mathcal{A}\}$ of intervals. (An interval $[a, b]$ is the set $\{x \in \mathcal{A} : a \leq x \leq b\}$.)

(b) On $\mathcal{I}$, define

   i. $[a,b] \vee [c,d] = [a \cup c, b \cup d]$

  ii. $[a,b] \wedge [c,d] = [a \cap c, b \cap d]$

 iii. $[a,b]^* = [b',b']$

Show that $\mathcal{I}$ is not complemented and that the truth tables of $\vee$ and $\wedge$ are precisely those of Lukasiewicz three-valued ones. Also show that if $[a \cap b, b' \cup a] = [c \cap d, d' \cup c]$, then $P(a|b) = P(c|d)$.

Chapter 5

BASIC CONNECTIVES

Consider a piece of information of the form "If (x is A and y is not B), then (z is C or z is D)". An approach to the translation of this type of knowledge is to model it as fuzzy sets. To translate completely the sentence above, we need to model the connectives "and", "or", and "not", as well as the conditional "If...then...". This combining of evidence, or "data fusion", is essential in building expert systems or in synthesizing controllers. But the connectives experts use are domain dependent— they vary from field to field. The connectives used in data fusion in medical science are different from those in geophysics. So there are many ways to model these connectives. The search for appropriate models for "and" has led to a class of connectives called "t-norms". Similarly, for modeling "or" there is a class called "t-conorms". In this chapter we will investigate ways for modeling basic connectives used in combining knowledge that comes in the form of fuzzy sets. These models may be viewed as extensions of the analogous connectives in classical two-valued logic. A model is obtained for each choice of such extensions, and one concern is with isomorphisms between the algebraic systems that arise.

5.1   t-norms

Consider first the connective "and". When A and B are ordinary subsets of a set U, then the table

A B	0	1
0	0	0
1	0	1

83

gives the truth evaluation of "A and B" in terms of the possible truth values 0 and 1 of A and B. The table just specifies a map $\wedge : \{0,1\} \times \{0,1\} \rightarrow \{0,1\}$. When A and B are fuzzy subsets of U, truth values are the members of the interval $[0,1]$, and we need to extend this map to a map $\wedge : [0,1] \times [0,1] \rightarrow [0,1]$. One such extension is given by $x \wedge y = \min\{x,y\}$. This mapping does agree with the table above when x and y belong to $\{0,1\}$. We make the following observations about $x \wedge y = \min\{x,y\}$.

- 1 acts as an identity. That is, $1 \wedge x = x$.

- $\wedge$ is commutative. That is, $x \wedge y = y \wedge x$.

- $\wedge$ is associative. That is, $x \wedge (y \wedge z) = (x \wedge y) \wedge z$.

- $\wedge$ is increasing in each argument. That is, if $v \leq w$ and $x \leq y$ then $v \wedge x \leq w \wedge y$.

Any binary operation

$$\triangle \colon [0,1] \times [0,1] \rightarrow [0,1]$$

satisfying these properties is a candidate for modeling the connective "and" in the fuzzy setting. Note that $\wedge$ is idempotent also, that is, $x \wedge x = x$, but we do not require this for modeling this connective. It turns out that such operations have already appeared in the theory of probabilistic metric spaces where they were related to the problem of extending geometric triangular inequalities to the probabilistic setting [91, 92]. They were termed "triangular norms", or "t-norms" for short. We will use these t-norms as a family of possible connectives for fuzzy intersection. Now, t-norms are binary operations on $[0,1]$ and a common practice is to denote them by T, and write $T(x,y)$. However, this notation can become awkward, especially in expressions such as $T(v, T(T(w,x), y))$. Also, in algebra, it is customary to write binary operations between their arguments, such as $x + y$, $x \wedge y$, and so on. So we are going to write t-norms as binary operations are generally written, between their arguments. One notation we will use for a t-norm will be $\triangle$, and we write $x \triangle y$. The expression $T(v, T(T(w,x), y))$, using the associative law, becomes simply

$$v \triangle ((w \triangle x) \triangle y) = v \triangle w \triangle x \triangle y.$$

Here is the formal definition.

Definition 5.1.1 *A binary operation* $\triangle : [0,1] \times [0,1] \rightarrow [0,1]$ *is a* **t-norm** *if it satisfies the following:*

1. $1 \triangle x = x$

2. $x \triangle y = y \triangle x$

3. $x \triangle (y \triangle z) = (x \triangle y) \triangle z$

4. If $w \le x$ and $y \le z$ then $w \triangle y \le x \triangle z$

The first, second, and fourth conditions give $0 \triangle x \le 0 \triangle 1 = 0$, and the associative law gives unambiguous meaning to expressions such as $u \triangle v \triangle w \triangle x \triangle y \triangle z$. Some examples follow.

- $x \triangle_0 y = \begin{cases} x \wedge y & \text{if } x \vee y = 1 \\ 0 & \text{otherwise} \end{cases}$

- $x \triangle_1 y = 0 \vee (x + y - 1)$

- $x \triangle_2 y = \frac{xy}{2 - (x + y - xy)}$

- $x \triangle_3 y = xy$

- $x \triangle_4 y = \frac{xy}{x + y - xy}$

- $x \triangle_5 y = x \wedge y$

The t-norm $\triangle_0$ is not continuous. The t-norms $\triangle_0$ and $\triangle_5$ are extremes.

Proposition 5.1.2 *If $\triangle$ is a t-norm, then for $x, y \in [0, 1]$,*

$$x \triangle_0 y \le x \triangle y \le x \triangle_5 y$$

The proof is easy and is an exercise at the end of this chapter.

The t-norm $\wedge$ is the only **idempotent** one, that is, the only t-norm $\triangle$ such that $x \triangle x = x$ for all x. For any t-norm $\triangle$, $x \triangle x$ is never greater than x. These facts are exercises. So for a t-norm $\triangle \ne \wedge$ there will be an element x such that $x \triangle x < x$.

Definition 5.1.3 *A t-norm $\triangle$ is **convex** if whenever $x \triangle y \le c \le x_1 \triangle y_1$, then there is an r between x and x_1 and an s between y and y_1 such that $c = r \triangle s$.*

For *t*-norms, the condition of convexity is equivalent to continuity. We refer to the condition as convex. This formulation has the advantage of being strictly order theoretic, allowing us to remain within the algebraic context of $\mathbb{I}$ as a lattice.

Corollary 5.1.4 *If a t-norm $\triangle$ is convex and $a < b$, then there is $c \in [a, 1]$ such that $a = b \triangle c$.*

Proof. $a \triangle b \leq 1 \triangle a = a < 1 \triangle b$, so by convexity, there is such a c.

■

Proposition 5.1.5 *A t-norm is convex if and only if it is continuous in each variable.*

Proof. If the t-norm is continuous in each variable, then it is convex, using the intermediate value theorem. Assume $\triangle$ is convex. Since $\triangle$ is monotone increasing in each variable, any point of discontinuity of $f(y) = x \triangle y$ is a jump. But convexity precludes this, and so $\triangle$ is continuous in each variable. ■

If instead of $[0, 1]$, we used a lattice, or even just a partially ordered set $\mathbb{L}$, then convexity makes sense, while there may be no notion of continuity on $\mathbb{L}$. Thus convexity, while equivalent to continuity on $[0, 1]$, applies to a much wider class of objects which might be used as fuzzy values.

Definition 5.1.6 *A t-norm $\triangle$ is **Archimedean** if it is convex, and for each $a, b \in (0, 1)$, there is a positive integer n such that*

$$a^{[n]} = \overbrace{a \triangle a \triangle \cdots \triangle a}^{n \ times} < b$$

In general we will write $a \triangle a = a^{[2]}$, $a \triangle a \triangle a = a^{[3]}$, and so on. We use $a^{[n]}$ instead of a^n for this t-norm power to distinguish it from a multiplied by itself n times.

The examples $\triangle_1, \triangle_2, \triangle_3$, and $\triangle_4$ are all Archimedean. For convex t-norms, the condition for Archimedean simplifies, as the corollary to the following proposition attests.

Proposition 5.1.7 *If a t-norm $\triangle$ is **Archimedean**, then for $a, b \in (0, 1)$, $a \triangle b < b$.*

Proof. If $\triangle$ is **Archimedean**, then for $a \in (0, 1)$, clearly $a \triangle a < a$ lest $a^{[n]} = a$ for all n. If $a < b$, then $a \triangle b \leq b \triangle b < b$. If $a > b$, then $a \triangle b \leq 1 \triangle b = b$. If $a \triangle b = b$, then $a^{[n]} \triangle b = b$ for all n, but for sufficiently large n, $a^{[n]} \leq b$, and $b = a^{[n]} \triangle b \leq b \triangle b$, an impossibility. ■

Corollary 5.1.8 *The following are equivalent for a convex t-norm $\triangle$.*

   1. $\triangle$ is Archimedean.

   2. $a \triangle a < a$ for all $a \in (0, 1)$.

Proof. Archimedean clearly implies the second condition. Assume that $a \triangle a < a$ for all $a \in (0, 1)$, and let $b \in (0, 1)$. Then $\bigwedge_n a^{[n+1]} = \bigwedge_n a^{[n]} = a \triangle \bigwedge_n a^{[n]}$, whence $\bigwedge_n a^{[n]} = 0$, and the corollary follows. ■

5.2 Generators of t-norms

Different situations demand different t-norms, and it is of some importance to be able to construct them. There is an easy way, at least for Archimedean t-norms. Let $0 \leq a < 1$ and let f be any **order isomorphism** from $[0, 1]$ to $[a, 1]$. This means that f is one-to-one and onto and $x \leq y$ if and only if $f(x) \leq f(y)$. They will be referred to simply as **isomorphisms** from $[0, 1]$ to $[a, 1]$. If $a = 0$, they are **automorphisms** of $[0, 1]$, that is elements of $Aut(\mathbb{I})$. Below is a picture of an order isomorphism f.

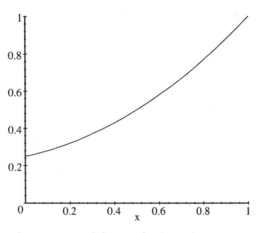

A generator f for an Archimedean t-norm

For an order isomorphism $f : [0, 1] \rightarrow [a, 1]$, and an Archimedean t-norm $\triangle$, define $\triangle_f$ by

$$x \triangle_f y = f^{-1}\left(\left(f(x) \triangle f(y)\right) \vee a\right)$$

Theorem 5.2.1 $\triangle_f$ *is an Archimedean t-norm.*

Proof. The proof is routine in all respects. From the formula, it is clear that $\triangle_f$ is commutative. Since f is increasing, so is $\triangle_f$, and

$$
\begin{aligned}
1 \triangle_f x &= f^{-1}(f(1) \triangle f(x) \vee a) \\
&= f^{-1}(f(x) \vee a) \\
&= f^{-1}(f(x)) \\
&= x
\end{aligned}
$$

For associativity, we need

$$x \triangle_f (y \triangle_f z) = (x \triangle_f y) \triangle_f z.$$

Using the fact that $a \leq f(x)$ for any x,

$$
\begin{aligned}
x \bigtriangleup_f (y \bigtriangleup_f z) &= x \bigtriangleup_f (f^{-1}(f(y) \bigtriangleup f(z) \vee a)) \\
&= f^{-1}(f(x) \bigtriangleup (f(f^{-1}(f(y) \bigtriangleup f(z) \vee a))) \vee a) \\
&= f^{-1}(f(x) \bigtriangleup ((f(y) \bigtriangleup f(z) \vee a)) \vee a) \\
&= f^{-1}(f(x) \bigtriangleup f(y) \bigtriangleup f(z) \vee a)
\end{aligned}
$$

and

$$
\begin{aligned}
(x \bigtriangleup_f y) \bigtriangleup_f z &= f^{-1}((f f^{-1}(f(x) \bigtriangleup f(y) \vee a)) \bigtriangleup f(z) \vee a) \\
&= f^{-1}((f(x) \bigtriangleup f(y) \vee a) \bigtriangleup f(z) \vee a) \\
&= f^{-1}(f(x) \bigtriangleup f(y) \bigtriangleup f(z) \vee a)
\end{aligned}
$$

■

Definition 5.2.2 *Let • denote the t-norm that is ordinary multiplication. If $\bigtriangleup = •_f$, then f is a **generator** of $\bigtriangleup$.*

The function f gives an Archimedean t-norm. The truly remarkable thing is that *every* Archimedean t-norm can be represented in this way. The proof of this fact is a bit technical, but perhaps worth including because of its fundamental nature. A principal reference is [59]. The theorem in essence goes back at least to [1]. There is a proof in [2] and some discussion in [92].

Theorem 5.2.3 *If $\bigtriangleup$ is an Archimedean t-norm then there is an $a \in [0,1)$ and an order isomorphism $f : [0,1] \to [a,1]$ such that*

$$
x \bigtriangleup y = f^{-1}(f(x)f(y) \vee a)
$$

for all $x, y \in [0,1]$. Also if g is an order isomorphism $[0,1] \to [b,1]$, then $x \bigtriangleup y = g^{-1}(g(x)g(y) \vee b)$ if and only if $f(x) = g(x)^r$ for some $r > 0$.

Proof. We outline the construction of a map f satisfying $f(x \bigtriangleup y) = f(x)f(y)$ for $x \bigtriangleup y \neq 0$. First we construct a sequence in $(0,1)$ and define the function on all finite products under $\bigtriangleup$ of the elements of this sequence. Then we extend this function to $[0,1]$ by continuity.

Let z be the largest element of $[0,1]$ satisfying $z \bigtriangleup z = 0$ and let $x_0 = \frac{1}{2}$. The function $x \mapsto x \bigtriangleup x$ is a one-to-one, onto, order preserving mapping from $[z,1]$ to $[0,1]$, so there is a (unique) number $x_1 \in [z,1]$ satisfying $x_1 \bigtriangleup x_1 = \frac{1}{2}$. Define inductively a sequence satisfying $x_n \bigtriangleup x_n = x_{n-1}$ for all positive integers n. This gets a strictly increasing sequence $\{x_i\}_{i=0}^{\infty}$ in the interval $[\frac{1}{2}, 1)$. The terms $\frac{1}{2} = x_0 < x_1 < x_2 < \dots$ form a bounded

infinite increasing sequence of points which must converge to some $y \leq 1$.
Since $x_n = x_{n+1} \vartriangle x_{n+1}$, this sequence also converges to $y \vartriangle y$. So
$y \vartriangle y = y$ and thus $y = 1$.

Extend this sequence by defining $x_n = x_{n+1} \vartriangle x_{n+1}$ for all negative
integers n. The sequence of points $x_0 \geq x_{-1} \geq x_{-2} \geq x_{-3} \geq \ldots$ converges
to some number $y \geq 0$ and since $x_{-(n+1)} = x_{-n} \vartriangle x_{-n}$, it also converges
to $y \vartriangle y$. Thus $y \vartriangle y = y$, whence $y = 0$. Note that the sequence
$x_0 \geq x_{-1} \geq x_{-2} \geq x_{-3} \geq \ldots$ may or may not be zero from some point
on, depending on the t-norm.

Now define a function f on the sequence $\{x_i : x_i \neq 0\}$ by

$$f(x_n) = \left(\frac{1}{2}\right)^{2^{-n}} \quad \text{if } x_n \neq 0$$

It is easy to check, using ordinary laws of exponents, that $f(x_n \vartriangle x_n) = f(x_n)f(x_n)$ for all integers n such that $x_n \vartriangle x_n \neq 0$.

Using induction, together with the property that the operation $\vartriangle$ is
strictly increasing in each variable, one can show that all nonzero products
(under $\vartriangle$) of x_i's can be written uniquely in the form

$$x_{i_1} \vartriangle x_{i_2} \vartriangle \cdots \vartriangle x_{i_n}$$

with $x_{i_1} < x_{i_2} < \cdots < x_{i_n}$. This property allows us to extend f to the
set of nonzero products of the x_i's by

$$f\left(x_{i_1} \vartriangle x_{i_2} \vartriangle \cdots \vartriangle x_{i_n}\right) = f(x_{i_1})f(x_{i_2}) \cdots f(x_{i_n})$$

The set of all points of the form $x_{i_1} \vartriangle x_{i_2} \vartriangle \cdots \vartriangle x_{i_n}$ is dense in
the unit interval. To see this, recall that we have already observed that
$\lim_{n \to -\infty} x_n = 0$ and $\lim_{n \to +\infty} x_n = 1$. For $a \in (0,1)$, we must have
$x_{n-1} < a < x_n$ for some integer $n = n_0$. If $a \neq x_{n_0}$ there is a smallest
positive integer n_1 such that $a < x_{n_0} \vartriangle x_{n_1}$. Continuing in this fashion
gives a sequence $x_{n_0} < x_{n_1} < \cdots < x_{n_i} < \ldots$ for which $\lim_{i \to +\infty} x_{n_0} \vartriangle$
$x_{n_1} \vartriangle \cdots \vartriangle x_{n_i} = a$. Now define

$$f(a) = \lim_{i \to +\infty} \prod_{k=0}^{i} f(x_{n_k})$$

To see that the function f maps the unit interval in the desired way
onto the interval $[f(0), 1]$, note that

$$\prod_{k=0}^{i} f(x_{n_k}) = \prod_{k=0}^{i} 2^{-2^{-n_k}} = 2^{-\sum_{k=0}^{i} 2^{-n_k}}$$

Since every real number z can be expressed in binary notation, that is,

$$z = \sum_{j=-\infty}^{m} a_j 2^j = \sum_{j=-m}^{\infty} a_j 2^{-j}$$

where $a_j \in \{0,1\}$, the set of numbers of the form $\sum_{k=0}^{i} 2^{-n_k}$ is dense in the set of real numbers. It is now easy to show that the function f is an isomorphism from $([0,1], \leq)$ to $([f(0), 1], \leq)$. Also $f(x \triangle y) = f(x) f(y)$ whenever $x \triangle y \neq 0$, so that $x \triangle y = f^{-1}(f(x) f(y))$ if $f(x) f(y) \geq f(0)$ and $x \triangle y = 0$ otherwise.

Suppose that a mapping $g : [0,1] \to [g(0), 1]$ gives the same t-norm as the f just constructed. Then for $r = \left(\ln \frac{1}{2} \right) / \left(\ln g \left(\frac{1}{2} \right) \right)$, $rg(\frac{1}{2}) = (g(\frac{1}{2}))^r = \frac{1}{2} = f(\frac{1}{2})$, and we see from the construction of f that $f(x_i) = g(x_i)^r$ and hence that $f(x) = g(x)^r$ for all $x \in [0,1]$. Conversely, it is easy to show that if $f(x) = g(x)^r$, then f and g give the same t-norm. ■

As mentioned, a function $f : [0,1] \to [a,1]$ such that

$$x \triangle y = f^{-1}(f(x) f(y) \vee f(0))$$

is called a **generator** of the t-norm $\triangle$. So every Archimedean t-norm has a generator, and we know when two generators f and g give the same t-norm, namely when $f(x) = g(x)^r$ for some $r > 0$. Now there are two kinds of generators: those $f \in Aut(\mathbb{I})$, that is, those f such that $f(0) = 0$, and those f such that $f(0) > 0$. Clearly if $f(x) = g(x)^r$ for some $r > 0$, then either both $f(0) = g(0) = 0$, or both $f(0)$ and $g(0)$ are positive. How are these two properties reflected in the behavior of the t-norms themselves?

Recall that $a \triangle a = a^{[2]}$, $a \triangle a \triangle a = a^{[3]}$, and so on.

Definition 5.2.4 *A t-norm $\triangle$ is **nilpotent** if for $a \neq 1$, $a^{[n]} = 0$ for some positive integer n, the n depending on a. A t-norm is **strict** if for $a \neq 0$, $a^{[n]} > 0$ for every positive integer n.*

It is easy to see that $a \triangle b = (a+b-1) \vee 0$ is nilpotent, and $a \triangle b = ab$ is strict.

Lemma 5.2.5 *Let f be a generator of the Archimedean t-norm $\triangle$. Then $a^{[n]} = f^{-1}(f(a)^n \vee f(0))$.*

The proof is left as an exercise. An immediate consequence is the following.

Corollary 5.2.6 *Let f be a generator of the Archimedean t-norm $\triangle$. Then $\triangle$ is nilpotent if and only if $f(0) > 0$, and $\triangle$ is strict if and only if $f(0) = 0$.*

This corollary could be rephrased simply to read "$\bullet_f$ is nilpotent if and only if $f(0) > 0$". Archimedean t-norms thus fall naturally into two classes: nilpotent ones and strict ones. Since we will be dealing almost exclusively with Archimedean t-norms, *the terms* nilpotent *and* strict *will mean in particular Archimedean ones.* We restate the previous theorem for the strict t-norm case.

Theorem 5.2.7 *The t-norm $\triangle$ is strict if and only if there is an automorphism $f \in Aut(\mathbb{I})$ such that $x \triangle y = f^{-1}(f(x)f(y))$. Another such automorphism g satisfies this condition if and only if $f(x) = g(x)^r$ for some $r > 0$.*

We know that two generators f and g give the same t-norm if and only if $f(x) = g(x)^r$ for some positive real number r. In particular, an Archimedean t-norm does not uniquely determine a generator for it. We need to sort out this situation.

Definition 5.2.8 *A **generating function** for an Archimedean t-norm is an order isomorphism $[0,1] \rightarrow [a,1]$, where $a \in [0,1)$. The set of all such functions for all such a is denoted $\mathbb{G}$.*

Generating functions may be composed: for $f, g \in \mathbb{G}$, fg is the function given by $(fg)(x) = f(g(x))$. Note that $Aut(\mathbb{I}) \subset \mathbb{G}$. It should be clear that the composition of generating functions is a generating function. Composition is associative and has an identity, namely the identity function on $[0,1]$. So it is a **monoid**. A monoid is a set with an associative binary operation $\circ$ that has an identity. A **submonoid** of a monoid is a subset that contains the identity and contains $x \circ y$ for any two elements of the subset. The unit interval together with any t-norm is a commutative monoid.

Let $\mathbb{R}^+$ be the set of positive real numbers. Now $\mathbb{R}^+$ is a group under ordinary multiplication of real numbers: this operation on $\mathbb{R}^+$ is associative, has an identity, and every element has an inverse. For each $r \in \mathbb{R}^+$, the mapping $[0,1] \rightarrow [0,1] : x \rightarrow x^r$ is in $Aut(\mathbb{I})$. We identify $\mathbb{R}^+$ with this subset of $Aut(\mathbb{I})$. Multiplication in $\mathbb{R}^+$ corresponds to composition of functions when $\mathbb{R}^+$ is viewed as this subset of $Aut(\mathbb{I})$. So $\mathbb{R}^+$ is a group under this composition, that is, is a **subgroup** of $Aut(\mathbb{I})$. Further, for $r \in \mathbb{R}^+$ and $f \in \mathbb{G}$, rf is the function given by $(rf)(x) = f(x)^r$. We have $\mathbb{R}^+ \subset Aut(\mathbb{I}) \subset \mathbb{G}$, with $\mathbb{G}$ a monoid and $\mathbb{R}^+$ and $Aut(\mathbb{I})$ subgroups of this monoid.

Declaring two generating functions equivalent if they generate the same t-norm is obviously an equivalence relation. The set of equivalence classes of this equivalence relation partitions $\mathbb{G}$. What are these equivalence classes? By the Theorem 5.2.3, they are the sets $\{\mathbb{R}^+ f : f \in \mathbb{G}\}$, where $\mathbb{R}^+ f = \{rf : r \in \mathbb{R}^+\}$ This proves the following proposition, but we give a strictly group theoretic proof that the $\mathbb{R}^+ f$ form a partition of $\mathbb{G}$.

Proposition 5.2.9 *For $f \in \mathbb{G}$, let $\mathbb{R}^+ f = \{rf : r \in \mathbb{R}^+\}$. The set $\{\mathbb{R}^+ f : f \in \mathbb{G}\}$ is a partition of $\mathbb{G}$. The generating functions f and g generate the same t-norm if and only if $\mathbb{R}^+ f = \mathbb{R}^+ g$.*

Proof. For each f, $\mathbb{R}^+ f$ is nonempty and contains f. If $h \in \mathbb{R}^+ f \cap \mathbb{R}^+ g$, then $h = rf = sg$ for some $r, s \in \mathbb{R}^+$. Thus $f = r^{-1}sg$, and so for any $t \in \mathbb{R}^+$, $tf = tr^{-1}sg \in \mathbb{R}^+ g$, whence $\mathbb{R}^+ f \subseteq \mathbb{R}^+ g$. By symmetry, $\mathbb{R}^+ g \subseteq \mathbb{R}^+ f$, so $\mathbb{R}^+ f = \mathbb{R}^+ g$. Thus the $\mathbb{R}^+ f$ form a partition of $\mathbb{G}$. The second half of the proposition follows immediately from Theorem 5.2.3. ∎

As an immediate consequence, we have the following theorem.

Theorem 5.2.10 *Let $\triangle_f$ be the Archimedean t-norm with generator f. Then $\triangle_f \to \mathbb{R}^+ f$ is a one-to-one correspondence between the set of Archimedean t-norms and the partition $\{\mathbb{R}^+ f : f \in \mathbb{G}\}$ of $\mathbb{G}$.*

For those $f \in Aut(\mathbb{I})$, we have $\mathbb{R}^+ f \subset Aut(\mathbb{I})$. Thus $\{\mathbb{R}^+ f : f \in Aut(\mathbb{I})\}$ partitions $Aut(\mathbb{I})$. In this situation, the $\mathbb{R}^+ f$ are right **cosets** of the subgroup $\mathbb{R}^+$ in the group $Aut(\mathbb{I})$.

Corollary 5.2.11 *Let $\bullet_f$ be the strict t-norm with generator f. Then $\bullet_f \to \mathbb{R}^+ f$ is a one-to-one correspondence between the set of strict t-norms and the right cosets of $\mathbb{R}^+$ in the group $Aut(\mathbb{I})$.*

Here is a problem that arises whenever a set is partitioned by equivalence classes: pick from each equivalence class a "canonical" element. That is, pick from each class a representative that has some particular feature. Sometimes such a representative is called a "canonical form". Such forms were considered in Section 4.6. We have that situation here. The monoid $\mathbb{G}$ is partitioned by the equivalence classes $\mathbb{R}^+ f$, where f ranges over $\mathbb{G}$. Every element rf of $\mathbb{R}^+ f$ gives the same Archimedean t-norm.

Lemma 5.2.12 *Let $a \in (0,1)$. An Archimedean t-norm has exactly one generator f such that $f(a) = a$.*

Proof. Let $f \in \mathbb{G}$. The lemma asserts that there is exactly one element rf in $\mathbb{R}^+ f$ such that $(rf)(a) = f(a)^r = a$. For $f(a) = b$, then there is exactly one $r \in \mathbb{R}^+$ such that $b^r = a$. Now $(rf)(a) = f(a)^r = b^r = a$. ∎

Proposition 5.2.13 *Let $a \in (0,1)$, and $\mathbb{G}_a = \{g \in \mathbb{G} : g(a) = a\}$. Then $\mathbb{G}_a$ is a submonoid of $\mathbb{G}$, and $g \to \bullet_g$ is a one-to-one correspondence between $\mathbb{G}_a$ and the set of Archimedean t-norms.*

Of course, those $g \in Aut(\mathbb{I})$ correspond to strict t-norms, and the rest to nilpotent ones. Also, $\mathbb{G}_a \cap Aut(\mathbb{I})$ which we denote by $Aut(\mathbb{I})_a$ is a subgroup of $Aut(\mathbb{I})$ whose intersection with $\mathbb{R}^+$ is the identity automorphism of $[0,1]$.

Corollary 5.2.14 $g \to \bullet_g$ *is a one-to-one correspondence between the subgroup $Aut(\mathbb{I})_a$ and the set of strict t-norms.*

For nilpotent t-norms, there is another way to pick a representative of each equivalence class of generators of that t-norm. If $\triangle$ is a nilpotent t-norm, and f any generator for it, then $f : [0,1] \to [b,1]$ for some $b \in (0,1)$. Let $a \in (0,1)$. Then there is an $r > 0$ such that $b^r = a$. Now, rf is an order isomorphism $[0,1] \to [b^r,1] = [a,1]$. Both f and rf are generators of $\triangle$, and the latter is a map $[0,1] \to [a,1]$, where we have chosen a arbitrarily in $(0,1)$. It should be clear that there is no other generator g of $\triangle$ such that $g(0) = a$.

Corollary 5.2.15 *Let $a \in (0,1)$. Then $f \to f^{-1}(f(x)f(y) \vee a)$ is a one-to-one correspondence between order isomorphisms $[0,1] \to [a,1]$ and nilpotent t-norms.*

It is easy to write down generators of Archimedean t-norms. Just any order isomorphism from $[0,1]$ to $[a,1]$ will do, and if $a = 0$, we get strict t-norms. But given such an order isomorphism f, the corresponding t-norm involves the inverse of f, and that might not be easy to compute. Also, given an Archimedean t-norm, there may be no way to get a generator for it short of the construction indicated in the proof of Theorem 5.2.3. But there are a number of explicit families of Archimedean t-norms known, along with their generators and inverses of those generators. We will see some of these later. Following are three well known t-norms along with their generators and inverses.

Example 5.2.16 The strict t-norm $x \triangle y = xy$ has generator $f(x) = x$, and so is particularly easy. The inverse of f is just f itself. Also note that for any $r > 0$, $f(x) = x^r$ will work. In this case, $f^{-1}(x) = x^{\frac{1}{r}}$ and

$$f^{-1}(f(x)f(y)) = (x^r y^r)^{\frac{1}{r}} = ((xy)^r)^{\frac{1}{r}} = xy$$

Example 5.2.17 $x \triangle y = \frac{xy}{x+y-xy}$ is a strict t-norm and has generator $f(x) = e^{-\frac{1-x}{x}}$, and $f^{-1}(x) = \frac{1}{1-\ln x}$ as is easily computed. We have

$$
\begin{aligned}
f^{-1}(f(x)f(y)) &= \frac{1}{1 - \ln(f(x)f(y))} \\[2mm]
&= \frac{1}{1 - \ln(e^{-\frac{1-x}{x}} e^{-\frac{1-y}{y}})} \\[2mm]
&= \left(1 + \frac{1-x}{x} + \frac{1-y}{y}\right)^{-1} \\[2mm]
&= \frac{xy}{x + y - xy}
\end{aligned}
$$

Example 5.2.18 $x \triangle y = (x + y - 1) \vee 0$ is a nilpotent t–norm. It has generator $f(x) = e^{x-1}$. Since $f(0) = e^{-1}$, f^{-1} is defined on $[e^{-1}, 1]$ and on that interval $f^{-1}(x) = 1 + \ln x$.

$$
\begin{aligned}
f^{-1}(f(x)f(y) \vee e^{-1}) &= f^{-1}(e^{x-1}e^{y-1} \vee e^{-1}) \\[2mm]
&= 1 + ((x + y - 2) \vee (-1)) \\[2mm]
&= (x + y - 1) \vee 0
\end{aligned}
$$

In the section on t-conorms, we will give additional examples of t-norms, together with t-conorms, their generators, and negations connecting them.

Historically, Archimedean t-norms have been represented by maps $g : [0, 1] \to [0, \infty]$, where g is a strictly decreasing function satisfying $0 < g(0) \leq \infty$ and $g(1) = 0$. In this case the binary operation satisfies

$$
g(x \triangle y) = (g(x) + g(y)) \wedge g(0)
$$

and since this minimum is in the range of g,

$$
x \triangle y = g^{-1}((g(x) + g(y)) \wedge g(0))
$$

Such functions g are called **additive generators** of the t-norm $\triangle$.

The following proposition shows that these two types of representations give the same t-norms. We use the multiplicative representation, because it allows us to remain within the context of the unit interval on which t-norms are defined.

Proposition 5.2.19 *Suppose $g : [0,1] \to [0,\infty]$ is a continuous, strictly decreasing function, with $0 < g(0) \leq \infty$ and $g(1) = 0$. Let*

$$
\begin{aligned}
f(x) &= e^{-g(x)} \\
x \bullet_f y &= f^{-1}(f(x)f(y) \vee f(0)) \\
x \triangle_{g^+} y &= g^{-1}((g(x) + g(y)) \wedge g(0))
\end{aligned}
$$

Then $f : [0,1] \to [f(0), 1]$ is order preserving, one-to-one and onto, and $\triangle_{g^+} = \bullet_f$.

Proof. It is easy to see that $f : [0,1] \to [f(0), 1]$ is such a mapping. Note that for x in the range of f, then $-\ln x$ is in the range of g, and $f^{-1}(x) = g^{-1}(-\ln x)$. For $f(x)f(y) \geq f(0)$, we have

$$
\begin{aligned}
x \triangle_f y &= f^{-1}(f(x)f(y)) \\
&= f^{-1}((e^{-g(x)})(e^{-g(y)})) \\
&= f^{-1}(e^{-(g(x)+g(y))}) \\
&= g^{-1}(-\ln(e^{-(g(x)+g(y))})) \\
&= g^{-1}(g(x) + g(y)) \\
&= x \triangle_{g^+} y
\end{aligned}
$$

For $f(x)f(y) < f(0)$, we have $x \triangle_f y = f^{-1}f(0) = 0$. Also $g(x) + g(y) > g(0)$ implies $x \triangle_g y = g^{-1}(g(0)) = 0$. ∎

We will not use additive generators, but the reader should be aware that they are used by some, but to our knowledge, never to an advantage over multiplicative ones. At the end of this chapter, there are some exercises about additive generators.

5.3 Isomorphisms of t-norms

We have a way to construct Archimedean t-norms: for $a \in [0,1)$ and an order isomorphism $f : [0,1] \to [a,1]$ define

$$
x \bullet_f y = f^{-1}(f(x)f(y) \vee a)
$$

In particular, if $a = 0$, then f is just an automorphism of $([0,1], \leq) = \mathbb{I}$. And Theorem 5.2.3 says that this gets them all. But this still leaves the question of when two t-norms are essentially the same. We sort that out in this section.

A word about notation is in order. We will use various symbols to denote t-norms, including $\triangle, \circ,$ and $\diamond$. We have previously used $f \circ g$ on occasion to denote composition of functions. However, standard practice for us is to use fg instead. The context will make clear the meaning. This is common in mathematics. The same symbol is often used to denote different operations. For example, $+$ is used to denote addition of matrices as well as of numbers. Also, we will have many occasions to write $ab \vee c$, where a, b, and c are numbers. This will always mean $(ab) \vee c$. But the meaning of $a + b \vee c$ is not so well established and would be written $(a + b) \vee c$ or $a + (b \vee c)$ depending on the meaning intended.

Let $\circ$ be a t-norm and consider the system $(\mathbb{I}, \circ)$. This system is simply $\mathbb{I}$ with an additional structure on it, namely the operation $\circ$. Let $\diamond$ be another t-norm on $\mathbb{I}$. The following definition makes precise the notion of the systems $(\mathbb{I}, \circ)$ and $(\mathbb{I}, \diamond)$ being structurally the same.

Definition 5.3.1 *Let $\circ$ and $\diamond$ be t-norms. The systems $(\mathbb{I}, \circ)$ and $(\mathbb{I}, \diamond)$ are **isomorphic** if there is an element $h \in Aut(\mathbb{I})$ such that $h(x \circ y) = h(x) \diamond h(y)$. We write $(\mathbb{I}, \circ) \approx (\mathbb{I}, \diamond)$. The mapping h is an **isomorphism**.*

This means that the systems $([0, 1], \leq, \circ)$ and $([0, 1], \leq, \diamond)$ are isomorphic in the sense of universal algebra: there is a one-to-one map from $[0, 1]$ onto $[0, 1]$ that preserves the operations and relations involved. If $(\mathbb{I}, \circ) \approx (\mathbb{I}, \diamond)$, we also say that the t-norms $\circ$ and $\diamond$ are isomorphic.

Isomorphism between t-norms is an equivalence relation and so partitions t-norms into equivalence classes. The t-norm min is rather special. A t-norm $\circ$ is **idempotent** if $a \circ a = a$ for all $a \in [0, 1]$. If $\circ$ is idempotent, then for $a \leq b$, $a = a \circ a \leq a \circ b \leq a \circ 1 = a$, so $\circ = \min$. Thus min is the *only* idempotent t-norm. Further, it should be clear that the only t-norm isomorphic to min is min itself. It is in an equivalence class all by itself.

An isomorphism of a system with itself is called an **automorphism**. It is easy to show that the set of automorphisms of $(\mathbb{I}, \circ)$ is a subgroup of $Aut(\mathbb{I})$. Thus, with each t-norm $\circ$, there is a group associated with it, namely the **automorphism group**

$$Aut(\mathbb{I}, \circ) = \{f \in Aut(\mathbb{I}) : f(x \circ y) = f(x) \circ f(y)\}$$

This is also called the automorphism group of the t-norm $\circ$. For the t-norm $a \wedge b = \min\{a, b\}$, it is clear that $Aut(\mathbb{I}, \wedge) = Aut(\mathbb{I})$.

By an **isomorphism** from a group G to a group H we mean a one-to-one onto map $\varphi : G \to H$ such that $\varphi(xy) = \varphi(x)\varphi(y)$ for all x and y in G. In groups, it is customary to write "product" of x and y simply as xy. If H is a subgroup of a group G, and $g \in G$, then $g^{-1}Hg = \{g^{-1}hg : h \in H\}$ is a subgroup of G. This subgroup is said to be **conjugate to** H, or a **conjugate of** H. The map $h \to g^{-1}hg$ is an isomorphism from H to its conjugate $g^{-1}Hg$.

Theorem 5.3.2 *If two t-norms are isomorphic then their automorphism groups are conjugate.*

Proof. Suppose that $\circ$ and $\diamond$ are isomorphic. Then there is an isomorphism $f : (\mathbb{I}, \circ) \to (\mathbb{I}, \diamond)$. The map $g \to f^{-1}gf$ is an isomorphism from $Aut(\mathbb{I}, \diamond)$ to $Aut(\mathbb{I}, \circ)$, so $f^{-1}Aut(\mathbb{I}, \diamond)f = Aut(\mathbb{I}, \circ)$. ∎

Restating Theorem 5.2.7, we have

Theorem 5.3.3 *The Archimedean t-norm $\circ$ is strict if and only if it is isomorphic to multiplication, that is, if and only if there is an element $f \in Aut(\mathbb{I})$ such that $f(x \circ y) = f(x)f(y)$. Another element $g \in Aut(\mathbb{I})$ satisfies this condition if and only if $f = rg$ for some $r > 0$.*

So a generator of a strict t-norm $\circ$ is just an isomorphism from $Aut(\mathbb{I}, \circ)$ to $Aut(\mathbb{I}, \bullet)$. (The symbol $\bullet$ stands for the t-norm $x \bullet y = xy$, that is, ordinary multiplication of real numbers.)

Corollary 5.3.4 *For any strict t-norm $\circ$, $Aut(\mathbb{I}, \circ) \approx Aut(\mathbb{I}, \bullet)$.*

Corollary 5.3.5 *For any two strict t-norms $\circ$ and $\diamond$, $Aut(\mathbb{I}, \circ) \approx Aut(\mathbb{I}, \diamond)$.*

We spell out exactly what the isomorphisms from $Aut(\mathbb{I}, \circ)$ to $Aut(\mathbb{I}, \diamond)$ are.

Theorem 5.3.6 *Let $\bullet_f$ and $\bullet_g$ be strict t-norms. Then $h : (\mathbb{I}, \bullet_f) \to (\mathbb{I}, \bullet_g)$ is an isomorphism if and only if $g^{-1}rf = h$ for some $r > 0$. That is, the set of isomorphisms from $(\mathbb{I}, \bullet_f)$ to $(\mathbb{I}, \bullet_g)$ is the set*

$$g^{-1}\mathbb{R}^+ f = \{g^{-1}rf : r \in \mathbb{R}^+\}.$$

Proof. An isomorphism $h : (\mathbb{I}, \bullet_f) \to (\mathbb{I}, \bullet_g)$ gets an isomorphism $gh : (\mathbb{I}, \bullet_f) \to (\mathbb{I}, \cdot)$ which must be rf for some $r \in \mathbb{R}^+$. So $h = g^{-1}rf$. For any r, $g^{-1}rf$ is an isomorphism.

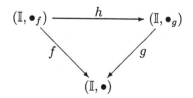

∎

Corollary 5.3.7 $Aut(\mathbb{I}, \bullet_f) = f^{-1}\mathbb{R}^+ f \approx \mathbb{R}^+$.

Proof. The set of automorphisms of $(\mathbb{I}, \bullet_f)$ is $f^{-1}\mathbb{R}^+ f$. It is a subgroup of $Aut(\mathbb{I})$, and is isomorphic to $\mathbb{R}^+$ via the mapping $f^{-1}rf \to r$. ∎

Corollary 5.3.8 $Aut(\mathbb{I}, \bullet) = \mathbb{R}^+$.

In the case of strict t-norms we have that $Aut(\mathbb{I}, \bullet_f) = f^{-1}\mathbb{R}^+ f \subseteq Aut(\mathbb{I})$. It turns out that these are the only convex t-norms with such automorphism groups.

Proposition 5.3.9 *Let* $\circ$ *be a convex t-norm. Then* $Aut(\mathbb{I}, \circ) = f^{-1}\mathbb{R}^+ f$ *for some* $f \in Aut(\mathbb{I})$ *if and only if* $\circ$ *is a strict t-norm.*

Proof. Suppose that $Aut(\mathbb{I}, \circ) = f^{-1}\mathbb{R}^+ f$ and suppose that for some $a \in (0,1)$, $a \circ a = a$. Then for any element $b \in (0,1)$, there is an element $g \in Aut(\mathbb{I}, \circ)$ such that $g(a) = b$, namely $g = f^{-1}rf$ where $r = \ln f(b) / \ln f(a)$. Thus

$$b \circ b = g(a) \circ g(a) = g(a \circ a) = g(a) = b,$$

so that $\circ$ is idempotent. But the only idempotent t-norm is min, and $Aut(\mathbb{I}, \min) = Aut(\mathbb{I}) \neq f^{-1}\mathbb{R}^+ f$. Thus $a \circ a < a$ for all $a \in (0,1)$, and $\circ$ is Archimedean. The t-norm is not nilpotent, by the preceding corollary, and thus it is strict. ∎

We cannot conclude that the function f in the proposition is a generator of $\circ$. Corollary 5.3.7 says that $Aut(\mathbb{I}, \bullet_f) = f^{-1}\mathbb{R}^+ f$, where $\bullet_f$ is the t-norm with generator f. But as we will see later, it is possible that $f^{-1}\mathbb{R}^+ f = g^{-1}\mathbb{R}^+ g$ without g being a generator of $\bullet_f$.

Now we look at isomorphisms between nilpotent t-norms. There are two basic facts : any two are isomorphic, and each has a trivial automorphism group.

Theorem 5.3.10 *Let* $\bullet_f$ *and* $\bullet_g$ *be nilpotent. Let* $r \in \mathbb{R}^+$ *with* $g(0) = (f(0))^r$. *Then* $g^{-1}rf$ *is the unique isomorphism from* $(\mathbb{I}, \bullet_f)$ *to* $(\mathbb{I}, \bullet_f)$.

Proof. First note that $g^{-1}rf \in Aut(\mathbb{I})$. To show that $g^{-1}rf$ is an isomorphism from $(\mathbb{I}, \bullet_f)$ to $(\mathbb{I}, \bullet_{f})$, we need to show that $g^{-1}rf(a \bullet_f b) = g^{-1}rf(a) \bullet_g g^{-1}rf(b)$.

$$
\begin{aligned}
g^{-1}rf(a \bullet_f b) &= g^{-1}rff^{-1}(f(a)f(b) \vee f(0)) \\
&= g^{-1}r(f(a)f(b) \vee f(0))
\end{aligned}
$$

$$
\begin{aligned}
g^{-1}rf(a) \bullet_g g^{-1}rf(b) &= g^{-1}(gg^{-1}rf(a)gg^{-1}rf(b) \vee g(0)) \\
&= g^{-1}(rf(a)rf(b) \vee g(0)) \\
&= g^{-1}((f(a)f(b))^r \vee f(0)^r) \\
&= g^{-1}r(f(a)f(b) \vee f(0))
\end{aligned}
$$

Suppose that $\varphi : (\mathbb{I}, \bullet_f) \to (\mathbb{I}, \bullet_g)$ is an isomorphism. Then

$$\varphi f^{-1}(f(a)f(b) \vee f(0)) = g^{-1}((g\varphi)(a)(g\varphi)(b) \vee g(0))$$

Thus

$$
\begin{aligned}
f^{-1}(f(a)f(b) \vee f(0)) &= \varphi^{-1} g^{-1}((g\varphi)(a)(g\varphi)(b) \vee g(0)) \\
&= (g\varphi)^{-1}((g\varphi)(a)(g\varphi)(b) \vee g\varphi(0))
\end{aligned}
$$

Since f and $g\varphi$ generate the same nilpotent t-norm, $g\varphi = rf$, so $\varphi = g^{-1}rf$, and $g(0) = (f(0))^r$ as asserted. ∎

Corollary 5.3.11 *If $\circ$ is a nilpotent t-norm, then $Aut(\mathbb{I}, \circ) = \{1\}$.*

By Theorem 5.3.10, there is exactly one isomorphism between any two nilpotent t-norms. The particular nilpotent t-norm $x \blacktriangle y = (x + y - 1) \vee 0$ is called **Lukasiewicz's t-norm**. It is perhaps the simplest looking nilpotent t-norm, just as multiplication is the simplest strict t-norm. In any case, if $\triangle$ is a nilpotent t-norm, it is isomorphic to $\blacktriangle$. There is an element f of $Aut(\mathbb{I})$ such that $f(x \circ y) = f(x) \blacktriangle f(y)$. In the strict case, a generator was an automorphism of $Aut(\mathbb{I})$. In the nilpotent case, a generator was not such an automorphism. But now we know that every nilpotent t-norm $\diamond$ comes about as $x \circ y = f^{-1}(f(x) \blacktriangle f(y))$ for a *unique* $f \in Aut(\mathbb{I})$. Such an f is an **L-generator** of the nilpotent t-norm $\circ$. It is just the unique isomorphism of $\circ$ with $\blacktriangle$. *We will consistently use $\blacktriangle$ to denote the Lukasiewicz t-norm and $\blacktriangle_f$ to denote the nilpotent t-norm $f^{-1}(f(x) \blacktriangle f(y))$. So $f(x \blacktriangle_f y) = f(x) \blacktriangle f(y)$, and this is for every automorphism $f \in Aut(\mathbb{I})$.*

Theorem 5.3.12 *For $f \in Aut(\mathbb{I})$, let $x \blacktriangle_f y = f^{-1}(f(x) \blacktriangle f(y))$. Then $f \to \blacktriangle_f$ is a one-to-one correspondence between $Aut(\mathbb{I})$ and the set of nilpotent t-norms.*

This one-to-one correspondence between $Aut(\mathbb{I})$ and the set of nilpotent t-norms gives the latter set the structure of a group, namely the binary operation defined by $\blacktriangle_f \blacktriangle_g = \blacktriangle_{fg}$. This is a triviality. Given *any* group G and *any* set S and a one-to-one correspondence $\phi : G \to S$, then S becomes a group under the binary operation $st = \phi(\phi^{-1}(s)\phi^{-1}(t))$. However, our correspondence suggests a way to get "natural" sets of nilpotent t-norms. For a subgroup G of $Aut(\mathbb{I})$, what is the corresponding set of nilpotent t-norms? One must compute $f^{-1}((f(x)+f(y)-1)\vee 0)$ for every $f \in G$. This may be difficult, the difficulty generally being in computing f^{-1}. There is one easy case, namely that of the subgroup $\mathbb{R}^+$. For any $r \in \mathbb{R}^+$, we have the nilpotent t-norm $r^{-1}((r(x)+r(y)-1) \vee 0)$, or if you will, $((x^r + y^r - 1) \vee 0)^{1/r}$. When $r = 1$, we get Lukasiewicz's t-norm, as should be.

Here is the formula for passing between generators and L-generators of nilpotent t-norms.

Proposition 5.3.13 *Let f be a generator of a nilpotent t-norm $\triangle$. Then the L-generator of $\triangle$ is*

$$g(x) = 1 - \frac{\ln f(x)}{\ln f(0)}$$

The inverse of g is

$$g^{-1}(x) = f^{-1}(f(0)^{1-x})$$

Proof. A generator for $\blacktriangle$ is $h(x) = e^{x-1}$, and $h^{-1}(x) = \ln(x) + 1$. From Theorem 5.3.10, the unique isomorphism from $\triangle$ to $\blacktriangle$ is

$$
\begin{aligned}
g(x) &= h^{-1} f(x)^{\frac{\ln h(0)}{\ln f(0)}} \\
&= \frac{\ln h(0)}{\ln f(0)} \ln f(x) + 1 \\
&= \frac{-1}{\ln f(0)} \ln f(x) + 1 \\
&= 1 - \frac{\ln f(x)}{\ln f(0)}.
\end{aligned}
$$

Calculating g^{-1} is easy. ∎

Note that $g(x)$ is independent of which f is picked, since any other generator is rf for some $r \in \mathbb{R}^+$. Also, if $f(0) = e^{-1}$, then $g(x) = 1 + \ln f(x)$.

5.4   Negations

The complement A' of a fuzzy set A has been defined by $A'(x) = 1 - A(x)$. This is the same as following A by the function $[0, 1] \to [0, 1] : x \to 1 - x$. This latter function is an **involution** of the lattice $\mathbb{I} = ([0, 1], \leq)$. That is, it is order reversing and applying it twice gives the identity map. In fuzzy set theory, such a map $\eta : [0, 1] \to [0, 1]$ is also called a **strong negation**. A strong negation η satisfies:

(i) $\eta(0) = 1$, $\eta(1) = 0$.

(ii) η is nonincreasing.

(iii) $\eta(\eta(x)) = x$.

A map satisfying only the first two conditions is a **negation**. It is clear that there are many of them: any nonincreasing map that starts at 1 and goes to 0. Such simple maps as $\eta(x) = 1$ if $x = 1$ and $= 0$ otherwise, and $\eta(x) = 0$ if $x = 0$ and $= 1$ otherwise, are negations. But they are of little interest to us here. *We will restrict our attention to involutions, and refer to them simply as* **negations**. The reader should be warned that in other areas, including logic programming, constructive mathematics, and mathematical logic, strong negation has meanings other than involution.

We will use α for the particular negation $x \to 1 - x$, and η to denote a negation in general. Other commonly used notations are N and $'$.

The negation α is not in $Aut(\mathbb{I})$ since it reverses order rather than preserves it. A mapping $f : [0,1] \to [0,1]$ that is one-to-one and onto, and such that $f(x) \geq f(y)$ if and only if $x \leq y$ is an antiautomorphism of $\mathbb{I}$. The set of all automorphisms and antiautomorphisms is denoted $Map(\mathbb{I})$, and is a group under the operation of composition of functions. $Aut(\mathbb{I})$ is a subgroup of it. The composition of two antiautomorphisms is an automorphism, the inverse of an antiautomorphism is and antiautomorphism, and the composition of an automorphism and an antiautomorphism is an antiautomorphism. All this is easy to verify.

An element $f \in Map(\mathbb{I})$ has **order** n if $f^n = 1$, and n is the smallest such positive integer. If no such integer exists, the element has **infinite** order. All the elements of $Aut(\mathbb{I})$ have infinite order except 1 which has order 1. All antiautomorphisms are either of order two or of infinite order. Antiautomorphisms of order 2 are called **involutions**. So negations and involutions are the same, and α is just a particular involution. Following are the graphs of the involutions $\alpha(x) = 1 - x$ and $\eta(x) = (1-x)/(1+5x)$.

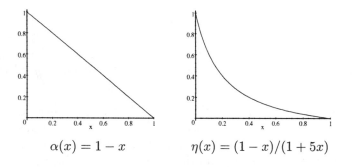

$$\alpha(x) = 1 - x \qquad \eta(x) = (1 - x)/(1 + 5x)$$

In any group, an element of the form $f^{-1}gf$ is a **conjugate** of g. In $Map(\mathbb{I})$, conjugates of automorphisms are automorphisms, conjugates of antiautomorphisms are antiautomorphisms, and conjugates of negations are negations. So an element of the form $f^{-1}\alpha f$ is also a negation.

If g is an element of a group G, then the set $\{f \in G : fg = gf\}$ is the **centralizer** of g. So this centralizer is just all those elements of G that commute with g. It is a subgroup of G. We will be interested in a subgroup of the centralizer of α, namely those elements in $Aut(\mathbb{I})$ that commute with α. This subgroup we denote $Z(\alpha)$. It is the subgroup $\{f \in Aut(\mathbb{I}) : f\alpha = \alpha f\}$. It is the **centralizer of α in** $Aut(\mathbb{I})$.

We will prove results analogous to those for t-norms in the previous sections. For example, every strict t-norm was constructed from the strict t-norm multiplication and an automorphism f of $\mathbb{I}$. In the same vein, every negation comes from the particular negation α and an automorphism. Here is the theorem.

Theorem 5.4.1 *Let β be a negation in $Map(\mathbb{I})$. Let*

$$f(x) = \frac{\alpha\beta(x) + x}{2}$$

Then $f \in Aut(\mathbb{I})$, and $\beta = f^{-1}\alpha f$. Furthermore, $g^{-1}\alpha g = \beta$ if and only if $gf^{-1} \in Z(\alpha)$.

Proof. Since f is the average of the two automorphisms $\alpha\beta$ and the identity, it is an automorphism.

$$\begin{aligned} f\beta(x) &= \frac{\alpha\beta(\beta(x)) + \beta(x)}{2} \\ &= \frac{1 - x + \beta(x)}{2} \end{aligned}$$

and

$$\begin{aligned} \alpha f(x) &= 1 - \frac{\alpha\beta(x) + x}{2} \\ &= \frac{1 + \beta(x) - x}{2} \end{aligned}$$

so we have $f\beta = \alpha f$, so $\beta = f^{-1}\alpha f$. Now, $g^{-1}\alpha g = f^{-1}\alpha f$ if and only if $gf^{-1}\alpha fg^{-1} = \alpha$ if and only if $gf^{-1}\alpha = \alpha gf^{-1}$ if and only if $gf^{-1} \in Z(\alpha)$. ∎

An automorphism f such that $\beta = f^{-1}\alpha f$ is a **generator** of β. In general, if η is a negation, we denote $f^{-1}\eta f$ by η_f. This means that f is a generator of α_f. So every negation has a generator, and we know when two elements of $Aut(\mathbb{I})$ generate the same negation. This theorem seems to be due to Trillas [97] who takes as generators functions from $[0, 1]$ to $[0, \infty]$.

One consequence of this theorem is that two elements f and g of $Aut(\mathbb{I})$ determine the same negation, that is, that $g^{-1}\alpha g = f^{-1}\alpha f$, if and only if $gf^{-1} \in Z(\alpha)$. But this is the same as $Z(\alpha)f = Z(\alpha)g$. Hence we have the following theorem.

Theorem 5.4.2 *The map $\alpha_f \to Z(\alpha)f$ is a one-to-one correspondence between the negations of $\mathbb{I}$ and the set of right cosets in $Aut(\mathbb{I})$ of the centralizer $Z(\alpha)$ of α.*

Actually, we can give a method for constructing all elements of $Z(\alpha)$.

Proposition 5.4.3 $Z(\alpha) = \{\frac{\alpha f\alpha + f}{2} : f \in Aut(\mathbb{I})\}$.

Proof. If $g \in Z(\alpha)$, then

$$\frac{\alpha g\alpha + g}{2} = \frac{g\alpha\alpha + g}{2} = \frac{g + g}{2} = g$$

so g has the right form. For the other inclusion, we have the equations

$$\left(\alpha\frac{\alpha f\alpha + f}{2}\right)(x) = 1 - \frac{\alpha f\alpha(x) + f(x)}{2}$$
$$= \frac{2 - \alpha f\alpha(x) - f(x)}{2}$$
$$= \frac{1 - \alpha f\alpha(x) + 1 - f(x)}{2}$$
$$= \frac{f\alpha(x) + \alpha f(x)}{2}$$
$$\left(\frac{\alpha f\alpha + f}{2}\alpha\right)(x) = \frac{\alpha f\alpha\alpha(x) + f\alpha(x)}{2}$$
$$= \frac{\alpha f(x) + f\alpha(x)}{2}$$

∎

This proposition gives a map

$$Aut(\mathbb{I}) \to Z(\alpha) : g \to \frac{\alpha g\alpha + g}{2}$$

This map fixes $Z(\alpha)$ elementwise. It is not a group homomorphism.

Consider two systems $(\mathbb{I}, \beta)$ and $(\mathbb{I}, \gamma)$ where β and γ are negations. They are **isomorphic** if there is a map $h \in Aut(\mathbb{I})$ with $h(\beta(x)) = \gamma(h(x))$, that is if $h\beta = \gamma h$, or equivalently if $\beta = h^{-1}\gamma h$. Let f and g be generators of β and γ, respectively. If h is an isomorphism, then $hf^{-1}\alpha f = g^{-1}\alpha gh$ which means that

$$f^{-1}\alpha f = h^{-1}g^{-1}\alpha gh = (gh)^{-1}\alpha gh$$

Therefore, f and gh generate the same negation, and so $zf = gh$ for some $z \in Z(\alpha)$. Thus $h \in g^{-1}Z(\alpha)f$. It is easy to check that elements of $g^{-1}Z(\alpha)f$ are isomorphisms $(\mathbb{I}, \beta) \rightarrow (\mathbb{I}, \gamma)$. We have the following theorem.

Theorem 5.4.4 *The set of isomorphisms from $(\mathbb{I}, \alpha_f)$ to $(\mathbb{I}, \alpha_g)$ is the set $g^{-1}Z(\alpha)f = \{g^{-1}zf : z \in Z(\alpha)\}$. In particular, $g^{-1}f$ is an isomorphism from $(\mathbb{I}, \alpha_f)$ to $(\mathbb{I}, \alpha_g)$.*

Note that $Z(\alpha)$ plays a role for negations analogous to that of $\mathbb{R}^+$ for strict t-norms. If we call two negations β and γ **isomorphic** if $(\mathbb{I}, \beta) \approx (\mathbb{I}, \gamma)$, then the previous theorem says in particular that any two negations are isomorphic. We have the following special cases.

Corollary 5.4.5 *The set of isomorphisms from $(\mathbb{I}, \alpha_f)$ to $(\mathbb{I}, \alpha)$ is the right coset $Z(\alpha)f$ of $Z(\alpha)$. In particular, the generator f of a negation β is an isomorphism from $(\mathbb{I}, \beta)$ to $(\mathbb{I}, \alpha)$.*

Noting that $f^{-1}Z(\alpha)f = Z(\beta)$, we have

Corollary 5.4.6 $Aut(\mathbb{I}, \alpha_f) = f^{-1}Z(\alpha)f = Z(\alpha_f)$, *and in particular,* $Aut(\mathbb{I}, \alpha) = Z(\alpha)$.

Since $z \rightarrow f^{-1}zf$ is an isomorphism from the group $Z(\alpha) = Aut(\mathbb{I}, \alpha)$ to $f^{-1}Z(\alpha)f = Aut(\mathbb{I}, \beta)$, we get

Corollary 5.4.7 *For any two negations β and γ, $Aut(\mathbb{I}, \beta) \approx Aut(\mathbb{I}, \gamma)$.*

Of course this last corollary follows also because the two systems $(\mathbb{I}, \beta)$ and $(\mathbb{I}, \gamma)$ are isomorphic. The main thrust of all this is that furnishing $\mathbb{I}$ with any negation yields a system isomorphic to that gotten by furnishing $\mathbb{I}$ with the negation $\alpha : x \rightarrow 1 - x$.

5.5 Nilpotent t-norms and negations

We get *all* negations as conjugates of α, but there are other ways to construct them. Nilpotent t-norms give rise to negations in various ways, and we look at three of those. In a Boolean algebra, $\bigvee\{y : x \wedge y = 0\}$ exists and is the complement of x, and sending x to this element is a negation in that Boolean algebra. Since $[0, 1]$ is a complete lattice, we can certainly perform the same construction for any binary operation $\triangle$ on $[0, 1]$, and in particular with respect to a nilpotent t-norm. For a nilpotent t-norm, this does turn out to be a negation, and what that negation is in terms of a generator and an L-generator for the t-norm is spelled out in the following theorem.

Theorem 5.5.1 *Let $\triangle$ be a nilpotent t-norm, f be a generator of $\triangle$, and g be the L-generator of $\triangle$. Let*

$$\eta_\triangle(x) \;=\; \bigvee\{y : y \triangle x = 0\}$$

$$N_f(x) \;=\; f^{-1}\left(\frac{f(0)}{f(x)}\right)$$

$$\alpha_g(x) \;=\; g^{-1}\alpha g$$

Then $\eta_\triangle(x) = N_f(x) = \alpha_g(x)$ and this function is a negation. In particular, $\eta_f(x)$ is independent of the particular generator f of $\triangle$.

Proof. Let g be an L-generator of $\triangle$. Then

$$
\begin{aligned}
\eta_\triangle(x) \;&=\; \bigvee\{y : y \triangle x = 0\}\\
&=\; \bigvee\{y : g^{-1}((g(x) + g(y) - 1) \vee 0) = 0\}\\
&=\; \bigvee\{y : (g(x) + g(y) - 1) \le 0\}\\
&=\; \bigvee\{y : (g(x) + g(y) - 1) = 0\}
\end{aligned}
$$

Hence $g(y) = 1 - g(x)$, whence $y = g^{-1}\alpha g(x)$. Thus $\eta_\triangle(x) = \alpha_g(x)$, and since $g^{-1}\alpha g$ is a negation, so is $\eta_\triangle(x)$.

$$
\begin{aligned}
N_f(x) \triangle x \;&=\; f^{-1}\left(\frac{f(0)}{f(x)}\right) \triangle x\\
&=\; f^{-1}\left(f\left(f^{-1}\left(\frac{f(0)}{f(x)}\right)\right) f(x) \vee f(0)\right)\\
&=\; f^{-1}\left(\left(\frac{f(0)}{f(x)}\right) f(x) \vee f(0)\right)\\
&=\; f^{-1}\left(f(0) \vee f(0)\right)\\
&=\; 0
\end{aligned}
$$

Therefore $N_f(x) \le \eta_\triangle(x)$. If $y \triangle x = 0$, then $0 = f^{-1}\left(f(y)f(x) \vee f(0)\right)$ and so $f(x)f(y) \le f(0)$. Thus $y \le f^{-1}\left(\frac{f(0)}{f(x)}\right)$ and the desired equality holds, that is, $N_f(x) = \eta_\triangle(x)$, and in particular $N_f(x)$ is a negation. ∎

Definition 5.5.2 *The negation in Theorem 5.5.1 is the **natural negation** associated with that nilpotent t-norm.*

So there are three equivalent ways to get the natural negation of a nilpotent t-norm $\triangle$: from a generator, from its L-generator, and directly from the t-norm as $\eta_\triangle(x) = \bigvee\{y : y \triangle x = 0\}$. This is the **residual** of x with respect to $\triangle$. It is of no interest to make this construction for a strict t-norm. In that case, $\eta_\triangle(x) = 0$ if $x \ne 0$ and $\eta_\triangle(0) = 1$.

Every nilpotent t-norm is of the form $\blacktriangle_f$ for a unique $f \in Aut(\mathbb{I})$, so we get $\eta_{\blacktriangle_f}(x) = f^{-1}\alpha f$. This implies that distinct nilpotent t-norms can have the same associated natural negation since $f^{-1}\alpha f$ determines f only up to left multiplication by elements of $Z(\alpha)$. Restated, this says that two nilpotent t-norms have the same natural negation if and only if their L-generators f and g satisfy $Z(\alpha)f = Z(\alpha)g$, that is, determine the same right coset of $Z(\alpha)$ in $Aut(\mathbb{I})$.

We now determine conditions on generators of nilpotent t-norms for them to give the same natural negation.

Theorem 5.5.3 *Let f and g be generators of nilpotent t-norms with* $f(0) = g(0) = a$. *Then $N_f = N_g$ if and only if for $x \in [a, 1]$,*

$$fg^{-1}\left(\frac{a}{x}\right) = \frac{a}{fg^{-1}(x)}$$

Proof. We have $\eta_f = \eta_g$ if and only if for all x,

$$f^{-1}\left(\frac{f(0)}{f(x)}\right) = g^{-1}\left(\frac{g(0)}{g(x)}\right)$$

Replacing x by $g^{-1}(x)$, this is equivalent to

$$\left(\frac{a}{fg^{-1}(x)}\right) = fg^{-1}\left(\frac{a}{x}\right)$$

and the theorem is proved. ∎

So $N_g = N_f$ if and only if the automorphism $h = fg^{-1}$ of $([a, 1], \leq)$ satisfies the equation $h(\frac{a}{x}) = \frac{a}{h(x)}$. Such automorphisms are easy to describe. First notice that $h(\sqrt{a}) = \sqrt{a}$, so that h induces an automorphism of $([a, \sqrt{a}], \leq)$. Further, h is determined on all of $[a, 1]$ by its action on $[a, \sqrt{a}]$ by the condition $h(\frac{a}{x}) = \frac{a}{h(x)}$, or equivalently, $h(x) = \frac{a}{h(\frac{a}{x})}$. Now if h is *any* automorphism of $([a, \sqrt{a}], \leq)$ define h on $[\sqrt{a}, 1]$ by

$$h(x) = \frac{a}{h(\frac{a}{x})}$$

Then h becomes an automorphism of $([a, 1], \leq)$ satisfying

$$h\left(\frac{a}{x}\right) = \frac{a}{h(x)}$$

Let H_a be this set of automorphisms of $([a, 1], \leq)$ just described. There are lots of them: they are in one-to-one correspondence with the automorphisms of $([a, \sqrt{a}], \leq)$. Take any such automorphism and extend it to an automorphism of $([a, 1], \leq)$ as indicated and this gets an element of H_a.

The set H_a is a subgroup of the automorphism group of $([a, 1], \leq)$. *Two generators f and g of nilpotent t-norms with $f(0) = g(0) = a$ give the same negation if and only if $fg^{-1} \in H_a$. In particular, f and hf for any $h \in H_a$ give the same negation:*

$$
\begin{aligned}
(hf)^{-1}\left(\frac{hf(0)}{hf(x)}\right) &= f^{-1}h^{-1}\left(\frac{h(a)}{hf(x)}\right) \\
&= f^{-1}h^{-1}\left(\frac{a}{hf(x)}\right) \\
&= f^{-1}h^{-1}h\left(\frac{a}{f(x)}\right) \\
&= f^{-1}\left(\frac{a}{f(x)}\right) \\
&= f^{-1}\left(\frac{f(0)}{f(x)}\right)
\end{aligned}
$$

A particularly simple class of nilpotent t-norms are those which have straight lines as generators, as in the picture below. For $a \in (0, 1)$ they are the functions $[0, 1] \to [a, 1]$ given by the formula $f(x) = (1 - a)x + a$.

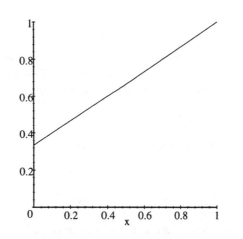

The generator $f(x) = (1 - a)x + a$, with $a = 1/3$

The inverse is $f^{-1}(x) = \frac{x-a}{1-a}$ and the resulting negation is

$$
\begin{aligned}
N_f(x) &= f^{-1}\left(\frac{f(0)}{f(x)}\right) \\
&= f^{-1}\left(\frac{a}{(1-a)x+a}\right) \\
&= \left(\frac{a}{(1-a)x+a} - a\right)/(1-a) \\
&= \frac{1-x}{1+\frac{1-a}{a}x}
\end{aligned}
$$

Since a ranges over $(0,1)$, $\frac{1-a}{a}$ ranges over $(0,\infty)$. A negation of the form

$$
\frac{1-x}{1+bx}
$$

with $b \in (-1,\infty)$ is called a **Sugeno negation**. Therefore the nilpotent t-norms with a linear generator give Sugeno negations, but only those with $b > 0$. Those negations with $b < 0$ come from t-conorms, which we will see in the next section.

5.6   t-conorms

The notion of t-norm plays the role of intersection, or in logical terms, "and". The dual of that notion is that of union, or "or". In the case of sets, union and intersection are related via complements. The well-known De Morgan formulas do that. They are

$$
\begin{aligned}
A \cup B &= (A' \cap B')' \\
A \cap B &= (A' \cup B')'
\end{aligned}
$$

where $'$ indicates complement. But in the fuzzy setting we have many "complements". Any negation plays such a role. For a binary operation $\triangle$ on $[0,1]$, we can define its **dual** with respect to **any** negation η, namely

$$
x \triangledown y = \eta(\eta(x) \triangle \eta(y))
$$

Since η is an involution, this last equation holds if and only if

$$
x \triangle y = \eta(\eta(x) \triangledown \eta(y))
$$

So if these equations hold, then we say that $\triangle$ and $\triangledown$ are **dual with respect to η**. In the case $\triangle$ is a t-norm, then $\triangledown$ is called a **t-conorm**. So a t-conorm is the dual of some t-norm with respect to some negation. The following theorem gives properties characterizing t-conorms.

Theorem 5.6.1 *A binary operation $\triangledown$ on $[0,1]$ is a t-conorm if and only if*

1. $0 \triangledown x = x$

2. $x \triangledown y = y \triangledown x$

3. $x \triangledown (y \triangledown z) = (x \triangledown y) \triangledown z$

4. If $w \le x$ and $y \le z$ then $w \triangledown y \le x \triangledown z$

Proof. Suppose that $\triangledown$ satisfies the conditions of the theorem. Let η be any negation, and let $x \triangle y = \eta(\eta(x) \triangledown \eta(y))$. Then $\triangle$ is a t-norm. For example,

$$
\begin{aligned}
(x \triangle y) \triangle z &= \eta(\eta(x) \triangledown \eta(y)) \triangle z \\
&= \eta[\eta\{\eta(\eta(x) \triangledown \eta(y))\} \triangledown \eta(z)] \\
&= \eta[(\eta(x) \triangledown \eta(y)) \triangledown \eta(z)] \\
&= \eta[\eta(x) \triangledown (\eta(y) \triangledown \eta(z))] \\
&= \eta[\eta(x) \triangledown \eta\{\eta(\eta(y) \triangledown \eta(z))\}] \\
&= x \triangle (y \triangle z)
\end{aligned}
$$

The rest of the verification that $\triangle$ is a t-norm is left for the reader. Conversely, if $x \triangledown y = \eta(\eta(x) \triangle \eta(y))$ for a t-norm $\triangle$ and a negation η, then it is easy to verify that $\triangledown$ satisfies the conditions of the theorem. ∎

The most frequently used negation is $x \to 1 - x$. If a t-norm and t-conorm are dual with respect to this negation, we will just say that they are **dual**. For a nilpotent t-norm, its **natural dual** is its dual with respect to its natural negation. For example, the natural dual of $\blacktriangle$ is $x \blacktriangledown y = (x + y) \wedge 1$, as is easily checked. This conorm is the **Lukasiewicz conorm**.

Definition 5.6.2 *A t-conorm is **Archimedean** if it is dual to a t-norm that is Archimedean, is **nilpotent** if it is dual to a nilpotent t-norm, and is **strict** if it is dual to a strict t-norm.*

A t-conorm $\triangledown$ is nilpotent if and only if for $x \in (0,1]$, $x^{[n]} = 1$ for some n, where $x^{[n]}$ means x conormed with itself n times. If a t-norm or t-conorm is Archimedean, then a dual of it with respect to any negation is Archimedean. Similar statements hold for nilpotent and strict.

If the t-norm $\triangle$ has a generator f, and η is a negation, then for the dual $\triangledown$ of $\triangle$ with respect to η we have

$$
\begin{aligned}
x \triangledown y &= \eta(\eta(x) \triangle \eta(y)) \\
&= \eta f^{-1}(f(\eta(x))(f\eta(y)) \vee f(0)) \\
&= (f\eta)^{-1}((f\eta)(x)(f\eta)(y) \vee (f\eta)(1))
\end{aligned}
$$

Thus Archimedean t-conorms have a representation by functions much in the same way as Archimedean t-norms do. The function $f\eta$ is a **cogenerator** for $\triangledown$. It is an order reversing, one-to-one mapping $[0,1] \to [b,1]$ for some $b \in [0,1)$. If g is such a mapping, then

$$x \triangledown y = g^{-1}(g(x)g(y) \vee g(1))$$

is an Archimedean t-conorm, is nilpotent if and only if $g(1) > 0$, and is strict if and only if $g(1) = 0$. These statements follow easily using the duality of t-conorms with t-norms. The conorm cogenerated by an order reversing, one-to-one mapping $g : [0,1] \to [b,1]$ will be denoted $\bullet_g$. So $\bullet_g$ is a t-norm if g is an isomorphism and a t-conorm if g is an anti-isomorphism.

If f is the L-generator of a nilpotent t-norm $\triangle$, and η is a negation, then

$$
\begin{aligned}
x \triangledown y &= \eta(\eta(x) \triangle \eta(y)) \\
&= \eta f^{-1}(f(\eta(x)) + (f\eta(y) - 1) \vee 0) \\
&= (f\eta)^{-1}((f\eta)(x) + (f\eta)(y) - 1) \vee 0) \\
&= (f\eta)^{-1}((f\eta)(x) \blacktriangle (f\eta)(y))
\end{aligned}
$$

Thus

$$f\eta(x \triangledown y) = (f\eta)(x) \blacktriangle (f\eta)(y)$$

Therefore, $f\eta$ is an anti-isomorphism from $\triangledown$ to $\blacktriangle$, and necessarily the only one. Such an anti-isomorphism is an **L-cogenerator** of the nilpotent t-conorm $\triangledown$.

There is a negation associated with nilpotent t-conorms. The construction is dual to the t-norm case. If $\triangledown$ is a nilpotent t-conorm and g is a cogenerator of it, the negation is

$$N_{\triangledown}(x) = g^{-1}\left(\frac{g(1)}{g(x)}\right)$$

Furthermore, $N(x) = \bigwedge\{y : x \triangledown y = 1\}$, dual to the t-norm case.

The proof of the following theorem is routine.

Theorem 5.6.3 *If f is a generator of a nilpotent t-norm $\triangle$ and β is a negation, then the negation associated with the t-conorm with cogenerator $f\beta$ is $\beta N_{\triangle}\beta$. That is, $N_{f\beta} = \beta N_f \beta$. In particular, if β is the natural negation of $\triangle$, then $N_{f\beta} = N_f$.*

The cogenerators of nilpotent t-conorms that are straight lines give Sugeno negations $(1-x)/(1+\lambda x)$ for $\lambda \in (-1,0)$. This is left as an exercise.

5.7 De Morgan systems

Let $\triangle$ be a t-norm and β a negation. Then $\triangledown$ defined by $x \triangledown y = \beta(\beta(x) \triangle \beta(y))$ is a t-conorm. If a t-norm and t-conorm are related in this way by the negation β, then $(\mathbb{I}, \triangle, \beta, \triangledown)$ is a **De Morgan system**, or a **De Morgan triple**, and the t-norm $\triangle$ and the t-conorm $\triangledown$ are said to be **dual to one another** via the negation β.

For notational reasons, we are going to adorn our operators with their generators. So a De Morgan system looks like $(\mathbb{I}, \bullet_f, \alpha_g, \bullet_h)$. Being a De Morgan system implies however that $\bullet_h = \bullet_{f\alpha_g}$. Now suppose that

$$q : (\mathbb{I}, \bullet_f, \alpha_g, \bullet_h) \to (\mathbb{I}, \bullet_u, \alpha_v, \bullet_w)$$

is an isomorphism. Then $q \in Aut(\mathbb{I})$ and the following hold.

$$\begin{aligned}
q(x \bullet_f y) &= q(x) \bullet_u q(y) \\
q(\alpha_g(x)) &= \alpha_v q(x) \\
q(x \bullet_h y) &= q(x) \bullet_w q(y)
\end{aligned}$$

But since $x \bullet_h y = \alpha_g(\alpha_g(x) \bullet_f \alpha_g(y))$ and $x \bullet_w y = \alpha_v(\alpha_v(x) \bullet_u \alpha_v(y))$, if the first two equations hold, then

$$\begin{aligned}
q(x \bullet_h y) &= q\left(\alpha_g(\alpha_g(x) \bullet_f \alpha_g(y))\right) \\
&= \alpha_v q((\alpha_g(x) \bullet_f \alpha_g(y))) \\
&= \alpha_v(q(\alpha_g(x)) \bullet_u q(\alpha_g(y))) \\
&= \alpha_v(\alpha_v(q(x)) \bullet_u (\alpha_v q(y))) \\
&= q(x) \bullet_w q(y)
\end{aligned}$$

Therefore to be an isomorphism, q need only be required to satisfy the first two conditions. That is, isomorphisms from $(\mathbb{I}, \bullet_f, \alpha_g, \bullet_h)$ to $(\mathbb{I}, \bullet_u, \alpha_v, \bullet_w)$ are the same as isomorphisms from $(\mathbb{I}, \bullet_f, \alpha_g)$ to $(\mathbb{I}, \bullet_u, \alpha_v)$. We will also call these systems **De Morgan systems**.

If the t-norm of a De Morgan system is strict, then so is its t-conorm, and if the t-norm is nilpotent, so is its t-conorm. It is convenient to separate the two cases—**strict De Morgan systems** and **nilpotent De Morgan systems**.

5.7.1 Strict De Morgan systems

In this section, all our De Morgan systems will be strict. To determine the isomorphisms q from $(\mathbb{I}, \bullet_f, \alpha_g)$ to $(\mathbb{I}, \bullet_u, \alpha_v)$, we just note that such a q must be an isomorphism from $(\mathbb{I}, \bullet_f)$ to $(\mathbb{I}, \bullet_u)$ and from $(\mathbb{I}, \alpha_g)$ to $(\mathbb{I}, \alpha_v)$. Therefore, from Theorems 5.3.6 and 5.4.4, we get the following theorem.

Theorem 5.7.1 *The set of isomorphisms from* $(\mathbb{I}, \bullet_f, \alpha_g)$ *to* $(\mathbb{I}, \bullet_u, \alpha_v)$
is the set

$$\left(u^{-1}\mathbb{R}^+f\right) \cap \left(v^{-1}Z(\alpha)g\right)$$

This intersection may be empty, of course. That is the case when the
equation $u^{-1}rf = v^{-1}zg$ has no solution for $r > 0$ and $z \in Z(\alpha)$. A
particular example of this is the case where $f = g = u = 1$, $v \notin Z(\alpha)$,
and $v\left(\frac{1}{2}\right) = \frac{1}{2}$. Then $r = v^{-1}z$ with $r > 0$ and $z \in Z(\alpha)$. But then

$$r\left(\frac{1}{2}\right) = v^{-1}z\left(\frac{1}{2}\right) = \left(\frac{1}{2}\right)^r = \frac{1}{2}$$

Thus $r = 1$, and so $v = z$. But $v \notin Z(\alpha)$. So there are De Morgan
systems $(\mathbb{I}, \bullet_f, \alpha_g)$ and $(\mathbb{I}, \bullet_u, \alpha_v)$ which are not isomorphic. When two
De Morgan systems are isomorphic, the isomorphism is unique. We need
a lemma.

Lemma 5.7.2 *For any f and $g \in Map(\mathbb{I})$,*

$$\left(f^{-1}\mathbb{R}^+f\right) \cap \left(g^{-1}Z(\alpha)g\right) = \{1\}$$

Proof. If $f^{-1}rf = g^{-1}zg$, then $gf^{-1}r = zgf^{-1}$. There is $x \in [0, 1]$
such that $gf^{-1}(x) = \frac{1}{2}$. For this x, $gf^{-1}r(x) = zgf^{-1}(x) = z\left(\frac{1}{2}\right) = \frac{1}{2}$,
and so $gf^{-1}(x^r) = \frac{1}{2}$. But $gf^{-1}(x) = \frac{1}{2}$, and since gf^{-1} is one-to-one,
$r = 1$ and the lemma follows. ∎

Theorem 5.7.3 $(\mathbb{I}, \bullet_f, \alpha_g) \approx (\mathbb{I}, \bullet_u, \alpha_v)$ *if and only if*

$$(\mathbb{I}, \bullet_u, \alpha_v) = (\mathbb{I}, \bullet_{fh}, \alpha_{gh})$$

*for some $h \in Aut(\mathbb{I})$, in which case h^{-1} is the only such isomorphism. In
particular,* $(\mathbb{I}, \bullet_f, \alpha_g) \approx (\mathbb{I}, \bullet, \alpha_{gf^{-1}})$.

Proof. It is easy to check that h^{-1} is an isomorphism from $(\mathbb{I}, \bullet_f, \alpha_g)$
to $(\mathbb{I}, \bullet_{fh}, \alpha_{gh})$. If k is such an isomorphism, then $k = u^{-1}rf = v^{-1}zg$
for some $r \in \mathbb{R}^+$ and $z \in Z(\alpha)$. Thus $u = rfk^{-1}$ and $v = zgk^{-1}$ and so
$(\mathbb{I}, \bullet_u, \alpha_v) = (\mathbb{I}, \bullet_{fk^{-1}}, \alpha_{gk^{-1}})$. If k were distinct from h^{-1}, then kh would
be a nontrivial automorphism of $(\mathbb{I}, \bullet_f, \alpha_g)$. But by the lemma, this is
impossible. ∎

One implication of this theorem, taking $f = g$, is that the theory of
the strict De Morgan systems $(\mathbb{I}, \bullet_f, \alpha_f)$ is the same as that of $(\mathbb{I}, \bullet, \alpha)$.
More generally this holds for $(\mathbb{I}, \bullet_f, \alpha_g)$ and $(\mathbb{I}, \bullet, \alpha_{gf^{-1}})$. This suggests
that in applications of strict De Morgan systems, one may as well take
the strict t-norm to be ordinary multiplication.

Corollary 5.7.4 $Aut((\mathbb{I}, \bullet_f, \alpha_g)) = \{1\}$.

Corollary 5.7.5 $(\mathbb{I}, \bullet, \beta) \approx (\mathbb{I}, \bullet, \gamma)$ *if and only if* $\gamma = r^{-1}\beta r$ *for some* $r \in \mathbb{R}^+$.

Taking $\beta = \alpha$ in this last Corollary, we see that $(\mathbb{I}, \bullet, \alpha) \approx (\mathbb{I}, \bullet, \gamma)$ if and only if $\gamma = r^{-1}\alpha r$ for some $r \in \mathbb{R}^+$. So strict De Morgan systems isomorphic to $(\mathbb{I}, \bullet, \alpha)$ are exactly those of the form $(\mathbb{I}, \bullet, \alpha_r)$ with $r \in \mathbb{R}^+$. Negations of the form $r^{-1}\alpha r$ are **Yager negations** [104]. Thus we have the following corollary.

Corollary 5.7.6 *The De Morgan systems* $(\mathbb{I}, \bullet, \beta)$ *which are isomorphic to* $(\mathbb{I}, \bullet, \alpha)$ *are precisely those with* β *a Yager negation.*

Here are some examples of strict De Morgan systems. It is easy to specify them. Just pick an $f \in Aut(\mathbb{I})$ and a negation η. Then $(\mathbb{I}, \bullet_f, \eta, \bullet_{f\eta})$ is a De Morgan system. Since the t-conorm $\bullet_{f\eta}$ is determined by $\bullet_f$ and η, one need specify only the t-norm and negation. But it may not be easy to compute the t-norms, negations, and t-conorms from given generators. In each of the examples below, we give t-norms, negations, t-conorms, and the generators of the t-norms. Also, each example will be a family of examples. Taking the same parameter(s) for the t-norm, t-conorm, and negation, gives a De Morgan system.

Dombi

t-norm
$$\frac{1}{1 + \left[(\frac{1-x}{x})^r + (\frac{1-y}{y})^r\right]^{\frac{1}{r}}} : r > 0$$

negation $\quad 1 - x$

t-conorm
$$\frac{1}{1 + \left[(\frac{1-x}{x})^{-r} + (\frac{1-y}{y})^{-r}\right]^{\frac{1}{-r}}} : r > 0$$

generator $\quad e^{-\left(\frac{1-x}{x}\right)^r} : r > 0$

Jane Doe #1-Hamacher

t-norm
$$\dfrac{1}{1+\left[(\frac{1-x}{x})^r+(\frac{1-y}{y})^r+a(\frac{1-x}{x})^r(\frac{1-y}{y})^r\right]^{\frac{1}{r}}} \quad : a>0,\ r>0$$

negation
$$\dfrac{1-x}{1+(a-1)x} \quad : a>0$$

t-conorm
$$\dfrac{1}{1+\left[(\frac{1-x}{x})^{-r}+(\frac{1-y}{y})^{-r}+a(\frac{1-x}{x})^{-r}(\frac{1-y}{y})^r\right]^{\frac{1}{-r}}} \quad : a>0,\ r>0$$

generator
$$\dfrac{x^r}{x^r+a(1-x)^r} \quad : a>0,\ r>0$$

Aczél-Alsina

t-norm
$$e^{-((-\ln x)^r+(-\ln y)^r)^{\frac{1}{r}}} \quad : r>0$$

negation
$$e^{\frac{1}{\ln x}}$$

t-conorm
$$e^{-\left((-\ln x)^{-r}+(-\ln y)^{-r}\right)^{\frac{1}{-r}}} \quad : r>0$$

generator
$$e^{-(-\ln x)^r} \quad : r>0$$

Frank

t-norm
$$\log_a\left[1+\dfrac{(a^x-1)(a^y-1)}{a-1}\right] \quad : a>0,\ a\neq 1$$

negation
$$1-x$$

t-conorm
$$1-\log_a\left[1+\dfrac{(a^{1-x}-1)(a^{1-y}-1)}{a-1}\right] \quad : a>0,\ a\neq 1$$

generator
$$\dfrac{a^x-1}{a-1} \quad : a>0,\ a\neq 1$$

Schweizer

t-norm $\left(\frac{1}{x^a} + \frac{1}{y^a} - 1\right)^{-\frac{1}{a}} : a > 0$

negation $1 - x$

t-conorm $1 - \left(\frac{1}{(1-x)^a} + \frac{1}{(1-y)^a} - 1\right)^{-\frac{1}{a}} : a > 0$

generator $e^{1-x^{-a}} : a > 0$

Jane Doe #2

t-norm $xye^{-a \ln x \ln y} : a > 0$

negation $1 - x$

t-conorm $1 - (1-x)(1-y)e^{a \ln(1-x) \ln(1-y)} : a > 0$

generator $\frac{1}{1 - a \ln x} : a > 0$

Jane Doe #3

t-norm $1 - (1 - (1 - (1-x)^a)(1 - (1-y)^a))^{\frac{1}{a}} : a > 0$

negation $1 - x$

t-conorm $(x^a + y^a - x^a y^a)^{\frac{1}{a}} : a > 0$

generator $1 - (1-x)^a : a > 0$

5.7.2 Nilpotent De Morgan systems

First we determine the isomorphisms between two nilpotent De Morgan systems $(\mathbb{I}, \bullet_f, \alpha_g)$ and $(\mathbb{I}, \bullet_u, \alpha_v)$. By Theorem 5.3.10, there is exactly one isomorphism from $(\mathbb{I}, \bullet_f)$ to $(\mathbb{I}, \bullet_u)$, namely $u^{-1}rf$ where $(rf)(0) = u(0)$. By Theorem 5.4.4, the isomorphisms from $(\mathbb{I}, \alpha_g)$ to $(\mathbb{I}, \alpha_v)$ are the maps $v^{-1}Z(\alpha)g$. So $u^{-1}rf$ must satisfy the condition that $v(u^{-1}rf)g^{-1} \in Z(\alpha)$. All this translates to the following.

Theorem 5.7.7 *For nilpotent De Morgan systems,* $(\mathbb{I}, \bullet_f, \alpha_g) \approx (\mathbb{I}, \bullet_u, \alpha_v)$ *if and only if*

$$(\mathbb{I}, \bullet_u, \alpha_v) = (\mathbb{I}, \bullet_{fh}, \alpha_{gh})$$

for some $h \in Aut(\mathbb{I})$, *in which case* h^{-1} *is the only such isomorphism.*

Note that in particular, taking $u(x) = e^{x-1}$ and $r = -\ln f(0)$

$$h^{-1} : (\mathbb{I}, \bullet_f, \alpha_g) \to (\mathbb{I}, \bullet_u, \alpha_{gf^{-1}ru}) = (\mathbb{I}, \blacktriangle, \alpha_v)$$

where $h^{-1}(x) = u^{-1}r^{-1}f = 1 + (\ln f(x))/\ln f(0)$, and $x \blacktriangle y = (x + y - 1) \vee 0$. Thus any nilpotent De Morgan system is isomorphic to one whose t-norm is the Lukasiewicz t-norm, just as any strict De Morgan system is isomorphic to one whose t-norm is multiplication.

It is a little simpler to use L-generators here. Recall that

$$x \blacktriangle_f y = f^{-1}(f(x) \blacktriangle f(y))$$

where $x \blacktriangle y = (x+y-1) \vee 0$. The only isomorphism from $(\mathbb{I}, \blacktriangle_f)$ to $(\mathbb{I}, \blacktriangle_u)$ is $u^{-1}f$, and so $(\mathbb{I}, \blacktriangle_f, \alpha_g) \approx (\mathbb{I}, \blacktriangle_u, \alpha_v)$ occurs only when $vu^{-1}fg^{-1} \in Z(\alpha)$. We get the same conclusion: $(\mathbb{I}, \blacktriangle_f, \alpha_g) \approx (\mathbb{I}, \blacktriangle_u, \alpha_v)$ if and only if $(\mathbb{I}, \blacktriangle_u, \alpha_v) = (\mathbb{I}, \blacktriangle_{fh}, \alpha_{gh})$.

There are some very special nilpotent De Morgan systems. Nilpotent t-norms have their natural negations, and for $\blacktriangle_f$ that natural negation is $f^{-1}\alpha f = \alpha_f$. We examine these De Morgan systems $(\mathbb{I}, \blacktriangle_f, \alpha_f)$.

Definition 5.7.8 *If* $\triangle$ *is a binary operation on a lattice* $\mathbb{L}$ *with* 0, *an element* x^* *in* $\mathbb{L}$ *is the* $\triangle$-***pseudocomplement*** *of an element* x *if* $x \triangle y = 0$ *exactly when* $y \le x^*$. *If every element* x *of* $\mathbb{L}$ *has a* $\triangle$-*pseudocomplement* x^*, *then* $*$ *is a* $\triangle$-*pseudocomplement for* $\mathbb{L}$.

If $\triangle$ is the meet operation $\wedge$ of the lattice, then $\wedge$-pseudocomplement is simply pseudocomplement as defined in Chapter 2. An element has at most one $\triangle$-pseudocomplement. For a nilpotent t-norm, $\triangle$, the natural negation $\eta_\triangle$ is a $\triangle$-pseudocomplement. Equivalently, α_f is a $\blacktriangle_f$-pseudocomplement. Pseudocomplements do not have to be negations, but these are. There are nilpotent De Morgan systems reminiscent of Stone algebras and Boolean algebras. We refer to [35] for a discussion of such De Morgan systems. We mention briefly here those reminiscent of Boolean algebras. These algebraic systems, which we call Boolean systems, are examples of MV-algebras, an important topic of study in fuzzy mathematics [15].

Definition 5.7.9 *The De Morgan system* $(\mathbb{I}, \triangle, \eta, \triangledown)$ *is a* ***Boolean system*** *if* η *is a* $\triangle$-*pseudocomplement and* $x \triangledown \eta(x) = 1$.

Theorem 5.7.10 *The De Morgan system* $(\mathbb{I}, \triangle, \eta, \triangledown)$ *is a Boolean system if and only if* $\triangle$ *is nilpotent and* η *is its natural negation. Thus any Boolean system is of the form* $(\mathbb{I}, \blacktriangle_f, \alpha_f, \blacktriangledown_{f\alpha_f})$.

Proof. If η is the natural negation of $\triangle$, then it is the $\triangle$-pseudo-complement, and $\eta(x \triangledown \eta(x)) = \eta(x) \triangle \eta\eta(x) = \eta(x) \triangle (x) = 0$, so $x \triangledown \eta(x) = 1$, and so $(\mathbb{I}, \triangle, \eta, \triangledown)$ is a Boolean system.

If $(\mathbb{I}, \triangle, \eta, \triangledown)$ is a Boolean system, then for $a \in (0,1)$, $a \triangle \eta(a) = 0$, whence $\triangle$ is nilpotent. ∎

A Boolean system is determined by its t-norm, but, as we know, many nilpotent t-norms have the same natural negation. The systems $(\mathbb{I}, \blacktriangle_f, \alpha_g, \blacktriangledown_{f\alpha_f})$ are De Morgan systems if and only if $\alpha_f = \alpha_g$, that is, if and only if $f = zg$ for some $z \in Z(\alpha)$. So in the Boolean systems $(\mathbb{I}, \blacktriangle_{zg}, \alpha_g, \blacktriangledown_{zgf\alpha_f})$, changing the z changes the t-norms and t-conorms, but does not change the negation. And $Z(\alpha)$ is an uncountable group, so there are uncountably many different Boolean systems with the same nilpotent t-norm and t-conorm. Still any two are isomorphic.

Theorem 5.7.11 *Any two Boolean systems are isomorphic.*

Proof. If $(\mathbb{I}, \blacktriangle_f, \alpha_f, \blacktriangledown_{f\alpha_f})$ and $(\mathbb{I}, \blacktriangle_g, \alpha_g, \blacktriangledown_{g\alpha_f})$ are Boolean systems, then fg^{-1} is an isomorphism between them, in fact the only one. ∎

A Boolean system is determined by its t-norm $\blacktriangle_f$ with unique $f \in Aut(\mathbb{I})$. Here are some examples of Boolean systems. We give, for each parameter, a t-norm, its natural negation, its t-conorm with respect to that natural negation, and the L-generator of the t-norm.

Schweizer-Sklar t-norm, Yager t-conorm

t-norm	$((x^a + y^a - 1) \vee 0)^{\frac{1}{a}} : a > 0$
natural negation	$(1 - x^a)^{\frac{1}{a}} : a > 0$
t-conorm	$((x^a + y^a))^{\frac{1}{a}} \wedge 1 : a > 0$
L-generators	$x^a : a > 0$

Yager t-norm, Schweizer-Sklar t-conorm

t-norm	$1 - (1 \wedge ((1-x)^a + (1-y)^a)^{\frac{1}{a}}) : a > 0$
natural negation	$1 - (1 - (1-x)^a)^{\frac{1}{a}} : a > 0$
t-conorm	$1 - (0 \vee ((1-x)^a + (1-y)^a)^{\frac{1}{a}}) : a > 0$
L-generators	$1 - (1-x)^a : a > 0$

Weber t-norm, Jane Doe #5 t-conorm

t-norm	$(a(x+y-1) - (a-1)xy) \vee 0 : a > 0, \ a \neq 1$
natural negation	$\dfrac{a(1-x)}{1 + a(1-x)} : a > 0, \ a \neq 1$
t-conorm	$(x + y - \frac{a-1}{a}xy) \wedge 1 : a > 0, \ a \neq 1$
L-generators	$1 - \log_a((a-1)(1-x) + 1) : a > 0, \ a \neq 1$

Jane Doe #5 t-norm, Weber t-conorm,

t-norm	$(\frac{1}{a}(x + y - 1 - (a-1)xy) \vee 0 : a > 0, \ a \neq 1$
natural negation	$\dfrac{1-x}{1 + (a-1)x} : a > 0, \ a \neq 1$
t-conorm	$(x + y + (a-1)xy) \wedge 1 : a > 0, \ a \neq 1$
L-generators	$\log_a((a-1)x + 1) : a > 0, \ a \neq 1$

The system $([0,1], \wedge, \vee, ')$, where $\wedge$, $\vee$, and $'$ are max, min, and $x' = 1 - x$ forms a De Morgan algebra in the usual lattice theoretic sense. If we replace $'$, which we have been denoting by α, by any other involution β, then the systems $([0,1], \wedge, \vee, ')$ and $([0,1], \wedge, \vee, \beta)$ are isomorphic. Isomorphisms between these algebras are exactly the isomorphisms between $(\mathbb{I}, \alpha)$ and $(\mathbb{I}, \beta)$. There are many such isomorphisms and these are spelled out in Theorem 5.4.4. This suggests that in applications of De Morgan systems where $\wedge$ and $\vee$ are taken for the t-norm and t-conorm, respectively, the negation may as well be $\alpha(x) = 1 - x$.

5.7.3 Non-uniqueness of negations in strict De Morgan systems

We have noted that a De Morgan system $(\mathbb{I}, \triangle, \beta, \triangledown)$ is determined by the system $(\mathbb{I}, \triangle, \beta)$. Of course, it is also determined by the system $(\mathbb{I}, \beta, \triangledown)$. Is it determined by $(\mathbb{I}, \triangle, \triangledown)$? How unique is the negation in a De Morgan system? The following lemma is straightforward and applies to both strict and nilpotent De Morgan systems.

Lemma 5.7.12 *If* $(\mathbb{I}, \triangle, \beta, \triangledown)$ *and* $(\mathbb{I}, \triangle, \gamma, \triangledown)$ *are De Morgan systems having the same t-norm and same t-conorm, then* $\gamma\beta \in Aut\,(\mathbb{I}, \triangle)$.

Proof. We need

$$\gamma\beta(a \triangle b) = \gamma\beta(a) \triangle \gamma\beta(b)$$

By hypothesis

$$\beta(\beta(a) \triangle \beta(b)) = \gamma(\gamma(a) \triangle \gamma(b))$$

Applying this to elements $\gamma(a)$ and $\gamma(b)$, we get

$$\beta(\beta\gamma(a) \triangle \beta\gamma(b)) = \gamma(a \triangle b)$$

and so

$$\beta\gamma(a) \triangle \beta\gamma(b) = \beta\gamma(a \triangle b)$$

as desired. ■

Corollary 5.7.13 *If* $(\mathbb{I}, \triangle, \beta, \triangledown)$ *and* $(\mathbb{I}, \triangle, \gamma, \triangledown)$ *are nilpotent De Morgan systems, then* $\beta = \gamma$.

The situation is more complicated in strict De Morgan systems. Suppose that $(\mathbb{I}, \triangle, \beta, \triangledown)$ and $(\mathbb{I}, \triangle, \gamma, \triangledown)$ are strict De Morgan systems and f is a generator of $\triangle$. We know from the lemma that $\gamma\beta$ is an automorphism of $(\mathbb{I}, \triangle)$. But automorphisms of $(\mathbb{I}, \triangle)$ are of the form $f^{-1}rf$ for $r > 0$. Thus $\gamma\beta = f^{-1}rf$ and $\beta = \gamma f^{-1}rf$. Now β is of order 2, and so is $f^{-1}\beta f$. Thus

$$\beta = \gamma f^{-1}rf = f^{-1}r^{-1}f\gamma$$

and

$$
\begin{aligned}
f\beta f^{-1} &= f\left(f^{-1}r^{-1}f\gamma\right)f^{-1}\\
&= r^{-1}f\gamma f^{-1} = f\gamma f^{-1}r = rf\beta f^{-1}r
\end{aligned}
$$

and so $rf\beta f^{-1}r = f\beta f^{-1}$. Let $\eta = f\beta f^{-1}$. Then η is an involution and $r\eta r = \eta$.

On the other hand, if η is an involution such that for some $r > 0$, $r\eta r = \eta$, then it is routine to check that for any t-norm $\circ$ with generator f, $f^{-1}\eta f$ and $f^{-1}\eta rf$ are negations which give the same t-conorm. Are there such involutions η? Yes, of course, with $r = 1$. But when $r = 1$, $\gamma\beta = f^{-1}rf = 1$ and $\gamma = \beta$. Are there such involutions with $r \neq 1$?

Let $\varepsilon(x) = e^{1/\ln x}$. Then $\varepsilon r = r^{-1}\varepsilon$ for any positive real number r. Moreover, for any $a > 0$, $a\varepsilon$ is an involution satisfying $ra\varepsilon r = a\varepsilon$. So $f^{-1}a\varepsilon f$ and $f^{-1}a\varepsilon rf$ are negations that give the same t-conorm. We get the following theorem.

Theorem 5.7.14 *Let $\triangle$ be a strict t-norm with generator f, $\varepsilon(x) = e^{1/\ln x}$, and let a and b be positive real numbers. Then $f^{-1}a\varepsilon f$ and $f^{-1}b\varepsilon f$ are negations that give the same t-conorm for $\triangle$.*

We look a moment at the case when the t-norm is multiplication. Suppose that the De Morgan system $(\mathbb{I}, \bullet, a\varepsilon)$ is isomorphic to $(\mathbb{I}, \bullet, \beta)$. By the lemma, $a\varepsilon\beta = r$ for some $r \in \mathbb{R}^+$. So $\beta = r^{-1}a\varepsilon$. We sum up.

Corollary 5.7.15 *The De Morgan systems $(\mathbb{I}, \bullet, a\varepsilon)$ for $a \in \mathbb{R}^+$ are all isomorphic. If $(\mathbb{I}, \bullet, a\varepsilon) \approx (\mathbb{I}, \bullet, \beta)$, then $\beta = r\varepsilon$ for some $r \in \mathbb{R}^+$.*

The negations $a\varepsilon$ satisfy $ra\varepsilon r = a\varepsilon$ for *all* $r \in \mathbb{R}^+$. There are no other negations with this property. However, there is a large family of negations β such that $r\beta r = \beta$ for a *fixed* $r \neq 1$ [32]. Thus for each one of these negations β, the negations $\gamma = f^{-1}\beta f$ and $\delta = f^{-1}\beta rf$ produce the same t-conorm from the t-norm generated by f. It seems likely that for some such negations the De Morgan systems $(\mathbb{I}, \bullet_f, f^{-1}\beta f)$ and $(\mathbb{I}, \bullet_f, f^{-1}\beta rf)$ are not isomorphic even though the t-conorms are the same. They will be, however, whenever $\sqrt{r}\beta\sqrt{r} = \beta$ also holds. Constructing and somehow classifying all such β seems not to have been done.

Let $\triangle$ and $\triangledown$ be a strict t-norm and strict t-conorm with generator f and cogenerator g, respectively. When does there exist a negation β such that $\triangle$ and $\triangledown$ are dual with respect to β? This means finding a negation β such that

$$\beta(f^{-1}(f(\beta(x))f(\beta(y)))) = g^{-1}(g(x)g(y))$$

This in turn means that $f\beta = rg$ for some $r > 0$. The existence of such a β is equivalent to the existence of a negation in the set $f^{-1}\mathbb{R}^+g$. There may be many or there may be no such negations.

5.8 Groups and t-norms

The group $\mathbb{A}$ of automorphisms of $\mathbb{I} = ([0,1], \leq)$ has played a fundamental role in our development of Archimedean t-norms. Every strict t-norm $\triangle$ comes from an element $f \in \mathbb{A}$ via $x \triangle y = f^{-1}(f(x)f(y))$, and every nilpotent one via $x \triangle y = f^{-1}((f(x) + f(y) - 1) \vee 0)$. For nilpotent ones, the f is unique. For strict ones, f is unique up to a left multiple by an element of $\mathbb{R}^+$. That is, f and rf, for $r \in \mathbb{R}^+$ give the same strict t-norm. The antiautomorphisms in the group $\mathbb{M}$ of all automorphisms and antiautomorphisms of $\mathbb{I}$ provided generators for t-conorms, and the involutions in $\mathbb{M}$ are the negations in De Morgan systems. The symmetries, that is, the automorphism groups of the algebraic systems arising from $\mathbb{I}$ with additional operations such as a t-norm, or a negation, or both, are of course subgroups of $\mathbb{A}$. Suffice it to say that these groups form a basis for Archimedean t-norm theory. In this section, we will investigate further some group theoretic aspects of t-norms. A particular point we want to make is that many of the examples of families of t-norms and t-conorms arise from simple combinations of a very few subgroups of $\mathbb{M}$.

5.8.1 The normalizer of $\mathbb{R}^+$

The multiplicative group $\mathbb{R}^+$ of positive real numbers is identified with a subgroup of $\mathbb{A}$ by $r(x) = x^r$ for each positive r. Multiplication of elements of $\mathbb{R}^+$ corresponds to composition of functions in $\mathbb{A}$. This subgroup $\mathbb{R}^+$ gained attention from the fact that the set of strict t-norms is in one-to-one correspondence with the set of right cosets of $\mathbb{R}^+$ in $\mathbb{A}$. This is the set $\{\mathbb{R}^+ f : f \in \mathbb{A}\}$, and it forms a partition of $\mathbb{A}$. Any two of the $\mathbb{R}^+ f$ are either equal or disjoint, and their union is all of $\mathbb{A}$. This is purely a group theoretic fact. If S is a subgroup of a group G, then $\{Sg : g \in G\}$ is a partition of G. Were $\mathbb{R}^+$ a normal subgroup of $\mathbb{A}$, that is, if $f^{-1}\mathbb{R}^+ f = \mathbb{R}^+$ for all $f \in \mathbb{A}$, then the set of these cosets would themselves form a group under the operation $(\mathbb{R}^+ f)(\mathbb{R}^+ g) = \mathbb{R}^+ fg$. This would put a group structure on the set of strict t-norms. But this is not the case. The group $\mathbb{R}^+$ is not a normal subgroup of $\mathbb{A}$. However, for any subgroup S of a group G, there is a unique largest subgroup $N(S)$ of G in which S is normal. This is the subgroup $\{g \in G : g^{-1}Sg = S\}$, and is called the **normalizer of S in G**. The first problem that arises is the identification of $N(\mathbb{R}^+)$. For this we will consider $\mathbb{R}^+$ as a subgroup of $\mathbb{M}$, so $N(\mathbb{R}^+)$ is the set $\{f \in \mathbb{M} : f^{-1}\mathbb{R}^+ f = \mathbb{R}^+\}$. There is a t-norm and t-conorm point to this. The elements of $N(\mathbb{R}^+)$ generate a set of t-norms and t-conorms which carry a group structure, and besides its own intrinsic interest, it turns out that many well-known families of t-norms and t-conorms arise rather directly from $N(\mathbb{R}^+)$.

We now proceed to the determination of $N(\mathbb{R}^+)$. If $f \in N(\mathbb{R}^+)$, then $f^{-1}\mathbb{R}^+f = \{f^{-1}rf : r \in \mathbb{R}^+\} = \mathbb{R}^+$, and in fact, $\varphi(r) = f^{-1}rf$ is an automorphism of $\mathbb{R}^+$. This just means that φ is one-to-one and onto and $\varphi(rs) = \varphi(r)\varphi(s)$. But it is more. The mapping φ is order preserving if f is an automorphism of $\mathbb{I}$, and order reversing if f is an antiautomorphism of $\mathbb{I}$. To check that φ is an automorphism of the group $\mathbb{R}^+$ is easy. To see that it preserves order if f is an automorphism of $\mathbb{I}$, suppose that $r < s$. For $\varphi(r) < \varphi(s)$, it suffices to show that for any $x \in [0,1]$, $x^{\varphi(r)} > x^{\varphi(s)}$. We have, using the fact that f^{-1} is an automorphism and hence preserves order,

$$
\begin{aligned}
x^{\varphi(r)} &= x^{f^{-1}rf} \\
&= f^{-1}rf(x) \\
&= f^{-1}(f(x))^r \\
&> f^{-1}(f(x))^s \\
&= x^{\varphi(s)}
\end{aligned}
$$

Similarly, if f is an antiautomorphism, then $\varphi(r) > \varphi(s)$, whenever $r < s$.

Now we need to determine the order preserving automorphisms and the order reversing automorphisms of $\mathbb{R}^+$.

Theorem 5.8.1 *The order preserving automorphisms of the multiplicative group $\mathbb{R}^+$ of positive real numbers are the maps $x \to x^t$ for positive real numbers t. The order reversing ones are the maps $x \to x^t$ for negative real numbers t.*

This theorem is well known. The proof is outlined in the exercises at the end of this chapter.

We are now ready to determine the normalizer $N(\mathbb{R}^+)$ of $\mathbb{R}^+$. Let $f \in N(\mathbb{R}^+)$. Then f^{-1} is also in the normalizer of $\mathbb{R}^+$, and $r \to frf^{-1}$ is an automorphism of $\mathbb{R}^+$. Thus $frf^{-1} = r^t$ for some nonzero real number t. Thus $fr = r^tf$, and this means that for $x \in [0,1]$, $f(x^r) = (f(x))^{r^t}$, and

$$
\begin{aligned}
f(x) &= f\left(\left(\frac{1}{e}\right)^{-\ln x}\right) \\
&= \left(f\left(\frac{1}{e}\right)\right)^{(-\ln x)^t}
\end{aligned}
$$

Writing $f(\frac{1}{e}) = e^{-a}$ for a positive a, we get that $f(x) = e^{-a(-\ln x)^t}$. It is easy to check that $t > 0$ gives automorphisms, $t < 0$ gives antiautomorphisms, and $t = -1$ gives negations, or involutions. Hence we have the following theorem.

Theorem 5.8.2 *The normalizer $N(\mathbb{R}^+)$ of $\mathbb{R}^+$ in $\mathbb{M}$ is given by*

$$N(\mathbb{R}^+) = \left\{ f \in \mathbb{M} : f(x) = e^{-a(-\ln x)^t}, a > 0, t \neq 0 \right\}$$

The function $f(x) = e^{-a(-\ln x)^t}$ is an automorphism of $\mathbb{I}$ if $t > 0$, an antiautomorphism if $t < 0$, and a negation if $t = -1$.

We now look more closely at the group structure of $N(\mathbb{R}^+)$. Let $\mathbb{R}^*$ denote the multiplicative group of nonzero real numbers. For $f(x) = e^{-a(-\ln x)^t}$ and $g(x) = e^{-a'(-\ln x)^{t'}}$,

$$
\begin{aligned}
(gf)(x) &= e^{-a'\left(-\ln\left(e^{-a(-\ln x)^t}\right)\right)^{t'}} \\
&= e^{-c'\left(c(-\ln x)^t\right)^{t'}} \\
&= e^{-c'(c)^{t'}(-\ln x)^{tt'}}
\end{aligned}
$$

Thus we have the following.

Corollary 5.8.3 *The normalizer $N(\mathbb{R}^+)$ of $\mathbb{R}^+$ in $\mathbb{M}$ is isomorphic to the group*

$$\{(c,t) : c > 0, t \neq 0\}$$

with multiplication given by

$$(c',t')(c,t) = \left(c'c^{t'}, t't\right).$$

The subgroup $\mathbb{R}^+$ corresponds to $\{(c,1) : c \in \mathbb{R}^+\}$ and the group $N(\mathbb{R}^+)/\mathbb{R}^+$ with $\{(1,t) : t \neq 0\}$. Thus the natural group structure carried by the set of norms and conorms with generators in $N(\mathbb{R}^+)$ is the multiplicative group $\mathbb{R}^*$ of the nonzero real numbers.

The group $N(\mathbb{R}^+)$ splits: $N(\mathbb{R}^+) = \mathbb{R}^+ \times \mathbb{T}$, with $\mathbb{R}^+$ normal and $\mathbb{T}$ isomorphic to $\mathbb{R}^*$.

To find the norms and conorms with generators in $N(\mathbb{R}^+)$, for $f \in N(\mathbb{R}^+)$, we must compute $f^{-1}(f(x)f(y))$. If

$$f(x) = e^{-c(-\ln x)^t}$$

then

$$f^{-1}(x) = e^{-\left(-\frac{\ln x}{c}\right)^{\frac{1}{t}}}$$

and

$$f^{-1}\left(f\left(x\right)f\left(y\right)\right) = f^{-1}\left(e^{-c(-\ln x)^t - c(-\ln y)^t}\right)$$

$$= e^{-\left(-\frac{-c(-\ln x)^t - c(-\ln y)^t}{c}\right)^{\frac{1}{t}}}$$

$$= e^{-\left((-\ln x)^t + (-\ln y)^t\right)^{\frac{1}{t}}}$$

Of course, the quantity c does not appear in the formula since the norm or conorm generated by f is independent of c. It is straightforward to check the items in the following corollary.

Corollary 5.8.4 *The t-norms with generators in $N(\mathbb{R}^+)$ are given by*

$$x \triangle y = e^{-\left((-\ln x)^t + (-\ln y)^t\right)^{\frac{1}{t}}}$$

with t positive. The t-conorms with generators in $N(\mathbb{R}^+)$ are given by

$$x \triangledown y = e^{-\left((-\ln x)^t + (-\ln y)^t\right)^{\frac{1}{t}}}$$

with t negative. Ordinary multiplication is the identity element of the group of t-norms and t-conorms. That is, for $t = 1$,

$$x \triangle y = e^{-(-\ln x - \ln y)} = xy.$$

The t-norms correspond to the positive elements of the group $\mathbb{R}^*$, given by its parameter t in $e^{-\left((-\ln x)^t + (-\ln y)^t\right)^{\frac{1}{t}}}$. Thus with each such t-norm with parameter t, there is associated the conorm with parameter $-t$. A **negation** is an element η of $\mathbb{M}$ of order 2. That is, η^2 is the identity automorphism. A t-norm $\triangle$ is **dual** to a t-conorm $\triangledown$ with respect to a negation η if

$$x \triangledown y = \eta(\eta(x) \triangle \eta(y))$$

It is an easy calculation to verify the following corollary.

Corollary 5.8.5 *The negations in $N(\mathbb{R}^+)$ are the elements $e^{-c(-\ln x)^{-1}} = e^{\frac{c}{\ln x}}$, that is, the elements in $N(\mathbb{R}^+)$ with parameter $t = -1$. For $t > 0$, the t-norm*

$$e^{-\left((-\ln x)^t + (-\ln y)^t\right)^{\frac{1}{t}}}$$

is dual to the t-conorm

$$e^{-\left((-\ln x)^{-t} + (-\ln y)^{-t}\right)^{\frac{1}{-t}}}$$

with respect to any of the negations $\eta(x) = e^{\frac{c}{\ln x}}$.

All the generators of the norms and conorms $e^{-((-\ln x)^t + (-\ln y)^t)^{\frac{1}{t}}}$ are in $N(\mathbb{R}^+)$. This is because these norms and conorms do have generators in $N(\mathbb{R}^+)$, namely $e^{-c(-\ln x)^t}$, with the positive t giving norms and the negative t giving conorms. Generators are unique up to composition with an element of $\mathbb{R}^+$, and since the group $N(\mathbb{R}^+)$ contains $\mathbb{R}^+$, our claim follows. Thus if a norm and conorm with generators in $N(\mathbb{R}^+)$ are dual, then they must be dual with respect to a negation in $N(\mathbb{R}^+)$. But for a generator $f(x) = e^{-c(-\ln x)^t}$ of a norm, and negation $e^{d/\ln x}$, we have $f\eta(x) = e^{(-c/d)(-\ln x)^{-t}}$. The following sums it up.

Corollary 5.8.6 *Let s and t be positive. The t-norm $e^{-((-\ln x)^s + (-\ln y)^s)^{\frac{1}{s}}}$ is dual to the t-conorm $e^{-((-\ln x)^{-t} + (-\ln y)^{-t})^{\frac{1}{-t}}}$ if and only if $s = t$, in which case they are dual with respect to precisely the negations $e^{\frac{c}{\ln x}}$ in $N(\mathbb{R}^+)$.*

5.8.2 Families of strict t-norms

We are going to express the set of generators of some well-known families of t-norms as simple combinations of just a few elements of $\mathbb{M}$, a couple of subgroups of $\mathbb{M}$, and one special subset of $\mathbb{A}$. These are the following:

- the subgroup $\mathbb{R}^+$ of $\mathbb{A}$;

- the subgroup $\mathbb{T} = \{e^{-(-\ln x)^t} : t \neq 0\}$ of $\mathbb{M}$;

- the element $\alpha(x) = 1 - x$ of $\mathbb{M}$;

- the element $f(x) = e^{-\frac{1-x}{x}}$ of $\mathbb{A}$;

- the set $\mathbb{F} = \{\frac{a^x - 1}{a - 1} : a > 0, a \neq 1\}$

Now we list some families of t-norms (and t-conorms in some cases) and express their sets of generators as promised.

1. The Hamacher family:

$$\text{t-norms} \qquad \left\{ \frac{xy}{x + y - xy + a(1 - x - y + xy)} : a > 0 \right\}$$

$$\text{generators} \qquad \left\{ \frac{x}{x + a(1 - x)} : a > 0 \right\} = f^{-1}\mathbb{R}^+ f$$

2. The Aczél-Alsina family:

t-norms $\quad \left\{ e^{-((-\ln x)^r + (-\ln y)^r)^{\frac{1}{r}}} : r > 0 \right\}$

t-conorms $\quad \left\{ e^{-((-\ln x)^r + (-\ln y)^r)^{\frac{1}{r}}} : r < 0 \right\}$

generators $\quad \left\{ e^{-(-\ln x)^r} ; r \neq 0 \right\} = \mathbb{T}$

Comments on this family: The only negation in this group of generators is the generator with $r = -1$, which gives the negation $e^{1/\ln x}$. So this group gives a family of De Morgan systems, namely the t-norm with parameter r, $r > 0$, the negation $e^{1/\ln x}$, and the t-conorm with parameter $-r$.

3. The Jane Doe #1 family:

t-norms $\quad \left\{ \dfrac{1}{1 + \left[(\frac{1-x}{x})^r + (\frac{1-y}{y})^r + (\frac{1-x}{x})^r (\frac{1-y}{y})^r \right]^{\frac{1}{r}}} : r > 0 \right\}$

t-conorms $\quad \left\{ \dfrac{1}{1 + \left[(\frac{1-x}{x})^r + (\frac{1-y}{y})^r + (\frac{1-x}{x})^r (\frac{1-y}{y})^r \right]^{\frac{1}{r}}} : r < 0 \right\}$

generators $\quad \left\{ \dfrac{x^r}{x^r + (1-x)^r} : r \neq 0 \right\} = f^{-1} \mathbb{T} f$

Comments on this family: The only negation in this group of generators is the generator with $r = -1$, which gives the negation α. So this group gives a family of De Morgan systems, namely the t-norm with parameter r, $r > 0$, the negation α, and the t-conorm with parameter $-r$.

4. The Jane Doe #3 family:

t-norms $\quad \left\{ 1 - (1 - (1 - (1-x)^a)(1 - (1-y)^a))^{\frac{1}{a}} : a > 0 \right\}$

generators $\quad \left\{ 1 - (1-x)^a : a > 0 \right\} = \alpha^{-1} \mathbb{R}^+ \alpha$

5. The Jane Doe #1-Hamacher family:

t-norms
$$\left\{ \frac{1}{1 + [(\frac{1-x}{x})^r + (\frac{1-y}{y})^r + a(\frac{1-x}{x})^r(\frac{1-y}{y})^r]^{\frac{1}{r}}} : a > 0, \ r > 0 \right\}$$

t-conorms
$$\left\{ \frac{1}{1 + [(\frac{1-x}{x})^r + (\frac{1-y}{y})^r + a(\frac{1-x}{x})^r(\frac{1-y}{y})^r]^{\frac{1}{r}}} : a > 0, \ r < 0 \right\}$$

generators
$$\left\{ \frac{x^r}{x^r + a(1-x)^r} : a > 0, \ r \neq 0 \right\} = f^{-1} N(\mathbb{R}^+) f$$

Comments on this family: The negations in this group of generators are those with parameters $a > 0$, $r = -1$, which are the negations $(1-x)/(1+(a-1)x)$. So this group gives a family of De Morgan systems, namely the t-norm with parameters $a > 0$, $r > 0$, the negation $(1-x)/(1+(a-1)x)$, and the t-conorm with parameters $a, -r$.

6. The Schweizer family:

t-norms
$$\left\{ (x^{-a} + y^{-a} - 1)^{-\frac{1}{a}} : a > 0 \right\}$$

generators
$$\left\{ e^{-(\frac{1-x^a}{x^a})} : a > 0 \right\} = f\mathbb{R}^+$$

7. The Jane Doe #2 family:

t-norms
$$\left\{ xye^{-a \ln x \ln y} : a > 0 \right\}$$

generators
$$\left\{ \frac{1}{1 - \ln x^a} : a > 0 \right\} = f^{-1}\mathbb{R}^+$$

8. The Dombi family:

$$\text{t-norms} \quad \left\{ \frac{1}{1 + \left(\left(\frac{1-x}{x} \right)^r + \left(\frac{1-y}{y} \right)^r \right)^{\frac{1}{r}}} : r > 0 \right\}$$

$$\text{t-conorms} \quad \left\{ \frac{1}{1 + \left(\left(\frac{1-x}{x} \right)^r + \left(\frac{1-y}{y} \right)^r \right)^{\frac{1}{r}}} : r < 0 \right\}$$

$$\text{generators} \quad \left\{ e^{-\left(\frac{1-x}{x} \right)^r} : r \neq 0 \right\} = \mathbb{T}f$$

Comments on this family: The coset $\mathbb{T}f$ has no negation in it. The t-norm with parameter r, $r > 0$, is dual to the t-conorm with parameter r, $r < 0$, with respect to the negation α. Denote the element $e^{-\left(\frac{1-x}{x} \right)^r}$ of $\mathbb{T}f$ by $t_r f$. Then element $f\alpha f^{-1} = e^{1/\ln x} = t_{-1} \in \mathbb{T}$. For a generator $t_r f$ of a t-norm in the Dombi family, $t_r f \alpha = t_r f \alpha f^{-1} f = t_r t_{-1} f \in \mathbb{T}f$.

9. The Frank family:

$$\text{t-norms} \quad \left\{ \log_a \left(1 + \frac{(a^x - 1)(a^y - 1)}{a - 1} \right) : a > 0, \, a \neq 1 \right\}$$

$$\text{generators} \quad \left\{ \frac{a^x - 1}{a - 1} : a > 0, \, a \neq 1 \right\} = \mathbb{F}$$

Comments on this family: The set of generators of this family does not seem to come from a group or a coset of a group of generators. It is the set of t-norms $\triangle$ satisfying the equation $x \triangle y + x \triangledown y = x + y$, where $\triangledown$ is the t-conorm dual to $\triangle$ with respect to the negation α. This family is discussed in Section 6.2.3.

5.8.3   Families of nilpotent t-norms

1. The Schweizer-Sklar family:

$$\text{t-norms} \quad \left\{ ((x^a + y^a - 1) \vee 0)^{\frac{1}{a}} : a > 0 \right\}$$

$$\text{L-generators} \quad \{ x^a : a > 0 \} = \mathbb{R}^+$$

2. The Yager family:

t-norms $\left\{(1 - ((1-x)^a + (1-y)^a)^{\frac{1}{a}}) \vee 0 : a > 0\right\}$

L-generators $\{1 - (1-x)^a : a > 0\} = \alpha\mathbb{R}^+\alpha$

3. The family Jane Doe #4:

t-norms $\left\{\left(\frac{1}{a}(x+y-1+(a-1)xy)\right) \vee 0 : a > 0, a \neq 1\right\}$

L-generators $\{\log_a((a-1)x+1) : a > 0, \ a \neq 1\} = \mathbb{F}^{-1}$

4. The Weber family:

t-norms $\{(a(x+y-1) - (a-1)xy) \vee 0 : a > 0\}$

L-generators $\{1 - \log_a((a-1)(1-x)+1) : a > 0\} = \alpha\mathbb{F}^{-1}\alpha$

5. The family Jane Doe #6:

t-norms $\left\{\log_a\left[\left(\frac{a^x + a^y - a - 1}{a-1} \vee 0\right)(a-1) + 1\right] : \right.$
$\left. a > 0, a \neq 1\right\}$

L-generators $\left\{\frac{a^x - 1}{a-1} : a > 0, a \neq 1\right\} = \mathbb{F}$

5.9 Interval-valued fuzzy sets

If $A: U \to [0,1]$ is a fuzzy subset of a set U, the value $A(x)$ for a particular x is typically associated with a degree of belief of some expert. An increasingly prevalent view is that this model is inadequate. Many believe that assigning an exact number to an expert's opinion is too restrictive, and that the assignment of an interval of values is more realistic. In this section, we will outline the basic framework of a model in which fuzzy values are intervals. This means developing a theory of the basic logical connectives for interval-valued fuzzy sets.

For ordinary fuzzy set theory, that which we have been considering so far, the basic structure on $[0,1]$ is its lattice structure, coming from its order $\leq$. The interval $[0, 1]$ is a complete lattice. Subsequent operations on $[0, 1]$ have been required to behave in special ways with respect to $\leq$. The basis of the theory we have discussed has come from the lattice $\mathbb{I} = ([0, 1], \leq)$.

Now consider fuzzy sets with interval values. The interval $[0, 1]$ is replaced by the set $\{(a, b) : a, b \in [0, 1], a \leq b\}$. The element (a, b) is just the *pair* with $a \leq b$. As we have seen, there is a notation for this set: $[0, 1]^{[2]}$. So if U is the universal set, then our new fuzzy sets are mappings $A : U \rightarrow [0, 1]^{[2]}$. Now comes the crucial question: with what structure should $[0, 1]^{[2]}$ be endowed? Again, lattice theory provides an answer. Use componentwise operations coming from the operations on $[0, 1]$. For example, $(a, b) \leq (c, d)$ if $a \leq c$ and $b \leq d$, which gives the usual lattice max and min operations

$$(a, b) \vee (c, d) \quad = \quad (a \vee c, b \vee d)$$
$$(a, b) \wedge (c, d) \quad = \quad (a \wedge c, b \wedge d)$$

The resulting structure, that is, $[0, 1]^{[2]}$ with this order, or equivalently, with these operations, is again a complete lattice. This complete lattice is denoted $\mathbb{I}^{[2]}$. We will use $\mathbb{I}^{[2]}$ as the basic building block for interval valued fuzzy set theory.

There is a natural negation on $\mathbb{I}^{[2]}$ given by $(a, b)' = (b', a')$ where $x' = 1 - x$. Since $\mathbb{I}^{[2]}$ comes equipped to provide the usual operations of sup and inf, we have immediately the De Morgan algebra $(\mathbb{I}^{[2]}, \wedge, \vee, ')$. This in turn yields a De Morgan algebra for the set of all interval valued fuzzy subsets of a set by

$$(A \wedge B)(s) \quad = \quad A(s) \wedge B(s)$$
$$(A \vee B)(s) \quad = \quad A(s) \vee B(s)$$
$$A'(s) \quad = \quad (A(s))'$$

5.9.1   t-norms

Our first problem here is that of *defining* t-norms. There is a natural embedding of $\mathbb{I}$ into $\mathbb{I}^{[2]}$, namely $a \rightarrow (a, a)$. This is how $\mathbb{I}^{[2]}$ generalizes $\mathbb{I}$. Instead of specifying a number a (identified with (a, a)) as a degree of belief, an expert specifies an interval (a, b) with $a \leq b$. So no matter how a t-norm is defined on $\mathbb{I}^{[2]}$, it should induce a t-norm on this copy of $\mathbb{I}$ in it.

A t-norm should be increasing in each variable, just as in the case for $\mathbb{I}$. On $\mathbb{I}$ this is equivalent to the conditions $a \triangle (b \vee c) = (a \triangle b) \vee (a \triangle c)$ and $a \triangle (b \wedge c) = (a \triangle b) \wedge (a \triangle c)$. But just increasing in each variable will not yield these distributive laws on $\mathbb{I}^{[2]}$. However, these distributive laws do imply increasing in each variable.

Now for the boundary conditions. Since t-norms are to generalize intersection, we certainly want $(a, b) \triangle (1, 1) = (a, b)$ for all $(a, b) \in \mathbb{I}^{[2]}$. It will follow that $(0, 0) \triangle (a, b) = (0, 0)$, but what about the element $(0, 1)$? How is it to behave? We require that $(0, 1) \triangle (a, b) = (0, b)$. There are some strong mathematical reasons for this, having to do with fixed points of automorphisms, but we choose to forego that discussion. We are led to the following definition.

Definition 5.9.1 *A commutative associative binary operation* $\triangle$ *on* $\mathbb{I}^{[2]}$ *is a **t-norm** if*

1. $(1, 1) \triangle (a, b) = (a, b)$

2. $x \triangle (y \vee z) = x \triangle y \vee x \triangle z$

3. $x \triangle (y \wedge z) = x \triangle y \wedge x \triangle z$

4. *The restriction of* $\triangle$ *to* $D = \{(a, a) : a \in [0, 1]\}$ *is a t-norm, identifying* D *with* $[0, 1]$

5. $(0, 1) \triangle (a, b) = (0, b)$

Several additional useful properties follow immediately.

Corollary 5.9.2 *The following hold for a t-norm* $\triangle$ *on* $\mathbb{I}^{[2]}$.

1. $\triangle$ *is increasing in each variable*

2. $(0, b) \triangle (c, d) = (0, e)$ *for some* $e \in [0, 1]$.

Proof. Suppose that $y \leq z$. Then

$$
\begin{aligned}
x \triangle y \vee x \triangle z &= x \triangle (y \vee z) \\
&= x \triangle z
\end{aligned}
$$

so $x \triangle y \leq x \triangle z$. For the second part,

$$(0, b) \triangle (c, d) \leq (0, 1) \triangle (c, d) = (0, d)$$

∎

The fundamental result about t-norms on $\mathbb{I}^{[2]}$ is the following theorem.

Theorem 5.9.3 *Every t-norm* $\triangle$ *on* $\mathbb{I}^{[2]}$ *is of the form*

$$(a, b) \triangle (c, d) = (a \diamond c, b \diamond d)$$

where $\diamond$ *is a t-norm on* $\mathbb{I}$.

Proof. Since the t-norm induces a t-norm on D, we have

$$(a, a) \triangle (c, c) = (a \diamond c, a \diamond c)$$

where $\diamond$ is a t-norm on $\mathbb{I}$. Now

$$
\begin{aligned}
(a, b) \triangle (c, d) &= (a, b) \triangle ((c, c) \vee (0, d)) \\
&= ((a, b) \triangle (c, c)) \vee ((a, b) \triangle (0, d)) \\
&= ((a, b) \triangle (c, c)) \vee (0, e)
\end{aligned}
$$

Therefore, the first component of $(a, b) \triangle (c, d)$ does not depend on d and similarly does not depend on b. Also $(a, b) \triangle (c, d)$ has its second component the same as

$$
\begin{aligned}
(a, b) \triangle (c, d) \triangle (0, 1) &= (a, b) \triangle (0, 1) \triangle (c, d) \triangle (0, 1) \\
&= (0, b) \triangle (0, d)
\end{aligned}
$$

Thus, the second component of $(a, b) \triangle (c, d)$ does not depend on a or c. So a t-norm on $\mathbb{I}^{[2]}$ acts componentwise. From $(a, a) \triangle (c, c) = (a \diamond c, a \diamond c)$ it follows that

$$(a, b) \triangle (c, d) = (a \diamond c, b \diamond d)$$

and the proof is complete. ∎

Now we define Archimedean, strict, and nilpotent t-norms on $\mathbb{I}^{[2]}$ just as for t-norms on $\mathbb{I}$. It is easy to see that in the notation of Theorem 5.9.3, a t-norm $\triangle$ is Archimedean, strict, or nilpotent if and only if the t-norm $\diamond$ is Archimedean, strict, or nilpotent, respectively. In effect, the theory of t-norms on $\mathbb{I}^{[2]}$ as we have defined them is reduced to the theory of t-norms on $\mathbb{I}$.

5.9.2   Negations and t-conorms

We will define t-conorms to be dual to t-norms, just as in the case of fuzzy set theory. This involves negations, which are certain antiautomorphisms. To determine these negations we must examine the set of automorphisms and antiautomorphisms of $\mathbb{I}^{[2]}$.

Definition 5.9.4 *An **automorphism** of $\mathbb{I}^{[2]}$ is a one-to-one map f from $\mathbb{I}^{[2]}$ onto itself such that $f(x) \leq f(y)$ if and only if $x \leq y$. An **antiautomorphism** is a one-to-one map f from $\mathbb{I}^{[2]}$ onto itself such that $f(x) \leq f(y)$ if and only if $x \geq y$.*

An antiautomorphism f such that $f(f(x)) = x$ is an **involution**, or **negation**. The map α given by $\alpha(a, b) = (1 - b, 1 - a)$ is a negation, as is $f^{-1}\alpha f$ for any automorphism f. It turns out that there are no others. Furthermore, if f is an automorphism, that is, a one-to-one map of $[0, 1]$ onto itself such that $f(x) \leq f(y)$ if and only if $x \leq y$, then $(a, b) \rightarrow (f(a), f(b))$ is an automorphism of $\mathbb{I}^{[2]}$. It turns out that there are no others. It should be clear that automorphisms take $(0, 0)$ and $(1, 1)$ to themselves. Antiautomorphisms interchange these two elements.

In the plane, $\mathbb{I}^{[2]}$ is the triangle pictured. Each leg is mapped onto itself by automorphisms.

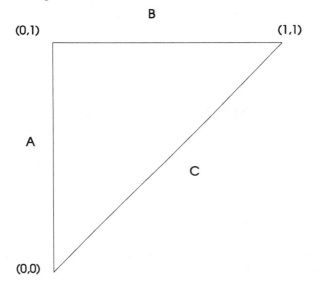

Lemma 5.9.5 *Let A, B, and C be as in the picture. If f is an automorphism, then $f(A) = A$, $f(B) = B$, and $f(C) = C$.*

Proof. Let f be an automorphism. Since $f(x) \leq f(y)$ if and only if $x \leq y$, it follows that

$$
\begin{aligned}
f(x \vee y) &= f(x) \vee f(y) \\
f(x \wedge y) &= f(x) \wedge f(y) \\
f(0, 0) &= (0, 0) \\
f(1, 1) &= (1, 1)
\end{aligned}
$$

Suppose that $f(a, b) = (c, c)$. Then

$$
\begin{aligned}
f(a, b) &= f((a, a) \vee ((b, b) \wedge (0, 1))) \\
&= f(a, a) \vee (f(b, b) \wedge f(0, 1)) \\
&= (c, c)
\end{aligned}
$$

No two elements strictly less than a diagonal element (c, c) can have join (c, c) and no two elements strictly greater than a diagonal element (c, c) can have meet (c, c). Since $f(a, a) \vee (f(b, b) \wedge f(0, 1)) = (c, c)$, we need only rule out $f(0, 1) = (c, c)$. So suppose that $f(0, 1) = (c, c)$. Then every element less than (c, c) is the image of an element less than $(0, 1)$. So for $0 < a < c$, $f(a, a) \geq (0, c)$. If $f(x, y) = (0, c)$, then $f(a, a) \geq f(x, y)$ for all $a > 0$, whence $(a, a) \geq (x, y)$ for all $a > 0$. It follows that $x = y = 0$. Thus $c = 0$, but $f(0, 0) = (0, 0)$. Therefore only elements from C go to C under an automorphism f. Since the inverse of an automorphism is an automorphism, $f(C) = C$ for all automorphisms f.

If $f(a, b) = (0, c)$, then

$$
\begin{aligned}
f(a, b) &= f(a, a) \vee f(0, b) \\
&= (d, d) \vee f(0, b) \\
&= (0, c)
\end{aligned}
$$

Therefore $d = 0$, implying that $a = 0$. Hence $f(A) = A$.

If $f(a, b) = (c, 1)$, then

$$
\begin{aligned}
f(a, b) &= f(b, b) \wedge f(a, 1) \\
&= (d, d) \wedge f(a, 1) \\
&= (c, 1)
\end{aligned}
$$

Thus $d = 1$, and so $b = 1$. It follows that $f(B) = B$. ∎

Theorem 5.9.6 *Every automorphism f of $\mathbb{I}^{[2]}$ is of the form $f(a, b) = (g(a), g(b))$, where g is an automorphism of $\mathbb{I}$.*

Proof. Since f is an automorphism of C, it induces an automorphism g of $\mathbb{I}$, namely $(g(a), g(a)) = f(a, a)$. Now $f(0, 1) = (0, 1)$ since $f(A) = A$ and $f(B) = B$. Thus

$$
\begin{aligned}
f(a, b) &= f(a, a) \vee (f(b, b) \wedge f(0, 1)) \\
&= (g(a), g(a)) \vee ((g(b), g(b)) \wedge (0, 1)) \\
&= (g(a), g(b))
\end{aligned}
$$

∎

So automorphisms of $\mathbb{I}^{[2]}$ are of the form $(a, b) \to (f(a), f(b))$ where f is an automorphism of $\mathbb{I}$. We will use f to denote both the automorphism of $\mathbb{I}$ and the corresponding automorphism of $\mathbb{I}^{[2]}$.

Let α be the antiautomorphism of $\mathbb{I}$ given by $\alpha(a) = 1 - a$. Then $(a, b) \to (\alpha(b), \alpha(a))$ is an antiautomorphism of $\mathbb{I}^{[2]}$ which we also denote by α. If g is an antiautomorphism of $\mathbb{I}^{[2]}$, then $g = \alpha f$ for the automorphism $f = \alpha g$. Now

$$
\begin{aligned}
g(a,b) &= \alpha f(a,b) \\
&= \alpha(f(a), f(b)) \\
&= (\alpha f(b), \alpha f(a)) \\
&= (g(b), g(a))
\end{aligned}
$$

We have

Corollary 5.9.7 *The antiautomorphisms of $\mathbb{I}^{[2]}$ are precisely the anti-automorphisms $(a,b) \to (g(b), g(a))$, where g is an antiautomorphism of $\mathbb{I}$.*

In particular, we have

Corollary 5.9.8 *The negations of $\mathbb{I}^{[2]}$ are precisely the negations $(a,b) \to (\eta(b), \eta(a))$ where η is an negation of $\mathbb{I}$.*

Corollary 5.9.9 *The negations of $\mathbb{I}^{[2]}$ are precisely the negations $(a,b) \to (f^{-1}\alpha f(b), f^{-1}\alpha f(a))$ where f is an automorphism of $\mathbb{I}$.*

Just as for $\mathbb{I}$, we define a t-conorm to be the dual of a t-norm with respect to some negation.

Definition 5.9.10 *Let $\triangle$ be a binary operation and η a negation on $\mathbb{I}^{[2]}$. The **dual of $\triangle$ with respect to** η is the binary operation $\triangledown$ given by*

$$
x \triangledown y = \eta(\eta(x) \triangle \eta(y))
$$

*If $\triangle$ is a t-norm, then $\triangledown$ is called a **t-conorm**.*

Theorem 5.9.11 *Every t-conorm $\triangledown$ on $\mathbb{I}^{[2]}$ is of the form*

$$
(a,b) \triangledown (c,d) = (a \diamond c, b \diamond d)
$$

where $\diamond$ is a t-conorm on $\mathbb{I}$.

The proof is a consequence of the definition of t-conorm and the corresponding theorems about t-norms and negations. The theory of t-norms, t-conorms, and negations on $\mathbb{I}^{[2]}$ has thus been reduced to that theory on $\mathbb{I}$.

The following theorem gives properties characterizing t-conorms.

Theorem 5.9.12 *A binary operation $\triangledown$ on $\mathbb{I}^{[2]}$ that is commutative and associative is a t-conorm if and only if*

1. $(0,0) \bigtriangledown (a,b) = (a,b)$

2. $x \bigtriangledown (y \vee z) = x \bigtriangledown y \vee x \bigtriangledown z$

3. $x \bigtriangledown (y \wedge z) = x \bigtriangledown y \wedge x \bigtriangledown z$

4. *The restriction of $\bigtriangledown$ to $D = \{(a,a) : a \in [0,1]\}$ is a t-conorm*

5. $(0,1) \bigtriangledown (a,b) = (0,b)$

Much fuzzy set theory can be extended to interval valued fuzzy sets, and we have just indicated the beginnings of such a theory for t-norms, t-conorms, and negations. The theory and applications of interval valued fuzzy set theory seems destined to become an important topic, but we will not pursue it further here.

5.10   Exercises

1. Verify that $\triangle_0$, $\triangle_1$, $\triangle_2$, $\triangle_3$, $\triangle_4$, $\triangle_5$ are indeed t-norms.

2. Show that if $\triangle$ is a t-norm, then

$$\triangle_0 \leq \triangle \leq \triangle_5$$

3. Show that $\wedge$ is the only idempotent t-norm.

4. Show that $\triangle_1$ is nilpotent.

5. Show that $\triangle_5$ is continuous, but not Archimedean. Show that $\triangle_1, \triangle_2, \triangle_3$, and $\triangle_4$ are Archimedean.

6. Show that

$$x \triangle y = 1 - \left(1 \wedge [(1-x)^2 + (1-y)^2]^{1/2}\right)$$

   is nilpotent.

7. Furnish the details of the proof of Proposition 5.1.5

8. Let f be a generator of an Archimedean t-norm $\triangle$. Show that

$$\overbrace{a \triangle a \triangle a \triangle \cdots \triangle a}^{n \; times} = f^{-1}(f(a)^n \vee f(0))$$

   Show that the t-norm is nilpotent if and only if $f(0) > 0$.

9. Let f and g be generators of nilpotent t-norms. Show that there is an $h \in Aut(\mathbb{I})$ such that $hf = g$.

10. Complete the proof of Theorem 5.2.9.

11. Prove that $Aut(\mathbb{I})_a$ is a subgroup of $Aut(\mathbb{I})$.

12. Prove that an Archimedean t-norm $\triangle$ is strict if and only if $x \triangle y$ is strictly increasing for $x, y \in (0, 1)$.

13. Prove that the only t-norm isomorphic to min is min itself.

14. Let $\circ$ be a t-norm. Show that the set of automorphisms of $(\mathbb{I}, \circ)$ is a subgroup of $Aut(\mathbb{I})$.

15. Show that if f is a one-to-one mapping of $[0, 1]$ onto itself, $\circ$ is an Archimedean t-norm, and $f(x \circ y) = f(x) \circ f(y)$, then $f \in Aut(\mathbb{I}, \circ)$.

16. Prove that if $\circ$ is a strict t-norm, then $f(x) = x \circ x$ is an automorphism and $g(x) = (1 - x) \circ (1 - x)$ is an antiautomorphism of $\mathbb{I}$.

17. Prove that $Map(\mathbb{I})$ is a group under composition of functions.

18. In $Map(\mathbb{I})$, prove that

    (a) conjugates of automorphisms are automorphisms,

    (b) conjugates of antiautomorphisms are antiautomorphisms, and

    (c) conjugates of negations are negations.

19. Prove that the centralizer of an element is a subgroup.

20. Show that the average $(f(x) + g(x))/2$ of two automorphisms of $\mathbb{I}$ is an automorphism of $\mathbb{I}$.

21. Prove that $Z(f^{-1}\alpha f) = f^{-1}Z(\alpha)f$, and so the centralizer $Z(\beta)$ of a negation $\beta = f^{-1}\alpha f$ is the group

$$f^{-1}Z(\alpha)f = f^{-1}\left\{ \frac{\alpha g \alpha + g}{2} : g \in Aut(\mathbb{I}) \right\} f$$

22. Show that the map

$$\Phi : Aut(\mathbb{I}) \to Z(\alpha) : g \to \frac{\alpha g \alpha + g}{2}$$

is not a group homomorphism. In fact, show that

$$\Phi(f)\Phi(g) = \Phi(f\Phi(g)) \neq \Phi(fg)$$

23. Let H be a subgroup of a group G, and let $g \in G$. Show that $h \to g^{-1}hg$ is an isomorphism from H to its conjugate $g^{-1}Hg$.

24. Show directly that if $f, g \in Aut(\mathbb{I})$, and

$$f^{-1}((f(x) + f(y) - 1) \vee 0) = g^{-1}((g(x) + g(y) - 1) \vee 0)$$

for all $x, y \in [0, 1]$, then $f = g$.

25. For $r \in \mathbb{R}^+$ and $x \in [0, 1]$, let $f_r(x) = \frac{x}{r - (r-1)x}$, and let $G = \{f_r : r \in \mathbb{R}^+\}$.

   (a) Prove that G is a subgroup of $Aut(\mathbb{I})$.
   (b) Prove that $r \to f_r$ is an isomorphism from the group $\mathbb{R}^+$ to G.
   (c) Compute the set $\{\blacktriangle_f : f \in G\}$ of nilpotent t-norms.
   (d) Compute the set $\{f^{-1}\alpha f : f \in G\}$ of negations.

26. Let $x \bigtriangleup y = \frac{xy}{2 + xy - (x+y)}$. Prove that

   (a) $\bigtriangleup$ is a strict t-norm.
   (b) $f(x) = \frac{x}{2-x}$ is a generator of $\bigtriangleup$.
   (c) $g(x) = \frac{1}{2} \log(\frac{2-x}{x})$ is an additive generator of $\bigtriangleup$.

27. Calculate the duals of $\bigtriangleup_0, \bigtriangleup_1, \bigtriangleup_2, \bigtriangleup_3, \bigtriangleup_4, \bigtriangleup_5$ with respect to α. Denote this dual of $\bigtriangleup_i$ by $\bigtriangledown_i$.

28. Show that

$$\bigtriangleup_0 \ \le \ \bigtriangleup_1 \le \bigtriangleup_2 \le \bigtriangleup_3 \le \bigtriangleup_4 \le \bigtriangleup_5$$

$$\bigtriangledown_5 \ \le \ \bigtriangledown_4 \le \bigtriangledown_3 \le \bigtriangledown_2 \le \bigtriangledown_1 \le \bigtriangledown_0$$

29. Show that if $\bigtriangledown$ is a t-conorm, then

$$\bigtriangledown_5 \le \bigtriangledown \le \bigtriangledown_0$$

30. For $p \ge 1$, let

$$x \bigtriangleup_p y \ = \ 1 - 1 \wedge [(1-x)^p + (1-y)^p]^{1/p}$$

$$x \bigtriangledown_p y \ = \ 1 \wedge (x^p + y^p)^{1/p}$$

   (a) Show that $x \bigtriangleup_p y$ is a nilpotent t-norm with additive generator $f_p(x) = (1-x)^p$, and for $y \in [0, 1]$,

$$f_p^{-1}(y) = 1 - y^{1/p}$$

(b) Show that $x \nabla_p y$ is a nilpotent t-conorm with additive generator $g_p(x) = x^p$, and for $y \in [0,1]$,

$$g_p^{-1}(x) = y^{1/p}$$

31. Let $\lambda > 1$.

(a) Show that

$$x \triangle_\lambda y = 0 \vee \frac{x+y-1+\lambda xy}{1+\lambda}$$

is a nilpotent t-norm with additive generator

$$f_\lambda(x) = 1 - \frac{\log(1+\lambda x)}{\log(1+\lambda)}$$

and for $y \in [0,1]$,

$$f_\lambda^{-1}(y) = \frac{1}{\lambda}[(1+\lambda)^{1-y} - 1]$$

(b) Show that $x \nabla_\lambda y = 1 \wedge (x+y+\lambda xy)$ is a nilpotent t-conorm with additive generator

$$g_\lambda(x) = \frac{\log(1+\lambda x)}{\log(1+\lambda)}$$

and for $y \in [0,1]$,

$$g_\lambda^{-1}(y) = \frac{1}{\lambda}[(1+\lambda)^y - 1]$$

32. For $p > 0$, and $x \in [0,1]$, let $f_p(x) = \frac{1}{p}(x^{-p} - 1)$.

(a) Verify that f_p is an additive generator for a t-norm.

(b) Find the inverse of f_p on the interval $[0, f_p(0)]$.

(c) What is the associated strict Archimedean t-norm?

33. Let $\triangle$ be an Archimedean t-norm. Let $x^{[2]} = x \triangle x$, $x^{[3]} = x \triangle x^{[2]}$, ... , $x^{[n]} = x \triangle x^{[n-1]}$. Prove that for $n \geq 1$, $f(x^{[n]}) = nf(x)$ for $x \in [0,1]$, where f is an additive generator of $\triangle$.

34. Show that a t-norm is Archimedean if and only if its dual with respect to α is Archimedean.

35. Show that a t-norm is strict if and only if its dual with respect to α is strict.

36. Prove that natural dual of ▲ is $x \blacktriangledown y = (x + y) \wedge 1$.

37. Show that the only isomorphism from $(\mathbb{I}, \blacktriangle_f)$ to $(\mathbb{I}, \blacktriangle_g)$ is $g^{-1}f$.

38. Show that any $\eta : [0, 1] \rightarrow [0, 1]$ such that $x \leq y$ implies $\eta(x) \geq \eta(y)$ and $\eta^2 = 1$ is one-to-one and onto and continuous, and thus is a negation.

39. Show that a negation has exactly one fixed point. That is, show that if η is a negation, then there is exactly one $x \in [0, 1]$ such that $\eta(x) = x$.

40. Let $\alpha(x) = 1 - x$, and let β be any map $[0, 1] \rightarrow [0, 1]$. Show that $\alpha\beta = \beta\alpha$ if and only if $\beta(x) + \beta(1 - x) = 1$.

41. Show that if a negation commutes with $\alpha(x) = 1 - x$, then its fixed point is $\frac{1}{2}$.

42. Show that $\eta(x) = (1 - x)/(1 + \lambda x)$, $\lambda > -1$ is a negation. Find its fixed point. (These are *Sugeno* negations.)

43. Prove that the cogenerators of nilpotent t-conorms that are straight lines give Sugeno negations $(1 - x)/(1 + \lambda x)$ for $\lambda \in (-1, 0)$.

44. Let $f \in Aut(\mathbb{I})$. Show that if η is a negation, then so is $f^{-1}\eta f$.

45. Show that if

$$\eta(x) = \frac{ax + b}{cx + d}$$

is a negation, then

$$\eta(x) = \frac{1 - x}{1 + \lambda x}$$

for some $\lambda > -1$.

46. Let $f(x) = e^{\frac{x-1}{x}}$ for $x \in [0, 1]$.

   (a) Verify that f is a generator for a t-norm and find that t-norm.

   (b) Let $g(x) = f(1 - x)$. Find the t-conorm associated with g.

47. Let g be an order reversing, one-to-one onto mapping $[0,1] \to [b,1]$ for some $b \in [0,1)$. Prove that

$$x \bigtriangledown y = g^{-1}(g(x)g(y) \vee g(1))$$

is an Archimedean t-conorm, is nilpotent if and only if $g(1) > 0$, and is strict if and only if $g(1) = 0$.

48. Let

$$(x \bigtriangleup y) \quad = \quad 0 \vee \frac{x+y-1+axy}{1+a}, \text{ for } a > -1$$

$$x \bigtriangledown y \quad = \quad 1 \wedge (x+y+axy), \text{ for } a > -1$$

(a) Verify that $\bigtriangleup$ and $\bigtriangledown$ are norms and conorms, respectively.
(b) Find a generator for $\bigtriangleup$.
(c) Find a cogenerator for $\bigtriangledown$.
(d) Find the negation associated with $\bigtriangleup$.
(e) When $a = 0$, show that the norm and conorm are $\bigtriangleup_1$ and $\bigtriangledown_1$, respectively.

49. Let

$$x \bigtriangleup y \quad = \quad \frac{1}{1 + \left[(\frac{1}{x} - 1)^a + (\frac{1}{y} - 1)^a \right]^{\frac{1}{a}}} \text{ for } a \geq 0$$

$$x \bigtriangledown y \quad = \quad \frac{1}{1 + \left[\frac{1}{x} - 1)^{-a} + (\frac{1}{y} - 1)^{-a} \right]^{\frac{-1}{a}}} \text{ for } a \geq 0$$

Show that these are norms and conorms, respectively. Find their generators and cogenerators and determine whether these norms and conorms are strict or nilpotent. Find the negation associated with the norm if it is nilpotent.

50. Let

$$x \bigtriangleup y \quad = \quad \frac{xy}{x \vee y \vee a} \text{ for } a \in (0,1)$$

$$x \bigtriangledown y \quad = \quad \frac{x+y-xy-(x \wedge y \wedge (1-a))}{(1-x) \vee (1-y) \vee a} \text{ for } a \in (0,1]$$

Show that these are norms and conorms, respectively. Find their generators and cogenerators and determine whether these norms and conorms are strict or nilpotent. Find the negation associated with the norm if it is nilpotent.

51. Prove Theorem 5.6.3.

52. Let $\mathbb{R}$ be the additive group of all real numers, and let f be an order preserving automorhism of $\mathbb{R}$. Suppose that $f(1) = r$.

   (a) Prove that $r > 0$.

   (b) Prove that $f(n) = rn$ for integers n.

   (c) Prove that $f(1/n) = r(1/n)$ for nonzero integers n.

   (d) Prove that $f(m/n) = r(m/n)$ for rational numbers m/n.

   (e) Prove that $f(x) = rx$ for $r \in \mathbb{R}$.

   (f) Prove that an order reversing automorphism of the group $\mathbb{R}$ is multiplication by a negative real number.

   (g) Prove that an order preserving or order reversing automorphism f of the multiplicative group $\mathbb{R}^+$ is given by $f(x) = x^t$ for a nonzero real number t.

53. Prove that if f is an antiautomrophism of $\mathbb{I}$, and f is in the normalizer of $\mathbb{R}^+$, then $r \to f^{-1}rf$ is an order reversing automorphism of the group $\mathbb{R}^+$.

54. Prove that in the normalizer

$$\left\{ e^{-a(-\ln x)^t} : a > 0, t \neq 0 \right\}$$

   of $\mathbb{R}^+$, the parameters $t > 0$ give automorphisms of $\mathbb{I}$, $t < 0$ give antiautomorphisms of $\mathbb{I}$, and $t = -1$ gives negations of $\mathbb{I}$.

55. Prove Corollary 5.8.5.

56. Prove Theorem 5.9.11.

57. Prove Theorem 5.9.12.

Chapter 6

ADDITIONAL TOPICS ON CONNECTIVES

This chapter is a continuation of the study of fuzzy connectives, and t-norms, t-conorms and negations will still play a basic role in the topics discussed. We begin with fuzzy implications, a topic which has applications in data fusion. The theory we develop for averaging operators puts into context several phenomena observed in Chapter 5, and sheds new light on the important family of Frank t-norms. The section on powers of t-norms provides a definition of the r-th power of a t-norm, and characterizes those automorphisms of the unit interval that are such powers. Powers of t-norms have played an important role in the development of t-norm theory, for example, in proving the existence of generators. The sensitivity of a connective is a measure of its robustness, and that section illustrates various such measures. Copulas are important objects in joint distribution theory in statistics, and have some connections with t-norms. This connection is explored briefly in the last section.

6.1   Fuzzy implications

In classical two-valued logic, the table for $\Rightarrow$ is taken to be

$\Rightarrow$	0	1
0	1	1
1	0	1

This may be expressed on $\{0, 1\}$ by the formula $(a \Rightarrow b) = a' \vee b$. It is a binary operation on the truth values $\{0, 1\}$. In fuzzy logic, our set of truth values is $[0, 1]$, and so material implication $\Rightarrow$ should be a binary operation on $[0, 1]$. Such operations should agree with the classical case for $\{0, 1\}$.

A **fuzzy implication** is a map

$$\Rightarrow: [0, 1] \times [0, 1] \to [0, 1]$$

satisfying

$\Rightarrow$	0	1
0	1	1
1	0	1

Here are some examples.

1. $(x \Rightarrow y) = \begin{cases} 1 & \text{if} \quad x \leq y \\ 0 & \text{if} \quad x > y \end{cases}$

2. $(x \Rightarrow y) = (1 - x + y) \wedge 1$

3. $(x \Rightarrow y) = (1 - x) \vee y$

The class of all possible fuzzy implications consists of all functions $\Rightarrow$ defined on the unit square with the given values above on the four corners. There are *three basic constructions* of fuzzy implications. They arise from three ways to express implication in the classical case. The following are equivalent for that case.

- $(x \Rightarrow y) = \bigvee \{z : x \wedge z \leq y\}$

- $(x \Rightarrow y) = x' \vee y$

- $(x \Rightarrow y) = x' \vee (x \wedge y)$

These three conditions make sense on $[0, 1]$ when a t-norm is used for $\wedge$, a t-conorm for $\vee$, and a negation for $'$. There is a second motivation for the first class of implications. When we represent the fuzzy conditional "If x is A, then y is B" by the fuzzy subset $R = (A \times B) \vee (A' \times V)$, we realize that R is the largest solution D of the inequality $D \wedge (A \times V) \leq U \times B$. For the first class we need only a t-norm. The resulting operators are called R-**implications**, where R stands for **residuated** lattice.

Definition 6.1.1 *An R-implication is a map*

$$\Rightarrow: [0, 1] \times [0, 1] \to [0, 1]$$

of the form

$$(x \Rightarrow y) = \bigvee \{z \in [0, 1] : x \triangle z \leq y\}$$

*where $\triangle$ is a t-norm. The function $\Rightarrow$ is referred to as the R-**implication** **associated with** $\triangle$.*

The following calculations show that an R-implication is an implication. Let $\triangle$ be any t-norm.

- $1 \Rightarrow 0 = \bigvee\{z \in [0,1] : 1 \triangle z \leq 0\} = 0$ since $1 \triangle z = z$

- $0 \Rightarrow 0 = \bigvee\{z \in [0,1] : 0 \triangle z \leq 0\} = 1$ since $0 \triangle z = 0$

- $0 \Rightarrow 1 = \bigvee\{z \in [0,1] : 0 \triangle z \leq 1\} = 1$ since $0 \triangle z \leq 1$

- $1 \Rightarrow 1 = \bigvee\{z \in [0,1] : 1 \triangle z \leq 1\} = 1$ since $1 \triangle z \leq 1$

For R-implications $\Rightarrow$, it is always the case that $(x \Rightarrow y) = 1$ for $x \leq y$ since $x \triangle 1 = x \leq y$. This is in agreement with classical logic: $a \Rightarrow b = a' \vee b = 1$ if $a \leq b$. Here are some examples.

- For $x \triangle y = x \wedge y$

$$(x \Rightarrow y) = \bigvee\{z : x \wedge z \leq y\}$$
$$= \begin{cases} 1 & \text{if } x \leq y \\ y & \text{if } x > y \end{cases}$$

- For $x \triangle y = xy$

$$(x \Rightarrow y) = \bigvee\{z : xz \leq y\}$$
$$= \begin{cases} 1 & \text{if } x \leq y \\ y/x & \text{if } 0 < y < x \end{cases}$$

- For $x \triangle y = 0 \vee (x + y - 1)$

$$(x \Rightarrow y) = \bigvee\{z : 0 \vee (x + z - 1) \leq y\}$$
$$= 1 \wedge (1 - x + y)$$

When $\triangle$ is an Archimedean t-norm, then there is a simple formula for $\Rightarrow$ in terms of a generator of $\triangle$.

Theorem 6.1.2 *Let $\triangle$ be an Archimedean t-norm and f a generator of $\triangle$. Then*

$$(x \Rightarrow y) = f^{-1}\left(\frac{f(y)}{f(x)} \wedge 1\right)$$

Proof. We just calculate:

$$
\begin{aligned}
(x \Rightarrow y) &= \bigvee\{z : x \triangle z \le y\} \\
&= \bigvee\{z : f^{-1}(f(x)f(z) \vee f(0)) \le y\} \\
&= \bigvee\{z : (f(x)f(z) \vee f(0)) \le f(y)\} \\
&= \bigvee\{z : f(x)f(z) \le f(y)\} \\
&= \bigvee\{z : f(z) \le \frac{f(y)}{f(x)}\} \\
&= \bigvee\left\{z : z \le f^{-1}\left(\frac{f(y)}{f(x)} \wedge 1\right)\right\} \\
&= f^{-1}\left(\frac{f(y)}{f(x)} \wedge 1\right)
\end{aligned}
$$

Of course, if $x \le y$, we see that $(x \Rightarrow y) = 1$, and otherwise

$$
(x \Rightarrow y) = f^{-1}\left(\frac{f(y)}{f(x)}\right)
$$

Also note that if $x \ge y$, then

$$
f(0) \le \frac{f(y)}{f(x)} \le 1,
$$

so that f^{-1} is defined on $f(y)/f(x)$. ∎

A simple example is the case $x \triangle y = xy$. A generator for $\triangle$ is $f(x) = x$ and we have

$$
\begin{aligned}
(x \Rightarrow y) &= f^{-1}\left(\frac{f(y)}{f(x)} \wedge 1\right) \\
&= \frac{y}{x} \wedge 1
\end{aligned}
$$

In other words, $(x \Rightarrow y) = 1$ if $x \le y$ and is y/x otherwise.

We turn now to the second way of constructing fuzzy implications in the spirit of Boolean logic, namely $a \Rightarrow b = a' \vee b$. We simply use a t-conorm $\triangledown$ for $\vee$ and a negation η for $'$. The analogous definition for a fuzzy implication is the following:

Definition 6.1.3 *A* $\triangledown$*-implication is a map* $\Rightarrow: [0,1] \times [0,1] \to [0,1]$ *of the form* $(x \Rightarrow y) = \eta(x) \triangledown y$, *where* $\triangledown$ *is a t-conorm and* η *is a negation.*

It is easy to check that such an $\Rightarrow$ is indeed an implication operator. Some examples follow.

- For $x \bigtriangledown y = x \vee y$ and $\eta(x) = 1 - x$,

$$(x \Rightarrow y) = (1 - x) \bigvee y$$

- For $x \bigtriangledown y = x + y - xy$ and $\eta(x) = 1 - x$,

$$
\begin{aligned}
(x \Rightarrow y) &= \eta(x) \bigtriangledown y \\
&= (1 - x) \bigtriangledown y \\
&= 1 - x + y - (1 - x)y \\
&= 1 - x + xy
\end{aligned}
$$

- For $x \bigtriangledown y = 1 \wedge (x + y)$, and $\eta(x) = 1 - x$,

$$(x \Rightarrow y) = 1 \bigwedge (1 - x + y)$$

This example is the one used by Lukasiewicz.

If the t-conorm involved is Archimedean, then it has a cogenerator g, and so

$$
\begin{aligned}
(x \Rightarrow y) &= \eta(x) \bigtriangledown y \\
&= g^{-1}(g(\eta(x))g(y) \bigvee g(1))
\end{aligned}
$$

For example, let $g(x) = e^{-x^2}$ and $\eta(x) = \sqrt{1 - x^2}$. Then an easy calculation shows that

$$
(x \Rightarrow y) = \begin{cases} 1 & \text{if } x \leq y \\ \sqrt{1 - x^2 + y^2} & \text{if } x > y \end{cases}
$$

Consider a rule R of the form "If x is A then y is B else y is C" when A, B, and C are subsets of Ω. This rule is translated into

$$(x, y) \in A \times B \text{ or } (x, y) \in A' \times C$$

When A, B, and C are fuzzy subsets of U, this rule is translated into a fuzzy subset of $U \times U$ like this:

$$
\begin{aligned}
R(x, y) &= [(A \times B)(x, y) \bigtriangledown (A' \times C)(x, y)] \\
&= [A(x) \bigtriangleup B(y)] \bigtriangledown [A'(x) \bigtriangleup C(y)]
\end{aligned}
$$

where $(A \times B)(x, y) = A(x) \bigtriangleup B(y)$ for some t-norm $\bigtriangleup$. Here $\bigtriangledown$ is a t-conorm, and $A'(x) = \eta(A(x))$ with η a negation.

In particular, for $C = U$, $C(x) = U(x) = 1$ for all $x \in U$, and we have $z \bigtriangleup 1 = z$, and so

$$R(x, y) = [A(x) \bigtriangleup B(y)] \bigtriangledown \eta(A(x))$$

This special form appears also in quantum logic where the implication operators among formulas are defined by

$$A \Rightarrow B = A' \vee (A \wedge B)$$

This is the third equivalent form of implications for Boolean logic, and these implications are called **Q-implications**, with **Q** coming from the word quantum.

Definition 6.1.4 *Let* $(\triangle, \triangledown, \eta)$ *be a De Morgan system. This means that* $\triangle$ *and* $\triangledown$ *are dual with respect to the negation* η, *which means that* $x \triangle y = \eta((\eta(x) \triangledown \eta(y)))$. *A* **Q-implication** *is a binary operation* $\Rightarrow$ *on* $[0, 1]$ *of the form*

$$(x \Rightarrow y) = \eta(x) \triangledown (x \triangle y)$$

It is trivial to show that $\Rightarrow$ is an implication. Two examples follow.

- For the De Morgan system

$$\begin{aligned}
x \triangle y &= x \wedge y \\
x \triangledown y &= x \vee y \\
\eta(x) &= 1 - x
\end{aligned}$$

the Q-implication is

$$(x \Rightarrow y) = (x \wedge y) \vee (1 - x)$$

- For the De Morgan system

$$\begin{aligned}
x \triangle y &= (x + y - 1) \vee 0 \\
x \triangledown y &= (x + y) \wedge 1 \\
\eta(x) &= 1 - x
\end{aligned}$$

the Q-implication is

$$\begin{aligned}
(x \Rightarrow y) &= (x \triangle y) \triangledown \eta(x) \\
&= [((x + y - 1) \vee 0) + (1 - x)] \wedge 1 \\
&= \begin{cases} y & \text{if } x + y - 1 \geq 0 \\ 1 - x & \text{if } x + y - 1 < 0 \end{cases} \\
&= (1 - x) \vee y
\end{aligned}$$

6.2   Averaging operators

Taking the average of two elements of the unit interval $[0,1]$ is a binary operation on $[0,1]$. This operation has come into play a couple of times in Section 5.4. Every negation β is a conjugate of the negation $\alpha(x) = 1-x$. That is, there is an $f \in Aut(\mathbb{I})$ such that $\beta = f^{-1}\alpha f$. One such f is given by the formula

$$f(x) = \frac{\alpha\beta(x) + x}{2}$$

which takes the average of the elements $\alpha\beta(x)$ and x of $[0,1]$. Another instance was in determining the centralizer $Z(\alpha) = \{g \in Aut(\mathbb{I}) : g\alpha = \alpha g\}$. It was shown that $g \in Z(\alpha)$ if and only if for some $f \in Aut(\mathbb{I})$,

$$g(x) = \left\{ \frac{\alpha f \alpha(x) + f(x)}{2} \right\}$$

Both these involve the involution α. Why should such group theoretic properties of α involve $+$ and division by 2, here the average of two real numbers? Viewing average as a binary operation on $[0,1]$, just what is the special relation between average and α? The same formulas do not work if α is replaced by some other negation. This section addresses this issue. We define "averaging operators" on the unit interval, and investigate especially their relation with negations and nilpotent t-norms. Also, averaging operators provide new insight into the important class of Frank t-norms.

The averaging operators we consider are not "weighted" averages in the usual sense, although they share some of the basic properties. These averaging operators can be thought of as "skewed" averages. They provide a continuous scaling of the unit interval that is not provided by the lattice structure. Our characterization and many other facts about averaging operators can be found in the references [55, 71, 4, 27, 3, 23, 5, 105, 67].

We use the following definition which is a variant of those in the references.

Definition 6.2.1 *An **averaging operator** on* $\mathbb{I}$ *is a binary operation* $\dotplus : \mathbb{I}^2 \to \mathbb{I}$ *satisfying for all* $x, y \in [0,1]$,

1. $x \dotplus y = y \dotplus x$ ($\dotplus$ *is commutative*).

2. $y < z$ *implies* $x \dotplus y < x \dotplus z$ ($\dotplus$ *is strictly increasing in each variable*).

3. $x \dotplus y \le c \le x \dotplus z$ *implies there exists* $w \in [y, z]$ *with* $x \dotplus w = c$ ($\dotplus$ *is convex, i.e., continuous*).

4. $x \dotplus x = x$ ($\dotplus$ *is idempotent*).

5. $(x \dotplus y) \dotplus (z \dotplus w) = (x \dotplus z) \dotplus (y \dotplus w)$ ($\dotplus$ *is* bisymmetric).

The following properties of an averaging operator are well-known.

Proposition 6.2.2 *Let $\dotplus$ be an averaging operator. Then for each $x, y \in [0, 1]$,*

1. $x \wedge y \leq x \dotplus y \leq x \vee y$—*that is, the average of x and y lies* between *x and y;*

2. *the function $A_x : \mathbb{I} \to [x \dotplus 0, x \dotplus 1] : y \longmapsto x \dotplus y$ is an isomorphism——that is, A_x is an increasing function that is both one-to-one, and onto.*

Proof. If $x \leq y$, then $x \wedge y = x = x \dotplus x \leq x \dotplus y \leq y \dotplus y = y = x \vee y$. Similarly, if $y \leq x$, $x \wedge y \leq x \dotplus y \leq x \vee y$. Clearly the function A_x is strictly increasing and, in particular, one-to-one. Suppose $x \dotplus 0 \leq c \leq x \dotplus 1$. Then by convexity, there is a number $w \in [0, 1]$ with $x \dotplus w = c$. Thus A_x is onto. ∎

The standard averaging operator is the arithmetic mean:

$$\mathrm{av}\,(x, y) = \frac{x + y}{2}.$$

Other examples include the power means and logarithmic means:

$$x \dotplus y = \left(\frac{x^a + y^a}{2} \right)^{\frac{1}{a}}, \ a > 0$$

$$x \dotplus y = \log_a \left(\frac{a^x + a^y}{2} \right), \ a > 0, \ a \neq 1$$

Indeed, for any automorphism or antiautomorphism γ of $\mathbb{I}$,

$$x \dotplus y = \gamma^{-1} \left(\frac{\gamma(x) + \gamma(y)}{2} \right) = \gamma^{-1}\,(\mathrm{av}\,(\gamma(x), \gamma(y)))$$

is an averaging operator.

The preceding example is universal—that is, given an averaging operator $\dotplus$, there is an automorphism γ of $\mathbb{I}$ that satisfies

$$\gamma\,(x \dotplus y) = \frac{\gamma(x) + \gamma(y)}{2}$$

for all $x, y \in [0, 1]$. Here is a brief outline of the proof. This automorphism can be defined inductively on the collection of elements of $[0, 1]$ that are generated by $\dotplus$ from 0 and 1. Such elements can be written uniquely in one of the forms

$$x = 0, \quad x = 1, \quad x = 0\dotplus1, \quad \text{or}$$
$$x = ((\cdots((0\dotplus1)\dotplus a_1)\dotplus\cdots)\dotplus a_{n-1})\dotplus a_n$$

for $a_1, \ldots, a_n \in \{0, 1\}$, $n \geq 1$, and γ is then defined inductively by

$$\gamma(0) = 0; \qquad \gamma(1) = 1;$$

$$\gamma(x\dotplus a) = \frac{\gamma(x) + a}{2} \text{ if } \gamma(x) \text{ is defined and } a \in \{0, 1\}$$

The function γ satisfies

$$\gamma((\cdots((0\dotplus x_1)\dotplus x_2)\dotplus\cdots)\dotplus x_n) = \sum_{k=1}^{n} \frac{1}{2^{n-k+1}} x_k$$

where $x_1, ..., x_n$ is any sequence of 0's and 1's. Now γ is a strictly increasing function on a dense subset of $\mathbb{I}$ and thus γ extends uniquely to an automorphism of $\mathbb{I}$ (see, for example, Aczel [3] page 287). Moreover, there were no choices made in the definition of γ on the dense subset. Thus we have the following theorem.

Theorem 6.2.3 *The automorphism γ defined above satisfies*

$$\gamma(x\dotplus y) = \frac{\gamma(x) + \gamma(y)}{2}$$

for all $x, y \in [0, 1]$. Thus every averaging operator on $[0, 1]$ is isomorphic to the usual averaging operator on $[0, 1]$—that is, the systems $(\mathbb{I}, \dotplus)$ and $(\mathbb{I}, \mathrm{av})$ are isomorphic as algebras. Moreover, γ is the only isomorphism between $(\mathbb{I}, \dotplus)$ and $(\mathbb{I}, \mathrm{av})$.

If $\dotplus$ is any averaging operator, then the algebra $(\mathbb{I}, \dotplus)$ is called a **mean system**. By the theorem, any two mean systems are isomorphic.

Corollary 6.2.4 *For any averaging operator $\dotplus$, the automorphism group of $(\mathbb{I}, \dotplus)$ has only one element.*

Proof. Suppose that f is an automorphism of $(\mathbb{I}, \dotplus)$. Then γf is an isomorphism of $(\mathbb{I}, \dotplus)$ with $(\mathbb{I}, \mathrm{av})$, so by the previous theorem, $\gamma f = \gamma$. Thus $f = \gamma\gamma^{-1} = 1$. ∎

When an averaging operator is given by the formula

$$x \dotplus y = \gamma^{-1}\left(\frac{\gamma(x) + \gamma(y)}{2}\right)$$

for an automorphism γ of $\mathbb{I}$, we will call γ a **generator** of the operator $\dot{+}$ and write $\dot{+} = \dot{+}_\gamma$. From the theorem above, the generator of an averaging operator is unique.

An averaging operator on the unit interval $[0,1]$ enables us to define an average of two automorhisms or of two antiautomorphisms of that interval. Such averages will play a role in what follows.

Theorem 6.2.5 *If f,g are automorphisms [antiautomorphisms] of $\mathbb{I}$, and $\dot{+}$ is an averaging operator on $\mathbb{I}$, then $f \dot{+} g$ defined by $(f \dot{+} g)(x) = f(x) \dot{+} g(x)$ is again an automorphism [antiautomorphism] of $\mathbb{I}$.*

Proof. Suppose f and g are automorphisms of $\mathbb{I}$. If $x < y$, then $f(x) < f(y)$ and $g(x) < g(y)$ imply that $f(x) \dot{+} g(x) < f(y) \dot{+} g(y)$ since $\dot{+}$ is strictly increasing in each variable. Thus the map $f \dot{+} g$ is strictly increasing. Also, $(f \dot{+} g)(0) = f(0) \dot{+} g(0) = 0 \dot{+} 0 = 0$ and $(f \dot{+} g)(1) = f(1) \dot{+} g(1) = 1 \dot{+} 1 = 1$. It remains to show that f maps $[0,1]$ onto $[0,1]$. Let $y \in [0,1]$. Then $f(0) \dot{+} g(0) = 0 \leq y \leq 1 = f(1) \dot{+} g(1)$. Let

$$u = \bigvee \{x \in [0,1] : f(x) \dot{+} g(x) \leq y\}$$
$$v = \bigwedge \{x \in [0,1] : f(x) \dot{+} g(x) \geq y\}$$

If $u < w < v$, then $f(w) \dot{+} g(w) > y$ and $f(w) \dot{+} g(w) < y$, an impossibility. Thus $u = v$ and $f(u) \dot{+} g(u) = y$. This completes the proof for automorphisms. Similar remarks hold in the case f and g are antiautomorphisms of $\mathbb{I}$. ∎

6.2.1 Averaging operators and negations

Now we turn to one of the main topics of this section, the relation between averaging operators and negations. We show that each averaging operator naturally determines a negation, with respect to which the averaging operator is self-dual. Also we put in the general context of averaging operators the facts mentioned earlier about conjugates and centralizers of negations.

Theorem 6.2.6 *For each averaging operator $\dot{+}$ on $\mathbb{I}$, the equation*

$$x \dot{+} \eta(x) = 0 \dot{+} 1$$

defines a negation $\eta = \eta_+$ on $\mathbb{I}$ with fixed point $0 \dot{+} 1$.

Proof. Since $x \dot{+} 0 = 0 \dot{+} x \leq 0 \dot{+} 1 \leq x \dot{+} 1$, by Proposition 6.2.2, for each $x \in [0,1]$ there is a number $y \in [0,1]$ such that $x \dot{+} y = 0 \dot{+} 1$, and since A_x is strictly increasing, there is only one such y for each x. Thus the equation defines a function $y = \eta(x)$. Clearly $\eta(0) = 1$ and $\eta(1) = 0$. Suppose $0 \le x < y \le 1$. We know $x \dotplus \eta(x) = y \dotplus \eta(y) = 0 \dotplus 1$. If $\eta(x) \le \eta(y)$, then $x \dotplus \eta(x) < y \dotplus \eta(x) \le y \dotplus \eta(y)$ which is not the case. Thus $\eta(x) > \eta(y)$ and η is a strictly decreasing function. Now $\eta(\eta(x))$ is defined by $\eta(x) \dotplus \eta(\eta(x)) = 0 \dotplus 1$. But also, $\eta(x) \dotplus x = x \dotplus \eta(x) = 0 \dotplus 1$. Thus, applying Proposition 6.2.2 to $\eta(x)$, we see that $\eta(\eta(x)) = x$. It follows that η is a negation. If x is the fixed point of η, then $x = x \dotplus x = x \dotplus \eta(x) = 0 \dotplus 1$. ∎

It is immediate that the negation defined by the usual average $(x + y)/2$ is α, or using the notation of the theorem, that $\eta_{av} = \alpha$. Indeed, $(x + \eta(x))/2 = (0 + 1)/2$ yields $\eta(x) = 1 - x$.

Theorem 6.2.7 *Every isomorphism between mean systems respects the natural negation—that is, is an isomorphism of mean systems with natural negation.*

Proof. Suppose $f : (\mathbb{I}, \dotplus_1) \to (\mathbb{I}, \dotplus_2)$ is an isomorphism. Then

$$
\begin{aligned}
f(x) \dotplus_2 f\left(\eta_{\dotplus_1}(x)\right) &= f\left(x \dotplus_1 \eta_{\dotplus_1}(x)\right) = f(0 \dotplus_1 1) \\
&= f(0) \dotplus_2 f(1) = 0 \dotplus_2 1
\end{aligned}
$$

Thus $f\left(\eta_{\dotplus_1}(x)\right) = \eta_{\dotplus_2}(f(x))$. ∎

For this reason, **mean systems with natural negation** $\left(\mathbb{I}, \dotplus, \eta_{\dotplus}\right)$ will be referred to simply as **mean systems**.

Corollary 6.2.8 *If γ is the generator of $\dotplus$, then $\eta_{\dotplus} = \gamma^{-1}\alpha\gamma$. Every negation is the natural negation of some averaging operator.*

Proof. In the theorem, let $\dotplus_1 = \dotplus_\gamma$ and $\dotplus_2 = av$. Then $f\left(\eta_{\dotplus_1}(x)\right) = \eta_{\dotplus_2}(f(x))$ says that $\gamma\eta_{\dotplus} = \alpha\gamma$, or $\eta_{\dotplus} = \gamma^{-1}\alpha\gamma$. Since every negation is a conjugate of α, and γ is an arbitrary automorphism, the second statement of the corollary follows. ∎

Recall that a negation of the form $\gamma^{-1}\alpha\gamma$ is said to be **generated** by γ, and is written as $\alpha_\gamma = \gamma^{-1}\alpha\gamma$.

The corollary gives a way to compute the negation η associated with an averaging operator—that is if the generator γ of the averaging operator is known. Just conjugate α by γ. Finding η directly from the equation $x \dotplus \eta(x) = 0 \dotplus 1$ may or may not be easy.

Example 6.2.9 Following are three examples of averaging operators and their negations. That the negations are as stated is left as an exercise.

1. For $x \mathbin{\dot{+}} y = \dfrac{x+y}{2}$, $\eta_+(x) = 1 - x$.

2. For $x \mathbin{\dot{+}} y = \left(\dfrac{x^a + y^a}{2}\right)^{\frac{1}{a}}$, $\eta_+(x) = (1 - x^a)^{\frac{1}{a}}$, $a > 0$.

3. For $x \mathbin{\dot{+}} y = \log_a\left(\dfrac{a^x + a^y}{2}\right)$, $\eta_{+_a}(x) = \log_a(1 + a - a^x)$, $a > 0$, $a \neq 0$.

If $\triangle$ is a t-norm and f is an automorphism of $\mathbb{I}$, then $x \triangle_f y = f^{-1}(f(x) \triangle f(y))$ is a t-norm. If f is an antiautomorphism, $\triangle_f$ is a t-conorm. In particular, if f is a negation, $\triangle_f$ is the t-conorm dual to $\triangle$ with respect to this negation. For an averaging operator $\dot{+}$ it follows readily that if f is either an automorphism or an antiautomorphism of $\mathbb{I}$, then $x \mathbin{\dot{+}_f} y = f^{-1}(f(x) \dot{+} f(y))$ is again an averaging operator. Now an averaging operator $\dot{+}$ has its natural negation η_+. The following theorem shows that $\dot{+}$ is self-dual with respect to its natural negation—that is,

$$x \dot{+} y = \eta_+\left(\eta_+(y) \dot{+} \eta_+(x)\right)$$

Theorem 6.2.10 *Let $\dot{+}$ be an averaging operator on $\mathbb{I}$. Then η_+ is an antiautomorphism of the system $(\mathbb{I}, \dot{+})$. Moreover, it is the only antiautomorphism of $(\mathbb{I}, \dot{+})$.*

Proof. Let $\eta = \eta_+$. Since η is an antiautomorphism of $\mathbb{I}$, we need only show that $\eta(x \dot{+} y) = \eta(y) \dot{+} \eta(x)$ for all $x, y \in [0, 1]$. Now $\eta(x \dot{+} y)$ is the unique value satisfying the equation $(x \dot{+} y) \dot{+} \eta(x \dot{+} y) = 0 \dot{+} 1$. But by bisymmetry,

$$
\begin{aligned}
(x \dot{+} y) \dot{+} (\eta(y) \dot{+} \eta(x)) &= (x \dot{+} \eta(x)) \dot{+} (y \dot{+} \eta(y)) \\
&= (0 \dot{+} 1) \dot{+} (0 \dot{+} 1) = 0 \dot{+} 1
\end{aligned}
$$

It follows that $\eta(x \dot{+} y) = \eta(y) \dot{+} \eta(x)$. The last statement follows from Corollary 6.2.4. ∎

The centralizer $Z(\alpha) = \{g \in Aut(\mathbb{I}) : g\alpha = \alpha g\}$ is the set of elements of the form

$$\frac{\alpha f \alpha + f}{2}$$

for $f \in Aut(\mathbb{I})$, that is, the average of the two automorphisms $\alpha f \alpha$ and f. This gives $Z(\alpha)$ in terms of α and the ordinary average. This is just an instance of a general phenomenon.

Theorem 6.2.11 *Let $\dotplus$ be an averaging operator on $\mathbb{I}$ and let η be the negation determined by the equation $x \dotplus \eta(x) = 0 \dotplus 1$. Then the centralizer $Z(\eta) = \{g \in Aut(\mathbb{I}) : g\eta = \eta g\}$ of η is the set of elements of the form*

$$\eta f \eta \dotplus f$$

for automorphisms f of $\mathbb{I}$. Moreover, if $f \in Z(\eta)$, then $\eta f \eta \dotplus f = f$.

Proof. To show $\eta f \eta \dotplus f$ is in the centralizer of η, we need to show that $(\eta f \eta \dotplus f)(\eta(x)) = \eta((\eta f \eta \dotplus f)(x))$. We prove this by showing that $(\eta f \eta \dotplus f)\eta$ satisfies the defining property for η—that is, that $(\eta f \eta \dotplus f)(x) \dotplus (\eta f \eta \dotplus f)\eta(x) = 0 \dotplus 1$. Now

$$(\eta f \eta \dotplus f)(\eta(x)) = \eta f \eta \eta(x) + f\eta(x) = \eta f(x) \dotplus f\eta(x)$$

and

$$\eta((\eta f \eta \dotplus f)(x)) = \eta(\eta f \eta(x) \dotplus f(x))$$

By bisymmetry,

$$
\begin{aligned}
& [\eta f \eta(x) \dotplus f(x)] \dotplus [\eta f(x) \dotplus f\eta(x)] \\
= {} & [\eta f \eta(x) \dotplus f\eta(x)] \dotplus [\eta f(x) \dotplus f(x)] \\
= {} & [0 \dotplus 1] \dotplus [0 \dotplus 1] \\
= {} & [0 \dotplus 1]
\end{aligned}
$$

Thus the expression $(\eta f \eta \dotplus f)(\eta(x)) = \eta f(x) \dotplus f\eta(x)$ satisfies the defining equality for $\eta(\eta f \eta(x) \dotplus f(x))$, and we conclude that

$$(\eta f \eta \dotplus f)\eta = \eta(\eta f \eta \dotplus f)$$

Clearly, if $f \in Z(\eta)$, then $\eta f \eta \dotplus f = f$. It follows that every element of $Z(\eta)$ is of the form $\eta f \eta \dotplus f$ for some automorphism f of $\mathbb{I}$. ∎

To find an element to conjugate α by to get the negation β, one constructs

$$f(x) = \frac{\alpha\beta(x) + x}{2}$$

and $\beta = f^{-1}\alpha f$. That is, conjugate α by the ordinary average of the automorphisms $\alpha\beta$ and the identity. Again, this is just an instance of a more general phenomenon concerning averaging operations and negations. If η is a negation, then $\beta = f^{-1}\eta f$ where f is the average $\alpha\eta$ and the identity with respect to any averaging operator whose natural negation is η.

Theorem 6.2.12 *For any negation η, all negations are conjugates of η by automorphisms of $\mathbb{I}$. More specifically, if β is a negation, then*

$$\beta = f^{-1}\eta f$$

for f the automorphism of $\mathbb{I}$ defined by

$$f(x) = \eta\beta(x) \dotplus x,$$

where $\dotplus$ is any averaging operator such that $\eta = \eta_{\dotplus}$. Moreover, $\beta = g^{-1}\eta g$ if and only if $gf^{-1} \in Z(\eta)$.

Proof. We observed earlier that every negation is the natural negation of an averaging operator. The map $\eta\beta \dotplus \mathrm{id}$ is an automorphism of $\mathbb{I}$ since the composition of two negations is an automorphism and the average of two automorphisms is an automorphism (Theorem 6.2.5). To show that $\beta = (\eta\beta \dotplus \mathrm{id})^{-1} \eta (\eta\beta \dotplus \mathrm{id})$, we show that $(\eta\beta \dotplus \mathrm{id})\beta = \eta(\eta\beta \dotplus \mathrm{id})$. For any $x \in [0,1]$,

$$(\eta\beta \dotplus \mathrm{id})\beta(x) = (\eta\beta\beta(x) \dotplus \beta(x)) = \eta(x) \dotplus \beta(x)$$

and

$$\eta(\eta\beta \dotplus \mathrm{id})(x) = \eta(\eta\beta(x) \dotplus x)$$

Now by bisymmetry

$$
\begin{aligned}
[\eta\beta(x) \dotplus x] \dotplus [\eta(x) \dotplus \beta(x)] &= [\eta\beta(x) \dotplus \beta(x)] \dotplus [\eta(x) \dotplus x] \\
&= [0 \dotplus 1] \dotplus [0 \dotplus 1] = [0 \dotplus 1]
\end{aligned}
$$

Thus, using the defining property of η, $\eta(x) \dotplus \beta(x) = \eta(\eta\beta(x) \dotplus x)$, or

$$(\eta\beta \dotplus \mathrm{id})\beta = \eta(\eta\beta \dotplus \mathrm{id})$$

as claimed. ∎

The next theorem follows easily.

Theorem 6.2.13 *Let $\dotplus$ be an averaging operator on $\mathbb{I}$, and let η be the negation determined by the equation $x \dotplus \eta(x) = 0 \dotplus 1$. The map*

$$\mathrm{Neg}\,(\mathbb{I}) \rightarrow \mathrm{Aut}\,(\mathbb{I})/Z(\eta) : \beta \longmapsto Z(\eta)(\eta\beta \dotplus \mathrm{id})$$

is a one-to-one correspondence between the negations of $\mathbb{I}$ and the set of right cosets of the centralizer $Z(\eta)$ of η.

6.2.2 Averaging operators and nilpotent t-norms

We begin with a review of some notation.

- $\eta_\triangle$ is the negation naturally associated with a nilpotent t-norm $\triangle$ by the condition

$$\eta_\triangle(x) = \bigvee \{y : x \triangle y = 0\}$$

  that is, $x \triangle y = 0$ if and only if $y \le \eta_\triangle(x)$. This was discussed in Section 5.5. See also [35].

- $\blacktriangle$ is the symbol used for the Lukasiewicz t-norm.

$$x \blacktriangle y = (x + y - 1) \vee 0$$

- $\blacktriangle_\gamma$ stands for the nilpotent t-norm

$$x \blacktriangle_\gamma y = \gamma^{-1}((\gamma(x) + \gamma(y) - 1) \vee 0)$$

  generated by the automorphism γ. The automorphism γ is the L-generator of $\blacktriangle_\gamma$.

- $\alpha_\gamma = \gamma^{-1}\alpha\gamma$, the negation generated by γ.

- $\dot{+}_\gamma$ is the averaging operator generated by γ, and is given by

$$x \dot{+}_\gamma y = \gamma^{-1}\left(\frac{\gamma(x) + \gamma(y)}{2}\right)$$

- $\eta_{\dot{+}}$ is the natural negation determined by the averaging operator $\dot{+}$, and is given by

$$x \dot{+} \eta_{\dot{+}}(x) = 0 \dot{+} 1$$

It was observed in Theorem 6.2.6 that the negation generated by γ is the same as the negation associated with the averaging operator $\dot{+}_\gamma$—that is, $\alpha_\gamma = \eta_{\dot{+}_\gamma}$. A similar relationship holds for the nilpotent t-norm $\blacktriangle_\gamma$.

Proposition 6.2.14 *For an automorphism γ of $\mathbb{I}$, the negations α_γ, $\eta_{\blacktriangle_\gamma}$, and $\eta_{\dot{+}_\gamma}$ coincide—that is,*

$$x \blacktriangle_\gamma y = 0 \quad \text{if and only if} \quad y \le \alpha_\gamma(x)$$

and

$$x \dot{+}_\gamma \eta_{\blacktriangle_\gamma}(x) = x \dot{+}_\gamma \alpha_\gamma(x) = 0 \dot{+}_\gamma 1$$

Proof. Since $x \blacktriangle_\gamma y = \gamma^{-1}\left((\gamma(x) + \gamma(y) - 1) \vee 0\right)$, we have $x \triangle_\gamma y = 0$ if and only if $\gamma(x) + \gamma(y) - 1 \leq 0$ if and only if $\gamma(y) \leq 1 - \gamma(x)$ if and only if $y \leq \gamma^{-1}(1 - \gamma(x)) = \alpha_\gamma(x)$. The last equation follows. ∎

We remark that this same negation is often represented in the form

$$\eta(x) = f^{-1}\left(\frac{f(0)}{f(x)}\right)$$

for a *multiplicative generator* f of the nilpotent t-norm. See [35], for example.

There are a number of different averaging operators that give the same negation, namely one for each automorphism in the centralizer of that negation. The same can be said for nilpotent t-norms. However, there is a closer connection between averaging operators and nilpotent t-norms than a common negation. Given an averaging operator, one can determine the particular nilpotent t-norm that has the same generator, and conversely, as shown in the following theorem. This correspondence is a natural one—that is, it does not depend on the generator.

Theorem 6.2.15 *The condition*

$$x \triangle y \leq z \quad \text{if and only if} \quad x \dotplus y \leq z \dotplus 1$$

determines a one-to-one correspondence between nilpotent t-norms and averaging operators, namely, given an averaging operator $\dotplus$, define $\triangle_{\dotplus}$ by

$$x \triangle_{\dotplus} y = \bigwedge\{z : x \dotplus y \leq z \dotplus 1\}$$

This correspondence preserves generators. That is, $\triangle_{\dotplus}$ and $\dotplus$ have the same generator.

Proof. By Theorem 6.2.3, we may assume that $\dotplus = \dotplus_\gamma$ for an automorphism γ of $\mathbb{I}$. Then

$$
\begin{aligned}
x \triangle_{\dotplus} y &= \bigwedge\{z : x \dotplus_\gamma y \leq z \dotplus_\gamma 1\}\\
&= \bigwedge\left\{z : \gamma^{-1}\left(\frac{\gamma(x) + \gamma(y)}{2}\right) \leq \gamma^{-1}\left(\frac{\gamma(z) + \gamma(1)}{2}\right)\right\}\\
&= \bigwedge\{z : \gamma(x) + \gamma(y) \leq \gamma(z) + 1\}\\
&= \bigwedge\{z : \gamma(x) + \gamma(y) - 1 \leq \gamma(z)\}\\
&= \bigwedge\{z : (\gamma(x) + \gamma(y) - 1) \vee 0 \leq \gamma(z)\}\\
&= \bigwedge\{z : \gamma^{-1}((\gamma(x) + \gamma(y) - 1) \vee 0) \leq z\}\\
&= \gamma^{-1}((\gamma(x) + \gamma(y) - 1) \vee 0)
\end{aligned}
$$

Thus, in particular, $x \triangle_{\dotplus} y$ is a nilpotent t-norm. Moreover, $\triangle_{\dotplus}$ has the same generator as $\dotplus$. Thus the one-to-one correspondence $\dotplus_\gamma \longleftrightarrow \blacktriangle_\gamma$ is the natural one defined in the statement of the theorem. ∎

To describe the inverse correspondence directly—that is, without reference to a generating function, given a nilpotent t-norm $\triangle$, define a binary operation $*_\triangle$ by

$$x *_\triangle y = \bigvee \{ z : z \triangle z \leq x \triangle y \}$$

and define $\dotplus_\triangle$ by

$$x \dotplus_\triangle y = (x *_\triangle y) \wedge \left(\eta_\triangle \left(\eta_\triangle \left(x \right) *_\triangle \eta_\triangle \left(y \right) \right) \right)$$

This definition relies on the fact that for an averaging operator $\dotplus$, $\eta_{\dotplus}$ is an antiautomorphism of the system $\left(\mathbb{I}, \dotplus \right)$, (Theorem 6.2.10) —and in particular, $\dotplus$ is self-dual relative to $\eta_{\dotplus}$:

$$x \dotplus y = \eta_{\dotplus} \left(\eta_{\dotplus} \left(x \right) \dotplus \eta_{\dotplus} \left(y \right) \right)$$

The situation with strict t-norms is somewhat more complicated. We explore that in the next section.

6.2.3   De Morgan systems with averaging operators

The family of t-norms $\triangle$ that satisfy the equation

$$(x \triangle y) + (x \triangledown y) = x + y$$

for $x \triangledown y = \alpha \left(\alpha \left(x \right) \triangle \alpha \left(x \right) \right)$, the t-conorm dual to $\triangle$ relative to $\alpha \left(x \right) = 1 - x$, are called **Frank t-norms** [26]. Frank showed that this is the one-parameter family of t-norms of the form

$$x \triangle_{F_a} y = \log_a \left[1 + \frac{(a^x - 1)(a^y - 1)}{a - 1} \right], \ a > 0, \ a \neq 1$$

with limiting values

$$x \triangle_{F_0} y = x \wedge y$$

$$x \triangle_{F_1} y = xy$$

$$x \triangle_{F_\infty} y = (x + y - 1) \vee 0$$

Note that all the Frank t-norms for $0 < a < \infty$ are strict. The strict Frank t-norms are generated by functions of the form

$$F_a \left(x \right) = \frac{a^x - 1}{a - 1}, \ a > 0, \ a \neq 1$$

$$F_1 \left(x \right) = x$$

The limiting cases for $a = 0$, $a = 1$, and $a = \infty$ give the t-norms $\wedge$, multiplication, and $\blacktriangle$, respectively.

A t-norm $\triangle$ is called **nearly Frank** [69] if there is an isomorphism $h : (\mathbb{I}, \triangle, \alpha) \to (\mathbb{I}, \triangle_F, \alpha)$ of De Morgan systems for some Frank t-norm $\triangle_F$—that is, for all $x \in [0, 1]$,

$$h(x \triangle y) = h(x) \triangle_F h(y)$$
$$h\alpha(x) = \alpha h(x)$$

Definition 6.2.16 *A system $(\mathbb{I}, \triangle, \eta, \triangledown, \dotplus)$ is a **Frank system** if $\triangle$ is a t-norm (nilpotent or strict or idempotent), η is a negation, $\triangledown$ is a t-conorm, $\dotplus$ is an averaging operator, and the identities*

(1) $x \triangledown y = \eta(\eta(x) \triangle \eta(y))$ [$(\mathbb{I}, \triangle, \eta, \triangledown)$ *is a deMorgan system.*]

(2) $x \dotplus \eta(x) = 0 \dotplus 1$ [$(\mathbb{I}, \eta, \dotplus)$ *is a mean system with $\eta = \eta_{\dotplus}$.*]

(3) $(x \triangle y) \dotplus (x \triangledown y) = x \dotplus y$ [*The **Frank equation** is satisfied.*]

*hold for all $x, y \in [0, 1]$. A Frank system will be called a **standard Frank system** if $\dotplus = \mathrm{av} = \dotplus_{\mathrm{id}}$.*

Note that in a standard Frank system $(\mathbb{I}, \triangle, \eta, \triangledown, \dotplus)$, $\triangle$ is a Frank t-norm (nilpotent or strict) and $\eta = \alpha$. Also note that if $\dotplus$ is generated by $h \in \mathrm{Aut}(\mathbb{I})$, the Frank equation is

$$h^{-1}\left(\frac{h(x \triangle y) + h(x \triangledown y)}{2}\right) = h^{-1}\left(\frac{h(x) + h(y)}{2}\right)$$

which is equivalent to

$$h(x \triangle y) + h(x \triangledown y) = h(x) + h(y)$$

If $(\mathbb{I}, \triangle, \eta, \triangledown, \dotplus)$ is a Frank system, we will say that $(\mathbb{I}, \triangle, \dotplus)$ **determines a Frank system**, since η is determined algebraically by $\dotplus$, and $\triangledown$ by η and $\triangle$.

Theorem 6.2.17 *The system $(\mathbb{I}, \triangle, \eta, \triangledown, \dotplus)$ is a Frank system if and only if it is isomorphic to a standard Frank system.*

Proof. Suppose $(\mathbb{I}, \triangle, \dotplus)$ determines a Frank system. There is an automorphism g of $\mathbb{I}$ such that $\dotplus = \dotplus_g$, and g is also an isomorphism of Frank systems

$$g : (\mathbb{I}, \triangle, \dotplus_g) \approx (\mathbb{I}, \triangle_{g^{-1}}, \dotplus_{id})$$

where $\dot{+}_{id} = \mathrm{av}$. Thus $\triangle_{g^{-1}}$ is a Frank t-norm. The converse is clear. ∎

Thus $\left(\mathbb{I}, \bullet_f, \dot{+}_g\right)$ determines a strict Frank system if and only if f and g are related by $g \in F_a^{-1}\mathbb{R}^+ f$ for some a. Note that for every strict Archimedean, convex t-norm $\triangle = \bullet_f$ there is a two-parameter family of Frank systems

$$F_a^{-1}rf : \left(\mathbb{I}, \bullet_f, \dot{+}_{F_a^{-1}rf}\right) \approx \left(\mathbb{I}, \bullet_{F_a}, \mathrm{av}\right)$$

and for every averaging operator $\dot{+} = \dot{+}_g$ there is a one-parameter family of strict Frank systems

$$g : \left(\mathbb{I}, \bullet_{F_a g}, \dot{+}_g\right) \approx \left(\mathbb{I}, \bullet_{F_a}, \mathrm{av}\right)$$

Also, for every nilpotent Archimedean, convex t-norm $\triangle = \blacktriangle_\gamma$ there is a unique Frank system

$$\gamma : \left(\mathbb{I}, \blacktriangle_\gamma, \dot{+}_\gamma\right) \approx \left(\mathbb{I}, \blacktriangle, \mathrm{av}\right)$$

Thus every system of the form $\left(\mathbb{I}, \triangle\right)$ or $\left(\mathbb{I}, \dot{+}\right)$ is part of one or more Frank systems. However, not every De Morgan system can be extended to a Frank system. The following theorem identifies those that can. Recall that a nilpotent De Morgan system is called a **Boolean system** [35] if the negation is the one naturally determined by the t-norm.

Theorem 6.2.18 *A De Morgan system with nilpotent t-norm can be extended to a Frank system if and only if the system is Boolean. A De Morgan system $(\mathbb{I}, \triangle, \eta)$ with strict t-norm $\triangle$ can be extended to a Frank system if and only if there exists $a \in \mathbb{R}^+$ such that for $f, g \in \mathrm{Aut}\,(\mathbb{I})$ with $\triangle = \bullet_f$ and $\eta = \alpha_g$*

$$F_a^{-1}\mathbb{R}^+ f \cap Z\,(\alpha)\, g \neq \varnothing$$

In this case,

$$F_a^{-1} f \cap Z\,(\alpha)\, g = \{h\}$$

and the Frank system is

$$\left(\mathbb{I}, \bullet_f, \alpha_g, \dot{+}_h\right) = \left(\mathbb{I}, \bullet_{F_a h}, \alpha_h, \dot{+}_h\right)$$

Moreover, there is at most one such a. The t-norm in the Frank system is nearly Frank if and only if g is in the centralizer of α.

Proof. If the t-norm in a Frank system is nilpotent, it is generated by the same automorphism as the averaging operator. Thus the negation is also generated by the same automorphism as the t-norm.

Consider the De Morgan system $(\mathbb{I}, \bullet_f, \alpha_g)$ with strict t-norm. Assume

$$F_a^{-1}\mathbb{R}^+ f \cap Z(\alpha)\, g \neq \varnothing.$$

Then by Theorems 5.7.1 and 5.7.3,

$$F_a^{-1}\mathbb{R}^+ f \cap Z(\alpha)\, g = \{h\}.$$

Thus for some $r \in \mathbb{R}^+$ and $k \in Z(\alpha)$

$$F_a^{-1}rf = kg = h.$$

Thus $rf = F_a h$, implying $\bullet_f = \bullet_{F_a h}$. Thus

$$\left(\mathbb{I}, \bullet_f, \alpha_g, \dot{+}_h\right) = \left(\mathbb{I}, \bullet_{F_a h}, \alpha_h, \dot{+}_h\right)$$

is a Frank system. If $F_b^{-1}\mathbb{R}^+ f \cap Z(\alpha)\, g = \{k\}$ for some $b \in \mathbb{R}^+$, then $\bullet_{F_b k} = \bullet_f = \bullet_{F_a h}$, implying that $\bullet_{F_b k h^{-1}} = \bullet_{F_a}$. But no t-norm is both Frank and nearly Frank [69] from which it follows that $kh^{-1} = 1$ and, from that, that $a = b$ [26].

Now suppose that $(\mathbb{I}, \bullet_f, \alpha_g)$ can be extended to the Frank system $\left(\mathbb{I}, \bullet_f, \alpha_g, \dot{+}_h\right)$. Then α_g is the negation for $\dot{+}_h$ so that $\alpha_g = \alpha_h$ and we have $hg^{-1} \in Z(\alpha)$, or $h = kg$ with $k \in Z(\alpha)$. Thus,

$$\left(\mathbb{I}, \bullet_f, \alpha_g, \dot{+}_h\right) \approx \left(\mathbb{I}, \bullet_{fh^{-1}}, \alpha, \dot{+}\right)$$

which is isomorphic to $\left(\mathbb{I}, \bullet_{F_a}, \alpha, \dot{+}\right)$ for some $a \in \mathbb{R}^+$. So $rf = F_a h$ for some $r, a \in \mathbb{R}^+$, or

$$F_a^{-1}rf = h = kg \in F_a^{-1}\mathbb{R}^+ f \cap Z(\alpha)\, g$$

■

The intersection $F_a^{-1}\mathbb{R}^+ f \cap Z(\alpha)\, g$ may be empty for all $a > 0$. That is the case when the equation $F_a^{-1}rf = hg$ has no solution for $r, a > 0$ and $h \in Z(\alpha)$. For a particular example of this, take $f = \mathrm{id}$, $g(x) = x$ for $0 \leq x \leq \frac{1}{2}$. Then $F_a^{-1}r = hg$ implies that $h(x) = F_a^{-1}r(x)$ for $0 \leq x \leq \frac{1}{2}$ and since $h \in Z(\alpha)$, $h(x) = 1 - F_a^{-1}r(1-x) = \alpha\left(F_a^{-1}r(\alpha(x))\right)$ for $\frac{1}{2} \leq x \leq 1$. But then

$$g(x) = \begin{cases} x & \text{if } 0 \leq x \leq \frac{1}{2} \\ \alpha r^{-1} F_a \alpha F_a^{-1} r(x) & \text{if } \frac{1}{2} \leq x \leq 1 \end{cases}$$

Now simply choose g that is not differentiable at some $x_0 \in \left(\frac{1}{2}, 1\right)$, and such an equality cannot hold for any choice of a and r. So there are De Morgan systems $(\mathbb{I}, \bullet_f, \alpha_g)$ that are not parts of Frank systems.

6.3   Powers of t-norms

For positive integers n, the n-th **power** of a t-norm $\triangle$ is

$$x^{[n]} = \overbrace{x \triangle x \triangle \cdots \triangle x}^{n \text{ times}}$$

We outline in this section the theory for arbitrary positive real powers for Archimedean t-norms. The details may be found in [98].

For a strict t-norm, the notion of n-th power naturally extends to n-th roots by defining $x^{\left[\frac{1}{n}\right]}$ to be the unique solution to $\left(x^{\left[\frac{1}{n}\right]}\right)^{[n]} = x$ for positive integers n. This leads to the definition of positive rational powers by setting $x^{\left[\frac{m}{n}\right]} = \left(x^{\left[\frac{1}{n}\right]}\right)^{[m]}$ for positive integers m and n. One can show that the value of $\left(x^{\left[\frac{1}{n}\right]}\right)^{[m]}$ is independent of the representation of the rational $\frac{m}{n}$, and continue to develop the notion of real powers directly as is done for real powers of real numbers. However, we will take advantage of the fact that all strict t-norms are isomorphic to multiplication where powers are already defined, observing that if f is an isomorphism $f : (\mathbb{I}, \triangle) \to (\mathbb{I}, \bullet)$, then for positive integers n, $f\left(x^{[n]}\right) = \left(f(x)\right)^n$, so that $x^{[n]} = f^{-1}\left(\left(f(x)\right)^n\right)$. We extend this to arbitrary positive real numbers, calling on the continuity of the isomorphism.

Definition 6.3.1 *Given a strict t-norm $\triangle$ and a positive real number r, the r-**th power of** $\triangle$ is defined to be the function $x^{[r]} = f^{-1}\left(\left(f(x)\right)^r\right) = f^{-1}rf(x)$, where f is any isomorphism of $\triangle$ with multiplication.*

The function $x^{[r]}$ is independent of the choice of isomorphism, since any other isomorphism is of the form sf for some positive real number s, and

$$\begin{aligned}
(sf)^{-1}\left(sf(x)^r\right) &= f^{-1}\left(s^{-1}\left(f(x)^{rs}\right)\right) \\
&= f^{-1}\left(\left(f(x)\right)^{rss^{-1}}\right) \\
&= f^{-1}\left(\left(f(x)\right)^r\right)
\end{aligned}$$

Theorem 6.3.2 *Given a strict t-norm $\triangle$, the powers of $\triangle$ satisfy the following:*

1. $\left(x^{[r]}\right)^{[s]} = x^{[rs]}$ *for all $r, s \in \mathbb{R}^+$.*

2. $x^{[r]} \triangle x^{[s]} = x^{[r+s]}$ *for all $r, s \in \mathbb{R}^+$.*

3. $\left(x^{\left[\frac{1}{n}\right]}\right)^{[n]} = x$ *for all $n \in \mathbb{Z}^+$.*

4. $x^{[r]} = \lim_{\frac{m}{n} \to r} \left(x^{\left[\frac{1}{n}\right]} \right)^{[m]}$ *where* $n, m \in \mathbb{Z}^{+}$.

The proof is left as an exercise. Again, the details are in [98].

It can be shown [98] that a function $\delta : [0, 1] \to [0, 1]$ is the rth power for some strict t-norm $\triangle$ and some $r \in \mathbb{R}^{+}$ if and only if $\delta \in \text{Aut}(\mathbb{I})$ and one of the following holds.

1. $r > 1$ and $\delta(x) < x$ for all $x \in (0, 1)$.

2. $r = 1$ and $\delta(x) = x$ for all $x \in (0, 1)$.

3. $r < 1$ and $\delta(x) > x$ for all $x \in (0, 1)$.

There are many strict t-norms that have the same r-th power for any given power. Two automorphisms $f, g \in Aut(\mathbb{I})$ generate strict t-norms having the same r-th power if and only if fg^{-1} is in the centralizer $Z(r) = \{z \in Aut(\mathbb{I}) : zr = rz\}$. Also, the precise form of the elements of $Z(r)$ has been worked out. However, if all the r-th powers are known, the strict t-norm is uniquely determined. This is a direct consequence of the proof of Theorem 6.2.3.

The powers of a general nilpotent t-norm depend on the powers of the Lukasiewicz t-norm, in the same sense that powers of strict t-norms depend on powers for multiplication. So we first consider the question of powers for the Lukasiewicz t-norm $x \blacktriangle y = (x + y - 1) \vee 0$.

Proposition 6.3.3 *Let n be a positive integer. Then*

$$\overbrace{x \blacktriangle x \blacktriangle \cdots \blacktriangle x}^{n \ times} = (nx - n + 1) \vee 0.$$

The proof is left as an exercise.

Now $x^{[n]} = (nx - n + 1) \vee 0$ is a continuous function mapping $[0, 1]$ onto $[0, 1]$, and restricts to an isomorphism $\left[\frac{n-1}{n}, 1\right] \approx [0, 1]$. Thus for each $x \in (0, 1]$ there is a unique solution to the equation $y^{[n]} = x$. Call this solution $x^{\left[\frac{1}{n}\right]}$.

Proposition 6.3.4 *Let n be a positive integer and $x \in (0, 1]$. Then $\frac{1}{n}x - \frac{1}{n} + 1$ is the unique solution to $y^{[n]} = x$.*

The proof is left as an exercise.

Definition 6.3.5 *For the Lukasiewicz t-norm, $x \in (0, 1]$, and positive real numbers r, the r-th power of x is defined to be $x^{[r]} = (rx - r + 1) \vee 0$.*

For any positive integer n, 0 has multiple n-th roots, namely the interval of numbers from 0 to the largest $x \in [0,1]$ satisfying $x^{[n]} = 0$. The usual convention is that $0^{[n]} = 0$. Below are some examples of powers for the Lukasiewicz t-norm.

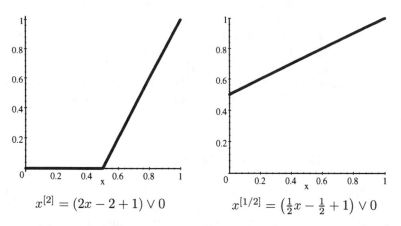

$$x^{[2]} = (2x - 2 + 1) \vee 0 \qquad\qquad x^{[1/2]} = \left(\tfrac{1}{2}x - \tfrac{1}{2} + 1\right) \vee 0$$

Proposition 6.3.6 *For positive real numbers r and s, and $x \in [0,1]$, the following hold.*

1. $\left(x^{[r]}\right)^{[s]} = x^{[rs]}$ if $x^{[r]} \neq 0$.

2. $x^{[r+s]} = x^{[r]} \blacktriangle x^{[s]}$.

The proof is left as an exercise.

From these propositions we see that for rationals $q = \frac{m}{n}$, $x^{\left[\frac{m}{n}\right]}$ is the m-th power of the n-th root of x and this definition depends only on the rational number $q = \frac{m}{n}$ and not on the particular representation of q as a quotient of integers.

Definition 6.3.7 *Given a nilpotent t-norm $\triangle$ and a positive real number r, the r-**th power** of $\triangle$ is defined to be the function*

$$x^{[r]} = f^{-1}\left(\left(r \cdot f(x) - r + 1\right) \vee 0\right)$$

where f is the L-generator of $\triangle$ and $r \cdot f(x)$ denotes the ordinary product of r and $f(x)$.

It can be shown that a nondecreasing function δ from $[0,1]$ to $[0,1]$ is the r-th power for some $r \in \mathbb{R}^+$ and some nilpotent t-norm $\triangle$ if and only if one of the following holds.

1. $r > 1$, $\delta(x) < x$ for all $x \in (0,1)$, and for some $a \in (0,1)$, δ is identically 0 on $[0,a]$ and induces an isomorphism $[a,1] \approx [0,1]$.

2. $r = 1$ and $\delta(x) = x$ for all $x \in (0, 1)$.

3. $r < 1$, $\delta(x) > x$ for all $x \in [0, 1)$, $\delta(0) = 0$ and for some $a \in (0, 1)$, δ induces an isomorphism $(0, 1] \approx (a, 1]$.

The same as for strict t-norms, there are many nilpotent t-norms that have the same r-th power for any given power. Two nilpotent t-norms with L-generators f and g have the same r-th power if and only if fg^{-1} is in a certain subgroup of $Aut(\mathbb{I})$ which is analogous to $Z(r)$ for the strict case. This subgroup is explicitly described in [98]. Again, if all the r-th powers of a nilpotent t-norm are known, the nilpotent t-norm is uniquely determined. Again, this is a direct consequence of the proof of Theorem 5.2.3.

6.4   Sensitivity of connectives

The modeling of fuzzy concepts through the assignment of membership functions as well as the choice of a fuzzy logic are subjective. This flexibility reflects the very nature of fuzziness. In specific applications, some choices must be made, depending of course on various factors. We illustrate some possible choices when robustness is an important factor. Here, by robustness we mean sensitivity of membership functions or of fuzzy logical connectives with respect to variations in their arguments. A logical connective is a mapping from either $[0, 1]^2$ or $[0, 1]$ into $[0, 1]$, as in the case of t-norms or negations, respectively. For any mapping $f : [0, 1]^n \to [0, 1]$, an extreme measure of sensitivity of f is as follows.

Definition 6.4.1 *For* $x = (x_1, x_2, ..., x_n)$, $y = (y_1, y_2, ..., y_n)$, *and* $\delta \in [0, 1]$ *let*

$$\rho_f(\delta) = \bigvee_{|x_i - y_i| \leq \delta} |f(x) - f(y)|$$

The function $\rho_f : [0, 1] \to [0, 1]$ *is an **extreme measure of sensitivity** of* f.

In order to compare different operations f, we proceed as in standard decision theory where $\rho_f(\delta)$ plays the role of the "risk" at δ of the procedure f. We say that f is **less sensitive** than g if for all δ, $\rho_f(\delta) \leq \rho_g(\delta)$, with strict inequality at some δ. Here are some examples.

Example 6.4.2 If $f(x, y) = x \wedge y$, then for $|x - u| \leq \delta$ and $|y - v| \leq \delta$, we have $x \leq u + \delta$ and $y \leq v + \delta$. So

$$x \wedge y \leq (u + \delta) \wedge (v + \delta) = (u \wedge v) + \delta$$

Similarly,

$$u \wedge v \leq x \wedge y) + \delta$$

so that $x \wedge y - u \wedge v \leq \delta$. Thus $\rho_\wedge(\delta) \leq \delta$. Taking $x = y = \delta$ and $u = v = 0$, we have $x \wedge y - u \wedge v = \delta$, so $\rho_\wedge(\delta) \geq \delta$. Thus $\rho_\wedge(\delta) = \delta$.

Example 6.4.3 If $f(x) = 1 - x$, then $|f(x) - f(y)| = |x - y| \leq \delta$, and $\rho_f(\delta)$ is attained at $x = \delta$, $y = 0$. Thus $\rho_f(\delta) = \delta$.

Example 6.4.4 If $f(x, y) = xy$, then $\rho_f(\delta) = 2\delta - \delta^2$.

Example 6.4.5 If $f(x, y) = x + y - xy$, then $\rho_f(\delta) = 2\delta - \delta^2$.

Example 6.4.6 If $f(x, y) = (x + y) \wedge 1$, then $\rho_f(\delta) = 2\delta \wedge 1$.

Example 6.4.7 If $f(x, y) = x \vee y$, then $\rho_f(\delta) = \delta$.

Proposition 6.4.8 *For any t-norm $\triangle$, it and its dual $\triangledown$ with respect to $\alpha(x) = 1 - x$ have the same sensitivity.*

The proof is left as an exercise.

Theorem 6.4.9 $x \wedge y$, $x \vee y$, *and* $\alpha(x) = 1 - x$ *are the least sensitive among all continuous t-norms, t-conorms, and negations, respectively.*

Proof. We showed in the first example above that $\rho_\wedge(\delta) = \delta$. If $\triangle$ is a t-norm, then

$$|1 \triangle 1 - (1 - \delta) \triangle (1 - \delta)| = |1 - (1 - \delta) \triangle (1 - \delta)|$$
$$\leq \rho_\triangle(\delta)$$

so $(1 - \delta) \geq (1 - \delta) \triangle (1 - \delta) > 1 - \rho_\triangle(\delta)$. Thus $\rho_\triangle(\delta) \geq \delta = \rho_\wedge(\delta)$.

Note $\wedge$ is the only t-norm $\triangle$ such that $\rho_\triangle(\delta) = \delta$. Indeed, for $\wedge \neq \triangle$, there are x, y such that $x \triangle y \neq x \wedge y$ and we may assume that $x \triangle y < x$. Now

$$|x \triangle 1 - 1 \triangle 1| = 1 - x \triangle y > 1 - x$$

so that $\rho_\triangle(1 - x) \neq 1 - x$. We leave the rest of the proof as exercises. ∎

An alternative to the measure above of extreme sensitivity of fuzzy logical connectives is a measure of **average** sensitivity. Let $f : [a, b] \to \mathbb{R}$. Assume that f has all derivatives and integrals necessary. A measure of the sensitivity of differentiable functions f at a point in $[a, b]$ is the square $f'(x)^2$ of its derivative at that point. Its average sensitivity would then be the "average" over all points in $[a, b]$ of $f'(x)^2$. This "average" is the quantity $\left(\int_a^b f'(x)^2 dx \right) / (b-a)$. More generally, if $f : [a,b]^n \to \mathbb{R}$, account needs to be taken of all the partial derivatives of f, and the quantity

$$\sum_{i=1}^n \left(\frac{\partial f}{\partial x_i} \right)^2$$

is an appropriate one. Its average sensitivity is then

$$S(f) = \frac{\int_{[a,b]^n} \sum_{i=1}^n \left(\frac{\partial f}{\partial x_i} \right)^2}{(b-a)^n}$$

In the case $[a,b] = [0,1]$, the case of interest to us, the **average sensitivity** of f becomes

$$S(f) = \int_{[0,1]^n} \sum_{i=1}^n \left(\frac{\partial f}{\partial x_i} \right)^2$$

Here are some examples for logical connectives on $[0,1]$.

Example 6.4.10 $S(\wedge) = S(\vee) = 1$.

Example 6.4.11 $S(\triangle) = \frac{2}{3}$ for $x \triangle y = xy$ and for $x \triangle y = x + y - xy$.

Example 6.4.12 $S(\triangle) = 1$ for $x \triangle y = (x+y) \wedge 1$ and for $x \triangle y = 0 \vee (x + y - 1)$.

Example 6.4.13 $S(\alpha) = 1$ for the negation $\alpha(x) = 1 - x$.

A t-norm and its dual with respect to $\alpha(x) = 1 - x$ have the same average sensitivity. The functions $\wedge$ and $\vee$ in the examples above are differentiable at all points in the unit square except for the line $x = y$, so there is no problem calculating the integrals involved. In certain situations, one may need to use more general notions of derivative.

Theorem 6.4.14 *The connectives* $x \triangle y = xy$, $x \triangledown y = x + y - xy$, *and* $\alpha(x) = 1 - x$ *have the smallest average sensitivity among t-norms, t-conorms, and negations, respectively.*

Proof. We need to show, for example, that $x \triangle y = xy$ minimizes

$$\int_0^1 \int_0^1 \left[\left(\frac{\partial \triangle}{\partial x} \right)^2 + \left(\frac{\partial \triangle}{\partial y} \right)^2 \right] dx dy$$

A standard fact from analysis is that $\triangle$ minimizes this expression if it satisfies the Laplace equation

$$\frac{\partial^2 \triangle}{\partial x^2} + \frac{\partial^2 \triangle}{\partial y^2} = 0$$

and of course it does. Similar arguments apply to the other two cases. ∎

As in the case of extreme measure of sensitivity, one can use the notion of average sensitivity to choose membership functions for fuzzy concepts. When facing a fuzzy concept such as a linguistic label, one might have a class of possible membership functions suitable for modeling the concept. A good choice could be the membership function within this class that minimizes average sensitivity.

6.5 Copulas and t-norms

The concept of t-norm is related to the familiar concept of joint cumulative distribution functions (CDF's) of random variables. Let $X = (X_1, X_2, \ldots, X_n)$ be an n-dimensional random vector defined on some probability space $(\Omega, \mathcal{A}, P)$. Suppose X takes values in the unit n-cube $[0, 1]^n$ and that the CDF F_j of each component X_j is $F_j(x) = x$. That is, each X_j is uniformly distributed over $[0, 1]$. Then the joint CDF F of X, namely

$$F(x_1, x_2, \ldots, x_n) = P(X_1 \leq x_1, X_2 \leq x_2, \ldots, X_n \leq x_n)$$

for $x = (x_1, x_2, \ldots, x_n) \in \mathbb{R}^n$ satisfies

1. $F_j(x_j) = F(1, \ldots, 1, x_j, 1, \ldots, 1) = x_j$ for $x_j \in [0, 1]$. (x_j is in the j-th place).

2. For any Borel subset B of $[0, 1]^n$, $P(X \in B) \geq 0$.

3. $F(x_1, x_2, \ldots, x_n) = 0$ if there is at least one j such that $x_j = 0$.

Axiomatically, a function $C : [0, 1]^n \to [0, 1]$ is called an **n-copula** if it satisfies

1. Each **margin** $C_j(x_j) = C(1, \ldots, 1, x_j, 1, \ldots, 1) = x_j$ for $x_j \in [0, 1]$,

2. C is **n-increasing**: for any Cartesian product $J = J_1 \times J_2 \times \ldots \times J_n$ of closed intervals $J_j = [d_j, e_j] \subseteq [0, 1]$,

$$Vol_C(J) = \sum_v sgn(v)C(v) \geq 0$$

the summation being over all vertices v of J. So if $v = (v_1, v_2, ..., v_n)$, then $v_j = d_j$ or e_j, and

$$sgn(v) = \left\{ \begin{array}{ll} 1 & \text{if} \quad v_j = d_j \text{ for an even number of } j's \\ -1 & \text{if} \quad v_j = d_j \text{ for an odd number of } j's \end{array} \right.$$

3. C is **grounded,** that is, $C(x_1, x_2, ..., x_n) = 0$ for all $(x_1, x_2, ..., x_n) \in [0, 1]^n$ such that $x_j = 0$ for at least one j.

Copulas are used in probability and statistics as follows. If F is a CDF on $\mathbb{R}^n$ with marginal CDF F_j's, then there exists an n-copula C, in general nonunique, such that for $(x_1, x_2, ..., x_n) \in \mathbb{R}^n$,

$$F(x_1, x_2, ..., x_n) = C(F_1(x_1), F_2(x_2), ..., F_n(x_n))$$

Of course, if the F_j are n one-dimensional CDF's, then the function F defined as above is a bona fide CDF on $\mathbb{R}^n$.

There is a good discussion of copulas and their relation to t-norms in [92]. We limit ourselves to some general comments. First, if $\triangle$ is a t-norm, then $T(x_1, x_2, ..., x_n) = x_1 \triangle x_2 \triangle ... \triangle x_n$ satisfies conditions 1 and 3 above in the definition of n-copula. So the relation between copulas and t-norms hinges partly on the second condition. But commutativity and associativity are issues. There are several pertinent theorems in [92], but they are rather technical. We state some facts that may be found there.

- The t-norms xy, $x \blacktriangle y$, and $x \wedge y$ are 2-copulas.

- $x \blacktriangle y \leq C(x, y) \leq x \wedge y$ for any 2-copula C.

- A 2-copula is a t-norm if and only if it is associative.

- A t-norm is a 2-copula if and only if it is 2-increasing.

- A t-norm $\triangle$ is a 2-copula if and only if for $a, b, c \in [0, 1]$, with $a \leq c$, it satisfies

$$c \triangle b - a \triangle b \leq c - a$$

6.6 Exercises

1. In the set of all subsets of a set show that

$$x' \cup y = \bigvee \{z : x \cap z \subseteq y\} = x' \cup (x \cap y)$$

2. Let $\triangle$ be a t-norm, ∇ a t-conorm, and η a negation. Verify that $(x \Rightarrow y) = (x \nabla \triangle y) \triangle \eta(x)$ is a fuzzy implication.

3. Is $(x \Rightarrow y) = x \wedge y$ an implication?

4. Calculate the Q-implications $(x \Rightarrow y)$ when

   (a) $x \triangle y = xy$, $x \nabla y = x + y - xy$, $\eta(x) = 1 - x$;
   (b) $x \triangle y = (x + y - 1) \vee 0$, $x \nabla y = (x + y \wedge 1)$, $\eta(x) = 1 - x$.

5. Let $\Rightarrow$ be a fuzzy implication. Show that

   $$(x \Rightarrow^* y) = ((1 - x) \Rightarrow (1 - y))$$

   is a fuzzy implication. Find the fuzzy implication in terms of $\triangle$, ∇, and η corresponding to the fuzzy relation $(A' \times V) \cup (U \times B)$.

6. Verify that

   $$(x \Rightarrow y) = \begin{cases} 1 & \text{if} \quad x < 1 \text{ or } y = 1 \\ 0 & \text{if} \qquad \text{otherwise} \end{cases}$$

   is a fuzzy implication.

7. Using $(x \Rightarrow y) = \eta(\bigwedge\{z : y \nabla z \geq x\})$, find $\Rightarrow$ when $\eta(x) = 1 - x$ and $x \nabla y = x + y - xy$.

8. Let $\triangle$ be a nilpotent t-norm. Show that the R-implication associated with $\triangle$ is the same as $(x \Rightarrow y) = \eta_\triangle[x \triangle \eta_\triangle(y)]$.

9. Let ∇ be a continuous t-norm and η a negation. Verify that the associated S-implication $(x \Rightarrow y) = \eta(x) \nabla y$ satisfies the following:

   (a) $(0 \Rightarrow x) = 1$ and $(1 \Rightarrow x) = x$ for all $x \in [0, 1]$.
   (b) $(x \Rightarrow (y \Rightarrow z)) = (y \Rightarrow (x \Rightarrow z))$.

10. For $\lambda > 0$, and $x \in [0, 1]$, let

    $$g(x) = \left[\frac{1 - x}{\lambda + (1 - \lambda)(1 - x)} \right]^{\frac{1}{\lambda}}$$

    (a) Verify that g is a generator of a t-conorm ∇.
    (b) Find the inverse of g.
    (c) Find the ∇-implication associated with ∇ and the negation $\eta(x) = (1 - x)/(1 + x)$.

11. Let $\triangledown$ be a t-conorm and η a negation.

   (a) Show that

   $$(x \Rightarrow y) = \eta\left(\bigwedge\{z \in [0,1] : z \triangledown y \geq x\}\right)$$

   is an implication.

   (b) Compute $\Rightarrow$ given in part (a) when

   i. $x \triangledown y = \begin{cases} x & \text{if } y = 0 \\ y & \text{if } x = 0 \\ 1 & \text{otherwise} \end{cases}$, $\eta(x) = \begin{cases} 1 & \text{if } x < 1 \\ 0 & \text{if } x = 1 \end{cases}$

   ii. $(x \triangledown y) = 1 \wedge (x + y)$, $\eta(x) = 1 - x$

   (c) Let f be a generator for a nilpotent t-norm. Show that

   $$(x \Rightarrow y) = f^{-1}\left(\frac{f(0)f(y)}{f(x)} \vee f(0)\right)$$

   is a fuzzy implication.

12. Show that a nilpotent t-conorm $\triangledown$ has a linear generator if and only if $\eta_\triangledown$ is a Sugeno negation $(1 - x)/1 - \lambda x)$ with $\lambda \in (-1, 0)$.

13. Show that $x \dotplus y = \left(\dfrac{x^a + y^a}{2}\right)^{\frac{1}{a}}$ is an averaging operator.

14. Show that $f(x) = (a^x - 1)/(a - 1)$ is a generator for the averaging operator $x \dotplus y = \log_a((a^x + a^y)/2)$, $a > 0$, $a \neq 1$.

15. Show that for $c > 0$ and $a \neq 0$,

   $$e^{-\frac{1}{c}\left(-\ln\left(\frac{e^{-c(-\ln x)^a} + e^{-c(-\ln y)^a}}{2}\right)\right)^{\frac{1}{a}}}$$

   is an averaging operator and find its generator.

16. Find the averaging operator whose generator is

   $$\frac{1}{1 + a\left(\frac{1-x}{x}\right)^p}$$

17. Show that if $\dotplus$ is an averaging operator and f and g are antiautomorphisms of $\mathbb{I}$, then $f \dotplus g$ is an antiautomorphism.

18. Verify that the negations of the averaging operators in Example 6.2.9 are as stated.

19. Let m, n, p, q be positive integers. Show that for a strict t-norm $\triangle$, $x \in [0, 1]$ there is a unique $y \in [0, 1]$ such that $y^{[n]} = x$. Denote this y by $x^{[\frac{1}{n}]}$. Show that $(x^{[\frac{1}{n}]})^{[m]} = (x^{[\frac{1}{q}]})^{[p]}$, and thus that $x^{[\frac{m}{n}]}$ is independent of the representation of the rational number m/n.

20. Prove Theorem 6.3.2.

21. Prove Proposition 6.3.3.

22. Prove Proposition 6.3.4.

23. Prove Proposition 6.3.6.

24. Show that if $\triangledown$ is dual to $\triangle$, then $\rho_\triangledown = \rho_\triangle$.

25. Verify that $x \triangle y = xy$ and $x \triangledown y = x + y - xy$ are equally sensitive.

26. Compute $\rho_f(\delta)$ for

    (a) $f(x, y) = xy$
    (b) $f(x, y) = (x + y) \wedge 1$
    (c) $f(x, y) = x + y - xy$
    (d) $f(x, y) = x \vee y$

27. Compute $S(f)$ in Examples 6.4.10 - 6.4.13 of Section 6.4.

28. Let $g(x, y) = 1 - f(1 - x, 1 - y)$. Show that $S(f) = S(g)$.

29. Let $w = (w_1, w_2, ..., w_n)$ with each $w_i \in [0, 1]$ and $\sum w_i = 1$. For $(x_1, x_2, ..., x_n) \in \mathbb{R}^n$, let $x_{(j)}$ be the j-th largest x_i, so that $x_{(i)} \geq x_{(2)} \geq ... \geq x_{(n)}$. Let

    $$F_w : \mathbb{R}^n \to \mathbb{R} : (x_1, x_2, ..., x_n) \to \sum w_j x_{(j)}$$

    F_w is an **ordered weighted averaging operator** (OWA) of dimension n.

    (a) Show that

    $$\min(x_1, x_2, ..., x_n) \leq F_w(x_1, x_2, ..., x_n) \leq \max(x_1, x_2, ..., x_n)$$

    (b) Determine F_w in the following cases:

    i. $w = (1, 0, ..., 0, 0)$

    ii. $w = (0, 0, ..., 0, 1)$

    iii. $w = (1/n, 0, ..., 0, 1/n)$

30. Show that a binary operation $\triangle$ on $[0, 1]$ is 2-increasing if and only if for all $0 \le x < y \le 1$ and $0 \le u < v \le 1$, the inequality

$$y \triangle u - x \triangle u \le y \triangle v - x \triangle v$$

    holds.

31. Show in detail that $C : [0, 1] \times [0, 1] \to [0, 1]$, defined by $C(x, y) = xy$ is a 2-copula.

32. Let $f : \mathbb{R} \to \mathbb{R}$. Show that f is 1-increasing if and only if f is nondecreasing.

33. Let $C : [0, 1]^2 \to [0, 1] : (x, y) \to x + y - xy$. Show that C is nondecreasing in each argument. Is C 2-increasing?

34. Let $f : [-1, 1]^2 \to [-1, 1] : (x, y) \to xy$. Show that f is 2-increasing. Is f nondecreasing in each argument?

35. Let $C : [0, 1]^2 \to [0, 1]$ be 2-increasing and grounded. Show that C is nondecreasing in each argument.

36. Let $U = \{u_1, u_2, ..., u_n\}$ be a finite set. A random set on U is a map S defined on a probability space $(\Omega, \mathcal{A}, P\}$ and taking values in the power set $\mathcal{P}(U)$. Its distribution is specified by its density function $f : \mathcal{P}(U) \to [0, 1] : B \to P\{\omega : S(\omega) = B\}$.

    (a) Consider the $\{0, 1\}$ valued random variables $X_j(\omega) = 1_{S(\omega)}(u_j)$, $j = 1, 2, ..., n$. Show that the distribution of S determines the joint distribution of $X = (X_1, X_2, ..., X_n)$, and vice versa.

    (b) Let A be a fuzzy subset of U. Let $S(A)$ denote the class of all random sets S on U such that $A(u_j) = P\{\omega : u_j \in S(\omega)\}$. Describe all possible distributions of S. (Hint: Use (a) and Sklar's Theorem.)

37. For $x = (x_1, x_2, ..., x_n)$ and $y = (y_1, y_2, ..., y_n)$, let $x \wedge y = (x_1 \wedge y_1, x_2 \wedge y_2, ..., x_n \wedge y_n)$, where $\wedge$ denoted min. Write $y \searrow x$ if each $y_j \searrow x_j$. A function $F : \mathbb{R}^k \to [0, 1]$ is a **cumulative distribution function** (CDF) if it satisfies

  (i) $\lim_{x_j \to -\infty} F(x_1, x_2, ..., x_n) = 0$ for each j, and

$\lim_{x_1 \to \infty, ..., x_n \to \infty} F(x_1, x_2, ..., x_n) = 1$.

(ii) F is right continuous, that is, if $y \searrow x$ then $F(y) \searrow F(x)$.

(iii) F is monotone decreasing, that is, if each $x_j \leq y_j$, then $F(x) \leq F(y)$.

(iv) For any k-dimensional rectangle $A = (x_1, y_1] \times (x_2, y_2] \times ... \times (x_k, y_k]$, and its set of vertices $V = \{x_1, y_1\} \times \{x_2, y_2\} \times ... \times \{x_k, y_k\}$, and for $sgn(v) = (-1)^{\tau(v)}$ where $\tau(v)$ is the number of x_j in $v \in V$.

$$\Delta_F(A) = \sum_{v \in V} sgn(v) F(v) \geq 0$$

(a) Show that (iii) is equivalent to (iv) when $k = 1$.

(b) Let $F : \mathbb{R}^2 \to [0, 1]$ be given by

$$F(x, y) = \begin{cases} 1 & \text{if} \quad s \geq 0, y \geq 0, x + y \geq 1 \\ 0 & \text{if} \quad \quad \quad \text{otherwise} \end{cases}$$

Show that F satisfies (i), (ii), and (iii) above. Does F satisfy (iv)?

(c) Let $F : \mathbb{R}^k \to [0, 1]$ such that for any $a, a_1, a_2, ..., a_n \in \mathbb{R}^k$,

$$F(a) \geq \sum_{\emptyset \neq I \subseteq \{1,2,....,n\}} (-1)^{|I|+1} F(\wedge_{b \in \{a, a_i : i \in I\}} b)$$

Show that such an F satisfies (iii) and (iv).

(d) Let $F : \mathbb{R}^2 \to [0, 1]$ be given by

$$F(x, y) = \begin{cases} 1 & \text{if} \quad s \geq 0, y \geq 0, \max(x, y \geq 1 \\ 0 & \text{if} \quad \quad \quad \text{otherwise} \end{cases}$$

Show that F is monotone nondecreasing. Does F satisfy the inequality in part (c)? (Hint: Consider $a = (x, y)$, $a_1 = (x_1, y_1)$, $a_2 = (x_2, y_2)$ with $0 \leq x_1 < 1 < x < x_2$ and $0 \leq y_2 < 1 < y < y_1$.)

38. Show that the following t-norms are 2-copulas.

   (a) $x \triangle y = xy$
   (b) $x \triangle y = (x + y - 1) \vee 0$
   (c) $x \triangle y = xy / (x + y - xy)$

39. Show that $\blacktriangle \leq \triangle$ for any t-norm $\triangle$ that is a 2-copula.

40. Show that Frank t-norms are 2-copulas.

Chapter 7

FUZZY RELATIONS

Relations, or associations among objects, are of fundamental importance in the analysis of real-world systems. Mathematically, the concept is a very general one and has been discussed earlier in Chapter 2. There are many kinds of relations: order relations, equivalence relations, and other relations with various important properties. Relations are ubiquitous in mathematics, and their generalizations to fuzzy theory are important. This chapter presents some of these generalizations with an emphasis on binary relations, especially fuzzy equivalence relations, which generalize ordinary equivalence relations.

7.1 Definitions and examples

Generalizing relations to fuzzy relations is easy. An **n-ary relation** is a subset R of the Cartesian product $U_1 \times U_2 \times ... \times U_n$ of n sets. The generalization to the fuzzy case is the natural one.

Definition 7.1.1 *An **n-ary fuzzy relation in a set** $V = U_1 \times U_2 \times ... \times U_n$ is a fuzzy subset R of V. If the sets U_i are identical, say U, then R is an **n-ary fuzzy relation on** U. A **2-ary** fuzzy relation is a **binary fuzzy relation**.*

So $R(u_1, u_2, ...u_n)$ represents the degree to which $(u_1, u_2, ...u_n)$ is compatible with the relation R. We shall deal mostly with binary fuzzy relations on a set U. Indeed, the motivation for fuzzy relations is to be able to quantify statements such as "John is much younger than Paul." For example, if R is a modeling of "much younger than", then $R(10, 15)$ is the degree to which 10 years old is considered as much younger than 15 years old. If R is a modeling of "is much greater than" on $U = (0, \infty)$, then R could be something like $R(x,y) = \max\{0, (x-y)/x\}$. If R is to represent "is close to" on the real numbers $\mathbb{R}$, then a reasonable function to use might be $R(x,y) = e^{-|x-y|}$.

The set $\mathcal{F}(U)$ of all fuzzy subsets of a set U gives rise to binary fuzzy relations R on $U \times \mathcal{F}(U)$ by $R(u, A) = A(u)$. That is, $R(u, A)$ is the degree to which the point u of U is compatible with the fuzzy set A of U.

Probability values are in $[0,1]$, so conditional probability evaluations can be also viewed as a binary fuzzy relation. If $(\Omega, \mathcal{A}, P)$ is a probability space, then P induces a binary fuzzy relation on $\mathcal{A} \times \mathcal{B}$, where $\mathcal{B} = \{b \in \mathcal{A} : P(b) \neq 0\}$, by $P(a,b) = P(a|b)$, the usual conditional probability.

7.2 Binary fuzzy relations

The generalization of relations to fuzzy relations is just that of going from subsets to fuzzy subsets. In fuzzifying various properties of relations, the standard procedure is to identify a relation with its indicator function, associate a property of a relation with properties of the indicator function of that relation, and translate those properties into properties of membership functions.

We consider now the special case of binary fuzzy relations on a set U. An ordinary binary relation on U is a subset of $U \times U$. Following are some important properties that these relations may have, and to generalize these concepts to the fuzzy case we need to state them in terms of their indicator functions. So let R be a relation on a set U. *We identify R with its indicator function.* Thus $R : U \times U \to \{0,1\}$ and the relation R on U is

1. **reflexive** if $R(x, x) = 1$;

2. **symmetric** if $R(x, y) = 1$ implies $R(y, x) = 1$;

3. **transitive** if $R(x, y) = R(y, z) = 1$ implies $R(x, z) = 1$;

4. **antisymmetric** if $R(x, y) = R(y, x) = 1$ implies $x = y$.

Many other properties of a relation are defined in terms of these. For example, R is an equivalence relation if it is reflexive, symmetric, and transitive.

There are many ways to translate these properties of a function from $U \times U \to \{0,1\}$ to a function $U \times U \to [0,1]$. But however translated, they must correspond to the properties for relations when restricted to that case. Let R be a fuzzy relation on a set U. Then R is

1. **reflexive** if $R(u, u) = 1$;

2. **symmetric** if $R(u, v) = R(v, u)$;

3. **transitive** if $R(u, w) \geq R(u, v) \wedge R(v, w)$;

4. **antisymmetric** if $R(u, v) > 0$ and $R(v, u) > 0$ imply $u = v$.

These properties do become the corresponding ones for relations when R takes values only in $\{0, 1\}$, as is easily seen. The value $R(u, v)$ represents the extent to which the elements u and v are similar, and certainly reflexivity is demanded. To be symmetric just says that v is exactly as similar to u as u is to v. Transitivity says that for all v, u is at least as similar to w as the minimum of the similarity of u and v and the similarity of v and w. This condition could be written as

$$\bigvee \{R(u, v) \wedge R(v, w) : v \in U\} \leq R(u, w)$$

We need a generalization of transitivity. The notion of transitivity above is **min-transitive** or **$\wedge$-transitive**. That is, $R(u, w) \geq R(u, v) \wedge R(v, w)$. We can replace $\wedge$ by any t-norm. So for a t-norm $\triangle$, the fuzzy relation R is **$\triangle$-transitive** if $R(u, w) \geq R(u, v) \triangle R(v, w)$ for all $v \in U$.

Definition 7.2.1 *A fuzzy relation R on a set U is a $\triangle$-fuzzy equivalence relation for the t-norm $\triangle$ if R is reflexive, symmetric, and $\triangle$-transitive. If $\triangle = \wedge$, then we say that R is a fuzzy equivalence relation.*

$\triangle$-fuzzy equivalence relations are fuzzy binary relations generalizing equivalence relations on sets. Intuitively, an equivalence relation on a set specifies when two elements are to be considered equal. For example, the relation "is the same age as" is an equivalence relation on any set of people. But "is about the same age as" is not as sharply defined and may demand modeling with a fuzzy equivalence relation.

Any equivalence relation is a $\triangle$-fuzzy equivalence relation for any t-norm $\triangle$. The definition is made that way. Exercises 13 and 15 give examples of $\triangle$-equivalence relations.

Fuzzy relations R are fuzzy sets, and so have α-cuts R_α. If R is a fuzzy relation on U, then

$$R_\alpha = R^{-1}(\uparrow\alpha) = R^{-1}([\alpha, 1]) = \{(u, v) : R(u, v) \geq \alpha\}$$

is a subset of $U \times U$, and so is a relation on U.

Theorem 7.2.2 *If R is a fuzzy relation on a set U, then R is a fuzzy equivalence relation if and only if each α-cut, R_α is an equivalence relation on U.*

Proof. Let R be a fuzzy equivalence relation. Then $(u, u) \in R_\alpha$ since $R(u, u) = 1 \geq \alpha$. If $(u, v) \in R_\alpha$, then $R(u, v) \geq \alpha$ so $R(v, u) \geq \alpha$ from symmetry, whence $(v, u) \in R_\alpha$. If (u, v) and $(v, w) \in R_\alpha$, then $R(u, w) \geq R(u, v) \wedge R(v, w) \geq \alpha$ so that $(u, w) \in R_\alpha$. Thus R_α is an equivalence relation. The other half of the proof is left as an exercise. ∎

This theorem does not hold for $\triangle$-fuzzy equivalence relations in general. (See Exercise 19, for example.)

So with each fuzzy equivalence relation R on a set U, there is associated a family of equivalence relations of U, namely the α-cuts R_α, one for each $\alpha \in [0, 1]$. Each of these equivalence relations induces a partition P_α of U, so we also have the associated family $\{P_\alpha : \alpha \in [0, 1]\}$ of partitions of U. Since the α-cuts of a fuzzy set determine that fuzzy set, then in particular this family of equivalence relations, or equivalently, this family of partitions, determines the fuzzy equivalence relation R. The set of equivalence relations $\mathcal{E}(U)$ on U is a complete lattice, and each R_α is an element of $\mathcal{E}(U)$. This is discussed in Section 2.2. In this complete lattice, we have for any subset $I \subseteq [0, 1]$,

$$
\begin{aligned}
\bigwedge_{\alpha \in I} R_\alpha &= \bigcap_{\alpha \in I} R_\alpha \\
&= \{(u, v) : R(u, v) \geq \alpha \text{ for all } \alpha \in I\} \\
&= \{(u, v) : R(u, v) \geq \bigvee_{\alpha \in I} \alpha \} \\
&= R_{\bigvee_{\alpha \in I} \alpha}
\end{aligned}
$$

Since the R_α form a chain, it is easy to see that $\bigvee_{\alpha \in I} R_\alpha = \bigcup_{\alpha \in I} R_\alpha$. Thus

$$
\begin{aligned}
\bigvee_{\alpha \in I} R_\alpha &= \bigcup_{\alpha \in I} R_\alpha \\
&= \{(u, v) : R(u, v) \geq \alpha \text{ for some } \alpha \in I\} \\
&\geq R_{\bigwedge_{\alpha \in I} \alpha}
\end{aligned}
$$

This last inequality is an equality if I is finite. (See Exercise 11.) Thus we have the following theorem.

Theorem 7.2.3 *Let R be a fuzzy equivalence relation on U. Then the map*

$$
[0, 1] \to \mathcal{E}(U) : \alpha \to R_\alpha
$$

is a lattice antihomomorphism.

Proof. The theorem just means that for any subset $I \subseteq [0, 1]$,

$$R_{\bigwedge_{\alpha \in I} \alpha} = \bigvee_{\alpha \in I} R_\alpha$$

$$R_{\bigvee_{\alpha \in I} \alpha} = \bigwedge_{\alpha \in I} R_\alpha$$

∎

So each fuzzy equivalence relation gives a family of partitions satisfying these last two equalities. The order relation of the set of partitions of a set is the one corresponding to the order relation on the associated equivalence relation, which is inclusion. This means that one partition is greater than another if its equivalence classes are unions of some equivalence classes of the other. The condition in the following definition is the characterizing condition for a family of partitions to be one corresponding to a fuzzy equivalence relation.

Definition 7.2.4 *A **partition tree** on a set U is a family $\{P_\alpha : \alpha \in [0, 1]\}$ of partitions of U such that $P_0 = U$, and for any subset I of $[0, 1]$, $P_{\bigvee_{\alpha \in I} \alpha} = \bigwedge_{\alpha \in I} P_\alpha$.*

One should note that when $I = \varnothing$, then $\bigvee_{\alpha \in I} \alpha = 0$ and $\bigwedge_{\alpha \in I} P_\alpha = U$.

Theorem 7.2.5 *Let S be the set of all fuzzy equivalence relations and $\mathcal{P}$ be the set of all partition trees on U. For $R \in S$, let P_α be the partition associated with the equivalence relation R_α. Then*

$$S \to \mathcal{P} : R \to \{P_\alpha : \alpha \in [0, 1]\}$$

is a bijection.

Proof. First, we need to show that this is actually a map, that is, that the family $\{P_\alpha : \alpha \in [0, 1]\}$ is a partition tree. But that was noted in the previous theorem. The map $S \to \mathcal{P}$ is one-to-one since the α-cuts of a fuzzy set determine that fuzzy set. We need only to show that every partition tree is the image of some fuzzy equivalence relation. Let $\{P_\alpha : \alpha \in [0, 1]\}$ be a partition tree. Define $R : U \times U \to [0, 1]$ by

$$R(u, v) = \bigvee \{\alpha : u \text{ and } v \text{ belong to the same member of } P_\alpha\}$$

The rest of the details are left as an exercise. ∎

Now we turn briefly to fuzzy partial orders. A relation on a set U is a partial order if it is reflexive, transitive, and antisymmetric. A fuzzy relation R on a set U is defined to be a **fuzzy partial order** if it is reflexive, transitive, and antisymmetric. A basic fact, similar to the case of fuzzy equivalence relations, is the following:

Theorem 7.2.6 *Let R be a fuzzy relation on a set U, and let $\{R_\alpha : \alpha \in [0,1]\}$ be its set of α-cuts. Then R is a fuzzy partial order on U if and only if each R_α is a partial order on U.*

Proof. Suppose R is a fuzzy partial order on U. We will prove transitivity. Suppose that (u,v) and (v,w) are in R_α. Then $R(u,v) \geq \alpha$ and $R(v,w) \geq \alpha$, whence $R(u,v) \wedge R(v,w) \geq \alpha$ and so $(u,w) \in R_\alpha$. The rest of the proof is left as an exercise. ∎

So with each fuzzy partial order on a set U, there is associated a family of partial orderings, namely the α-cuts of this relation. Those α-cuts determine the fuzzy partial order, of course, just as α-cuts determine any fuzzy set.

7.3   Operations on fuzzy relations

Let $f : U \to V$ and $g : V \to W$. If these functions are thought of as relations in $U \times V$ and $V \times W$, respectively, then the composition

$$
\begin{aligned}
g \circ f &= \{(u,w) \in U \times W : g(f(u)) = w\} \\
&= \{(u,w) \in U \times W : (u,v) \in f \text{ and } (v,w) \in g \text{ for some } v \in V\}
\end{aligned}
$$

This latter way of expressing this composition makes sense for arbitrary relations R and S in $U \times V$ and $V \times W$, respectively. This is how **composition of relations** is defined: $(u,w) \in R \circ S$ if and only if there is a $v \in V$ with $(u,v) \in R$ and $(v,w) \in S$. (Note that we reverse the order from that ordinarily used for functions. Were R and S functions, we would write $S \circ R$ for $R \circ S$.) With a view to generalizing to composition of fuzzy relations, identifying relations with their indicator functions, we may define $R \circ S$ by the equation

$$
(R \circ S)(u,w) = \bigvee_{v \in V} (R(u,v) \wedge S(v,w))
$$

Since $[0,1]$ is a complete lattice, this equation makes sense for fuzzy relations. Also in the fuzzy case, $\wedge$ may be replaced by any binary operation on $[0,1]$, and in particular by any t-norm. Here is the definition.

Definition 7.3.1 *Let R and S be fuzzy relations in $U \times V$ and $V \times W$, respectively, and let $\triangle$ be a t-norm. The **composition** $R \circ S$ of R and S with respect to $\triangle$ is the fuzzy relation on $U \times W$ with membership function*

$$
(R \circ S)(u,w) = \bigvee_v \{R(u,v) \triangle S(v,w)\}
$$

When $x \triangle y = x \wedge y$, $R \circ S$ is referred to as a **max-min composition**. When $x \triangle y = xy$, $R \circ S$ is a **max-product composition**. In the special case where the sets $U, V,$ and W are all finite, R and S can be represented in the form of matrices with entries in $[0, 1]$, and $R \circ S$ can be obtained as the matrix product of R and S provided that in this operation, addition is replaced by max and multiplication by $\triangle$. (See Exercise 3, for example.)

The composition of these relations is associative and distributes over sup and inf of fuzzy relations. For example, for any t-norm and fuzzy relations R, S, and T in $U \times V$, $V \times W$, and $W \times X$, respectively, we have

$$R \circ (S \circ T) = (R \circ S) \circ T \tag{7.1}$$

If R_i is a family of fuzzy relations on $U \times V$, then

$$\bigvee_i (R_i \circ S) = \left(\bigvee_i R_i \right) \circ S \tag{7.2}$$

$$\bigwedge_i (R_i \circ S) = \left(\bigwedge_i R_i \right) \circ S$$

and similarly in the other coordinate. These verifications are left as exercises.

The **direct image** of a fuzzy subset D of U under R is the fuzzy subset of V defined by

$$R(D)(v) = \bigvee_u \{R(u, v) \triangle D(u)\}$$

and the **inverse image** of a fuzzy subset E of V under R is the fuzzy subset of U defined by

$$R^{-1}(E)(u) = \bigvee_v \{R(u, v) \triangle E(v)\}$$

These concepts are extensions of corresponding concepts in the case of ordinary relations. When R is an ordinary subset of $U \times V$, and D is an ordinary subset of U, then $v \in R(D)$ if and only if there is a $u \in D$ such that $(u, v) \in R$. Thus $R(D)$ is the **projection** onto V of $R \cap (D \times V)$. Similarly, for $R^{-1}(E)$ consider the inverse relation R^{-1} on $V \times U$.

In general, let R be a fuzzy relation in $U_1 \times U_2 \times ... \times U_n$. Let $I = \{i_1, i_2, ..., i_k\}$ be a subset of $\{1, 2, ..., n\}$ with $i_1 < i_2 < ... < i_k$. The projection of R onto $U_{i_1} \times U_{i_2} \times ... \times U_{i_k}$ is defined to be the fuzzy subset of $U_{i_1} \times U_{i_2} \times ... \times U_{i_k}$ given by

$$S(u_{i_1}, u_{i_2}, ..., u_{i_k}) = \bigvee \{R(u_1, u_2 ..., u_n) : u_j \text{ with } j \notin I\}$$

For example, if R is a fuzzy relation on $U \times V \times W$, then the projection on $U \times V$ is given by $S(u, v) = \bigvee_{w \in W} R(u, v, w)$.

Let S be a fuzzy relation in $U \times V$. We define a fuzzy relation R in $U \times V \times W$ by $R(u, v, w) = S(u, v)$. Considering the set W as its indicator function, $R = S \times W$ since

$$(S \times W)(u, v, w) = \min\{S(u, v), W(w)\} = S(u, v)$$

The fuzzy set $S \times W$ is called the **cylindrical extension** of S. It is the largest fuzzy relation on $U \times V \times W$ having S as a projection. Indeed if T is a fuzzy relation with $T(u, v, w) > R(u, v, w)$, then the projection of T onto $U \times V$ has value at (u, v) at least $T(u, v, w) > R(u, v, w) = S(u, v)$.

7.4   Fuzzy partitions

An equivalence relation on a set gives a partition of that set, and vice versa. The analogy for fuzzy equivalence relations suggests properties for the notion of fuzzy partition. If $R : U \times U \to [0, 1]$ is a fuzzy relation on a set U, there is associated the family

$$\{R_u : U \to [0, 1] : v \to R(u, v)\}$$

of fuzzy subsets of U. If R were an equivalence relation, then R_u would be the equivalence class containing u, so this is an exact analog to the crisp case. When R is a fuzzy equivalence relation, we have

1. $R_u(u) = 1$ for each $u \in U$.

2. $R_u(v) = R_v(u)$ for all $u, v \in U$.

3. $R_u(v) \geq R_w(u) \wedge R_w(v)$ for all $u, v, w \in U$.

This suggests that a fuzzy partition of U could be defined as a family $\mathcal{P} = \{R_u : u \in U\}$ of fuzzy subsets of U satisfying these three properties. There is a clear one-to-one correspondence between the set of all such families and the set of all fuzzy equivalence relations on U. Note in particular that property 1 says that each of these fuzzy sets attains the value 1 for some $u \in U$. Call a fuzzy set A **normal** if $A(x) = 1$ for some x.

There is a different notion of a finite fuzzy partition that is of interest. A finite set $\{A_1, A_2, ..., A_n\}$ of nonempty subsets of a set U is a **partition** of U if

P1. $A_1 \cup A_2 \cup ... \cup A_n = U$ and

P2. $A_i \cap A_j = \varnothing$ if $i \neq j$.

This is equivalent to the condition $\sum_{i=1}^{n} \chi_{A_i}(x) = 1$ for all $x \in U$. In extending the notion of partition to fuzzy sets, we cannot simply use properties P1 and P2 since $\max\{A_i(x) : i = 1, ..., n\} = 1$ and $\min\{A_i(x), A_j(x)\} = 0$ for $i \neq j$ imply that $A_i(x) = 0$ or 1 and that the A_i are crisp sets. The condition of normality together with $\sum_{i=1}^{n} A_i(x) = 1$ does lead to a useful definition of finite partition for fuzzy sets.

Definition 7.4.1 *A **finite fuzzy partition** of a set U is a finite set of normal fuzzy subsets $\{A_1, A_2, ..., A_n\}$ of U such that $\sum_{i=1}^{n} A_i(x) = 1$ for all $x \in U$.*

This definition captures the meaning of properties P1 and P2 above in the following sense. Each x has a nonzero membership value for some A_i. If $A_i(x) = 1$ for some i, then it is 0 for all others. In fuzzy partitions, the degrees of membership of individuals in the various classes are measures of the intrinsic *heterogeneity* of individuals.

The need to consider fuzzy partitions is manifested in the design phase of fuzzy control, discussed in Chapter 13. Linguistic rules are of the form "If x_1 is $a_1, ..., x_n$ is a_n, then y is b", where the a_i are fuzzy subsets of the input spaces X_i, respectively, and b is a fuzzy subset of the output space Y, it is necessary to consider on each space involved an appropriate fuzzy partition of it, so that every input value can be classified according to the predetermined fuzzy partition for triggering corresponding rules. For example, for an interval of real numbers, a fuzzy partition could be "negative big", "negative medium", "negative small", "approximately zero", "positive small", "positive medium", and "positive big".

7.5 Fuzzy relations as Chu spaces

As mentioned in the previous section, fuzzy partitions of the input and output spaces are basic designs of fuzzy controllers. A fuzzy partition of a space X is used to extract information from elements of X, say from input data when X is the input space of a fuzzy controller. Specifically, let $\mathcal{A} = \{A_1, A_2, ... A_n\}$ be a fuzzy partition of X. Then elements of X are classified according to $\mathcal{A}$ via the evaluation relation $r : X \times \mathcal{A} \to [0, 1]$ given by $r(x, A_i) = A_i(x)$. Thus the triple $(X, r, \mathcal{A})$ is a **classification scheme**. That is, X is the class of things to be classified, $\mathcal{A}$ is a collection of properties used to classify elements of X, and r specifies the degree to which an element of X satisfies a property in $\mathcal{A}$. To be more general, the unit interval $[0, 1]$ can be replaced by some set K, for example a complete lattice. These general classification schemes are also known as

Chu spaces. See, for example, Barr [10]. Chu spaces offer a general framework for modeling phenomena such as concurrency in computer science, and information flow in distributed systems. We will elaborate on this concept with an example related to fuzzy logic.

A **K-Chu space** for a set K is a triple (X, r, A), where X and A are sets and $r : X \times A \to K$. Let $\mathcal{C}(K)$ denote the collection of all K-Chu spaces. These are the objects of a category, with a morphism from (X, r, A) to (Y, s, B) a pair (f, g) of maps, where $f : X \to Y$ and $g : B \to A$ that satisfy the adjointness condition

$$r(x, g(b)) = s(f(x), b)$$

for $x \in X$ and $b \in B$. That is, the following diagram commutes.

$$
\begin{array}{ccc}
X \times B & \xrightarrow{\; f \times 1 \;} & Y \times B \\[1em]
1 \times g \downarrow & & \downarrow s \\[1em]
X \times A & \xrightarrow[\; r \;]{} & K
\end{array}
$$

Composition of morphisms is defined as follows. If (f, g) and (u, v) are morphisms from (X, r, A) to (Y, s, B) and from (Y, s, B) to (Z, t, C), respectively, then the pair $(u \circ f, g \circ v)$ is a morphism from (X, r, A) to (Z, t, C). Indeed,

$$
\begin{aligned}
r(x, g \circ v(c)) &= r(x, g(v(c))) \\
&= s(f(x), v(c) \\
&= t(u(f(x)), c) \\
&= t(u \circ f(x), c)
\end{aligned}
$$

So composition of morphisms is given by $(f, g) \circ (u, v) = (u \circ f, g \circ v)$. Further, this composition of morphisms is associative, and the set of morphisms from any object to itself has an identity. With objects K-Chu spaces and morphisms as just defined, this yields a **Chu category**, denoted Chu(K). This situation defines a **category** in general: a class of objects $\mathcal{O}$ and for any two objects A and B in $\mathcal{O}$, a set $Mor(A, B)$, called the morphisms from A to B. There is a map $Mor(A, B) \times Mor(B, C) \to Mor(A, C)$, composition of morphisms, and this map is required to be associative, and each $Mor(A, A)$ must have an identity. The generic example is the class of sets with morphisms being maps from one set to another.

Here is an example of a Chu category. Let $K = [0, 1]$, and the objects be (X, r, A) where X is a set, $A = [0, 1]^X$, the set of all maps from X into

$[0, 1]$, and $r(x, a) = a(x)$, the evaluation map. Note that the elements of A are fuzzy subsets of X. We denote this category by $\mathcal{FUZZ}$.

Let (f, g) be a morphism from (X, r, A) to (Y, s, B). Then we have

$$
\begin{aligned}
r(x, g(b)) &= g(b)(x) \\
&= s(f(x), b) \\
&= b(f(x))
\end{aligned}
$$

Thus it suffices to specify $f : X \rightarrow Y$ since then $g : B \rightarrow A$ must be $g(b) = b \circ f$. This correspondence between sets and "evaluation fuzzy spaces" manifests itself through a functor F from the category of sets to $\mathcal{FUZZ}$. For a set X, let $F(X) = (X, r, [0, 1]^X) \in \mathcal{FUZZ}$. If $f : X \rightarrow Y$, let $F(f) = (f, \varphi(f))$ where $\varphi(f)(b) = b \circ f$. It is easy to check that $F(f)$ is a morphism $(X, r, [0, 1]^X) \rightarrow (Y, s, [0, 1]^Y)$ in $\mathcal{FUZZ}$, and that $F(g \circ f) = F(g) \circ F(f)$. This says essentially that $\mathcal{FUZZ}$ is equivalent to the category of sets, the category whose objects are sets and whose morphisms are ordinary maps.

This example can be varied in many ways. For example, take the Chu spaces to be triples $(X, r, [0, 1]^X)$ where X is a topological space and $[0, 1]^X$ is the set of all continuous maps from X into $[0, 1]$ with its usual topology. Morphisms are defined in the obvious way.

It seems that Chu spaces offer a general framework in which to describe various aspects of fuzzy set theory and its applications. For example, the relational approach to fuzzy control as described in 13.2 can be put into this framework. If X and Y are input and output spaces, respectively, of a control system, then a fuzzy system can be viewed as a map from $[0, 1]^X$ to $[0, 1]^Y$. Such a map can be defined from a fuzzy relation r on $X \times Y$ by

$$
r(a)(y) = \bigvee_{x \in X} (r(x, y) \wedge a(x))
$$

where $a \in [0, 1]^X$, $x \in X$ and $y \in Y$.

Morphisms between Chu spaces are novel for uncertainty analysis since they can be used to model interactions among concepts once these concepts are represented as Chu objects in a category.

7.6   Exercises

1. Let R and S be two $\wedge$-fuzzy equivalence relations on U. Show that $W : U \times U \rightarrow [0, 1] : (u, v) \rightarrow R(u, v) \wedge S(u, v\}$ is a $\wedge$-fuzzy equivalence relation on U.

2. Let

$$U = \{u_1, u_2, \ldots, u_n\}$$
$$V = \{v_1, v_2, \ldots, v_m\}$$
$$W = \{w_1, w_2, \ldots, w_k\}$$

Let R and S denote relations in $U \times V$ and $V \times W$, respectively, in the form of $n \times m$ and $m \times k$ matrices, respectively. Verify that $R \circ S$ is obtained by the product of the two matrices R and S when addition is replaced by $\vee$ and multiplication is replaced by $\wedge$.

3. Let

$$R = \begin{bmatrix} 0.1 & 0.0 & 0.6 \\ 0.2 & 0.3 & 1 \\ 0 & 0.4 & 0.5 \end{bmatrix} \quad \text{and} \quad S = \begin{bmatrix} 0.2 & 0.5 & 0 \\ 0.9 & 1 & 0.3 \\ 1 & 0.5 & 0.4 \end{bmatrix}$$

Find the max-min, and max-product composition of R and S.

4. Let R be a fuzzy relation in $U \times U$. Show that $R \circ R \le R$ if and only if R is transitive.

5. Let R be a fuzzy relation in $U \times U$. For $\alpha \in (0, 1]$, let

$$R_\alpha = \{(u, v) : R(u, v) \ge \alpha\}$$

That is, R_α is the α-cut of the fuzzy set R. Show that

   (a) R is symmetric if and only if the relations R_α are symmetric for all $\alpha \in (0, 1]$;

   (b) R is min transitive if and only if R_α are transitive for all $\alpha \in (0, 1]$.

6. Suppose that R and S are fuzzy equivalence relations on a set U. Is $R \circ S$ a fuzzy equivalence relation?

7. Show that if R and S are fuzzy equivalence relations on a set U, then so is $R \wedge S$. What about $R \vee S$?

8. Prove the other half of Theorem 7.2.2.

9. Complete the proof of Theorem 7.2.5.

10. Complete the proof of Theorem 7.2.6.

11. Let $\{P_\alpha : \alpha \in [0,1]\}$ be a partition tree. Show that for any subset I of $[0,1]$, $P_{\bigwedge_{\alpha \in I} \alpha} = \bigvee_{\alpha \in I} P_\alpha$.

12. Verify equations 7.1 and 7.2.

13. Let d be a metric on the set U. That is, $d : U \times U \to \mathbb{R}^+$ and satisfies

    (a) $d(x,y) = 0$ if and only if $x = y$

    (b) $d(x,y) = d(y,x)$

    (c) $d(x,y) \le d(x,z) + d(z,y)$

    Suppose that $d(x,y) \le 1$ for all $x,y \in U$. Define a fuzzy binary relation R on U by $R(x,y) = 1 - d(x,y)$. Show that R is a $\triangle$-fuzzy equivalence relation on U, where $\triangle$ is the t-norm

    $$x \triangle y = (x + y - 1) \vee 0$$

14. Let $\circ$ denote max-$\triangle$ composition of binary fuzzy relations on U, where $\triangle$ is any t-norm. Let R be a fuzzy binary relation on U which is $\triangle$-transitive and reflexive. Show that $R \circ R = R$.

15. Let R be an equivalence relation on a set X, and let $\triangle$ be a continuous t-norm on X. Verigy that R is a $\triangle$-fuzzy equivalence relation on X.

16. Let $\triangle$ be a continuous t-norm and let $R : [0,1] \times [0,1] \to [0,1]$ be defined by

    $$R(x,y) = \sup\{z \in [0,1] : z \triangle (x \vee y) \le x \wedge y\}$$

    Show that R is a $\triangle$-fuzzy equivalence on $[0,1]$.

17. Let R be a fuzzy subset of $[a,b] \times [a,b]$ with membership function $R(x,y) = e^{-|x-y|}$. Show that R is a product-fuzzy equivalence relation on $[a,b]$. That is, show that

    (a) $R(x,x) = 1$

    (b) $R(x,y) = R(y,x)$

    (c) $R(x,y) \ge R(x,z)R(z,y)$

18. Show that the α-cuts for R in the previous exercise do not all form equivalence relations.

19. Construct an example of a $\triangle$-fuzzy equivalence relation R on a finite set U so that for some α, R_α is not an equivalence relation on U.

20. Let R be a fuzzy equivalence relation on U. Define $d : U \times U \to [0,1]$ by $d(x,y) = 1 - R(x,y)$. Show that d is an **ultrametric.** That is, show that

    (a) $d(x,y) \geq 0$

    (b) $d(x,y) = d(y,x)$

    (c) $d(x,y) \leq d(x,z) \vee d(z,y)$

21. Let $\{A_1, A_2, ..., A_n\}$ be a fuzzy partition of a set U. Verify that if the A_i are crisp, then they are pairwise disjoint and have union U. Does a fuzzy set A together with its fuzzy complement A' form a fuzzy partition? Hint: think about normal fuzzy sets.

22. Are the following fuzzy partitions of $[0,10]$? Draw pictures of them.

    (a) $A_1(x) = x/10$, $A_2(x) = 1 - x/10$;

    (b) $A_1(x) = \begin{cases} 1 - x/5 & \text{if} \quad x < 5 \\ x/5 - 1 & \text{if} \quad 5 \leq x \end{cases}$,

    $A_2(x) = \begin{cases} x/5 & \text{if} \quad x < 5 \\ 2 - x/5 & \text{if} \quad 5 \leq x \end{cases}$;

    (c) $A_1(x) = \sin^2 x$, $A_2(x) = \cos^2 x$;

    (d) $A_1(x) = x/30$, $A_2(x) = x^2/100$, $A_3(x) = 1 - x/10$.

23. In the definition of morphisms of Chu categories, prove that composition of morphisms is indeed associative, and, that for each object, the set of morphisms of that object to itself has an identity under this composition.

24. In the definition of the functor F from the category of sets to the category $\mathcal{FUZZ}$, verify that $F(f)$ is a morphism and that $F(f \circ g) = F(f) \circ F(g)$.

Chapter 8

UNIVERSAL APPROXIMATION

This chapter is devoted to an example of system modeling in which fuzzy logic is put into effective use. We consider a system as an input-output map: $y = f(x)$. We assume that the internal structure of the system is unknown, but qualitative knowledge about its behavior is available, say, under the form of a collection of "If...then..." rules. The problem is to construct a mathematical description of the system, based upon available information, so that it will represent faithfully the "true" system. The construction process consists of translating linguistic rules into mathematical expressions using fuzzy sets and fuzzy logic and defuzzifying the combined fuzzy output. The systems so obtained are shown to be within a class of designs capable of approximating the "true" input-output relation to any degree of accuracy.

8.1 Fuzzy rule bases

A system in which aspects of fuzziness are involved is called a fuzzy system. Computer programs that emulate the decision making of human experts, where the knowledge available and the reasoning processes involve fuzzy concepts in a natural language, are fuzzy systems. Similarly, dynamical systems that are controlled by fuzzy controllers are fuzzy systems. A fuzzy controller itself is a fuzzy system since the control law is built using rules involving fuzzy concepts.

Consider a system where an input $x = (x_1, \ldots, x_n) \in \mathbb{R}^n$ will produce an output $y \in \mathbb{R}$. Suppose the relationship $y = f(x)$ is not known, but the behavior of the output with respect to input can be described as a collection of linguistic rules of the form

$$R_i : \quad \text{"If } x_1 \text{ is } A_{1i,} \ldots, x_n \text{ is } A_{ni} \quad \text{then } y \text{ is } B_i \text{"}, \quad i = 1, 2, \ldots, k$$

where the A's and B's are fuzzy sets. When an input $(x_1, \ldots, x_n)$ is observed, one needs to find an appropriate value for the output y.

In some areas of applications, for example fuzzy control, the interpretation of "If...then..." in the fuzzy rule base is slightly different from fuzzy implication of the previous chapter. Perhaps "If x is A then y is B" really means "If x is A then y is B, ELSE, if x is not A then y is 'undefined' ". This statement can be represented by the fuzzy relation

$$(A \times B) \cup (A' \times \varnothing)$$

where $\varnothing$ denotes "undefined", that is, $\varnothing(y) = 0$ for all y. The membership of this fuzzy relation is

$$
\begin{aligned}
(A \triangle B) \triangledown (N(A) \triangle \varnothing) &= (A \triangle B) \triangledown (N(A) \triangle 0) \\
&= (A \triangle B) \triangledown 0 \\
&= A \triangle B
\end{aligned}
$$

where $(A \triangle B)(x, y) = A(x) \triangle B(y)$, and similarly for the t-conorms. Thus in

$$R_i : \quad \text{"If } x_1 \text{ is } A_{1i}, \ldots, x_n \text{ is } A_{ni} \quad \text{then } y \text{ is } B_i \text{"}$$

each value y has a "rule weight" of

$$(\triangle_{1 \leq j \leq n} A_{ji}(x_i)) \triangle B_i(y)$$

From this interpretation, given a rule base of k rules, the property C, "appropriate values of output", is represented by a fuzzy subset of $\mathbb{R}$, depending on input $(x_1, \ldots, x_n)$, which can be written informally as

$$
\begin{aligned}
C(y) = \quad &\text{"If } x_1 \text{ is } A_{11}, \ldots, x_n \text{ is } A_{n1} \quad\quad \text{then } y \text{ is } B_1 \text{"}, \quad \text{or...} \\
&\ldots \text{ or "If } x_1 \text{ is } A_{1k}, \ldots, x_n \text{ is } A_{nk} \quad \text{then } y \text{ is } B_k \text{"}.
\end{aligned}
$$

Formally, with t-norm $\triangle$ and t-conorm $\triangledown$, we have

$$
\begin{aligned}
C(y) = \quad &\triangledown_{1 \leq i \leq k} [A_{1i}(x_1) \triangle A_{2i}(x_2) \triangle \\
&\cdots \triangle A_{ni}(x_n) \triangle B_i(y); \; i = 1, 2, \ldots, k]
\end{aligned}
$$

For example, if $a \triangle b = a \wedge b$ and $a \triangledown b = a \vee b$, then

$$C(y) = \bigvee_{1 \leq i \leq k} \{A_{1i}(x_1) \wedge \ldots \wedge A_{ni}(x_n) \wedge B_i(y)\}$$

The choice of $\triangle$ and $\triangledown$ might depend on problems at hand, or might be justified by some criteria for sensitivity analysis.

The rules R_i above are the result of modeling linguistic labels by fuzzy sets. This is done by choosing membership functions reflecting the semantics of the linguistic labels. Adjustment or tuning parameters of these membership functions may be necessary to get a faithful representation of the linguistic rules. We will not discuss these practical issues here.

From a theoretical viewpoint, assuming that we have a set of rules with specified membership functions and logical connectives, the problem is how to produce a single output corresponding to an input $x = (x_1, \ldots, x_n)$. Before treating the question, we consider some special cases.

Suppose the rules above are more precise in the sense that the B_i's are singletons, that is,

$$R_i: \quad \text{"If } x_1 \text{ is } A_{1i}, \ldots, x_n \text{ is } A_{ni} \quad \text{then } y = y_i\text{"}, \quad i = 1, 2, \ldots, k$$

In this case, $B_i = \{y_i\}$, so that

$$B_i(y) = \begin{cases} 1 & \text{if} \quad y = y_i \\ 0 & \text{if} \quad y \neq y_i \end{cases}$$

Therefore, for $y \neq y_i$, $i = 1, 2, \ldots k$,

$$\begin{aligned} A_{1i}(x_1) \triangle \ldots \triangle A_{ni}(x_n) \triangle B_i(y) &= A_{1i}(x_1) \triangle \ldots \triangle A_{ni}(x_n) \triangle 0 \\ &\leq 1 \triangle \ldots \triangle 1 \triangle 0 = 0 \end{aligned}$$

and hence

$$C(y) = 0 \triangledown \ldots \triangledown 0 = 0$$

For $y = y_j$, for some $j \in \{1, 2, \ldots, k\}$,

$$A_{1j}(x_1) \triangle \ldots \triangle A_{nj}(x_n) \triangle 1 = A_{1j}(x_1) \triangle \ldots \triangle A_{nj}(x_n)$$

so that

$$\begin{aligned} C(y_j) &= \triangledown[0, \ldots, 0, (A_{1j}(x_1) \triangle, \ldots \triangle, A_{nj}(x_n)), \triangle 0 \ldots \triangle 0] \\ &= A_{1j}(x_1) \triangle \ldots \triangle A_{nj}(x_n). \end{aligned}$$

One can view $C(y_j)$ as a weight for y_j (or a normalized weight $w_j = C(y_j)/\sum_{i=1}^{k} C(y_i)$). A plausible single output value could be a weighted average

$$y^* = \sum_{i=1}^{k} w_i y_i$$

As another example, consider the case where all fuzzy concepts involved express a form of measurement errors around some plausible values. In this case, Gaussian-type membership functions seem appropriate, so we can take

$$A_{ij}(z) = \exp\left\{-\frac{1}{2}\frac{(z - a_{ij})^2}{\sigma_{ij}^2}\right\}$$

$$B_j(z) = \exp\left\{-\frac{1}{2}\frac{(z - b_j)^2}{\beta_j^2}\right\}$$

For each rule R_j, b_j is a plausible candidate for the rule output, but with some degree depending upon $A_{ij}(x_i)$, $i = 1, \ldots, n$.

For example, choosing $a \triangle b = ab$ (product inference), the normalized weight of the output b_j of the rule R_j is

$$w_j = \frac{\prod_{i=1}^{n} A_{ij}(x_i) B_j(y)}{\sum_{j=1}^{k} \prod_{i=1}^{n} A_{ij}(x_i)}$$

so that the overall output of the system, as a weighted average, is

$$y^* = \sum_{j=1}^{k} w_j y_j$$

8.2   Design methodologies

Consider again the fuzzy set (output)

$$C(y) = \nabla_{1 \leq i \leq k}[A_{1i}(x_1) \triangle \ldots \triangle A_{ni}(x_n) \triangle B_i(y)]$$

In order to obtain a single value for the output corresponding to an input $x = (x_1, \ldots, x_n)$, we need to transform the membership function $C(y)$ into a real number $\mathcal{D}(C)$. That is, we have to **defuzzify** the fuzzy set

C. Such a transformation is called a **defuzzification procedure**. Some defuzzification procedures are **centroid defuzzification**

$$\mathcal{D}(C) = \frac{\int y C(y) dy}{\int C(y) dy}$$

and **center-of-maximum defuzzification**

$$\mathcal{D}(C) = \frac{m_- + m_+}{2}$$

where

$$m_- = \inf\{y : C(y) = \max_z C(z)\}$$

and

$$m_+ = \sup\{y : C(y) = \max_z C(z)\}$$

A chosen defuzzification procedure is used to produce an output for the system, namely

$$y^*(x_1, \dots, x_n) = \mathcal{D}(C)$$

where $C(y) = \triangledown[A_{ij} \triangle \cdots \triangle A_{ni}(x_n) \triangle B_i(y); \ i = 1, \dots, k]$. This is known as a "combine-then-defuzzify" strategy, that is, we first combine all the rules using fuzzy connectives to obtain an overall fuzzy set, and then defuzzify this fuzzy set by some chosen defuzzification procedure.

Note that the two examples in Section 8.1 present another strategy, namely "defuzzify-then-combine". Indeed, for each rule R_i, one first defuzzifies the rule output, say, by selecting a plausible output value y_i for the rule, and then combine all these y_i's via a discrete version of the centroid defuzzification procedure (weighted average) applied to the fuzzy set (y_i, w_i), $i = 1, \dots, k$.

Again, like the case of fuzzy logical connectives, the choice of defuzzification procedures might depend on the problems at hand. However, it is possible to make this choice "optimal" if we have some performance criteria.

So far we have described an inference design, based on a fuzzy rule base, without a rationale. Of course, a design is only useful only if it leads to a "good" representation of the input-output relation. Now, the "true" input-output relation $y = f(x)$ is unknown, and only some information about f is available, namely in the form of a fuzzy rule base. The previous designs are thus only approximations of f. Therefore, by a "good" representation, we mean a "good" approximation of f.

For a **design methodology**, that is, a choice of membership functions for the A's and B's, of logical connectives $\triangle$, $\triangledown$, and negation, and defuzzification procedure $\mathcal{D}$, we produce an input-output map

$$f^* : (x_1, \cdots, x_n) \to y^*$$

We need to investigate to what extent f^* will be a good approximation of f. Basically, this is a problem in the theory of approximation of functions. The following is a little technical, but it is necessary to spell out the main idea. The design above will lead to a good approximation if for any $\varepsilon > 0$, one can find an f^* such that $||f - f^*|| < \varepsilon$, where $|| \cdot ||$ denotes a distance between f and f^*. The following section contains necessary mathematical tools for investigating this problem.

8.3   Some mathematical background

In order to formulate and to prove some typical results about the approximation capability of fuzzy systems, we need some information about metric spaces. A metric space is a set with a distance on it. Here is the formal definition.

Definition 8.3.1 *Let X be a set. A map $d : X \times X \to \mathbb{R}$ is a **distance**, or a **metric** if*

1. *$d(x, y) \geq 0$ and $d(x, y) = 0$ if and only if $x = y$,*

2. *$d(x, y) = d(y, x)$, and*

3. *$d(x, y) \leq d(x, z) + d(y, z)$.*

   *A set together with such a function on it is a **metric space**. Condition 3 is the **triangle inequality**.*

Some examples follow.

- $\mathbb{R}$ with $d(x, y) = |x - y|$.

- $\mathbb{R}^n$ with $d((x_1, x_2, ..., x_n), (y_1, y_2, ..., y_n)) = \sqrt{\sum_{i=1}^{n}(x_i - y_i)^2}$. This is the ordinary distance on $\mathbb{R}^n$.

- $\mathbb{R}^n$ with $d((x_1, x_2, ..., x_n), (y_1, y_2, ..., y_n)) = \sup_{i=1}^{n}\{|x_i - y_i|\}$.

- $\mathbb{R}^n$ with $d((x_1, x_2, ..., x_n), (y_1, y_2, ..., y_n)) = \sum_{i=1}^{n}|x_i - y_i|$.

- For an interval $[a, b]$ of real numbers, let $C([a, b])$ be the set of all continuous functions $[a, b] \to \mathbb{R}$. For $f, g \in C([a, b])$ let $d(f, g) = \sup_{x \in [a,b]} |f(x) - g(x)|$. This is called the **sup-norm** on $C([a, b])$.

- If (X, d) is a metric space, then $e(x, y) = d(x, y)/(d(x, y)+1)$ makes X into a metric space.

- If X is any set, then $d(x, y) = 1$ if $x \neq y$ and $d(x, x) = 0$ makes X into a metric space.

The verifications that these are indeed metric spaces are left as exercises.

Let (X, d) and (Y, e) be metric spaces. A function $f : X \to Y$ is **continuous at a point** $x \in X$ if for any $\epsilon > 0$ there is a $\delta > 0$ such that $e(f(x), f(y)) < \epsilon$ whenever $d(x, y) < \delta$. The function f is **continuous** if it is continuous at every $x \in X$. If for each $\epsilon > 0$ there is a δ such that *for all* $x, y \in X$, $e(f(x), f(y)) < \epsilon$ whenever $d(x, y) < \delta$, then f is **uniformly continuous** on X.

A subset A of X is **open** if for each $a \in A$ there is $\epsilon > 0$ such that $B(a, \epsilon) = \{x : d(x, a) < \epsilon\} \subseteq A$. The complement of an open set is called **closed**. A point x is an **accumulation point** of A if for every $\epsilon > 0$, $B(x, \epsilon) \cap A \neq \varnothing$. The **closure** of A, denoted $\overline{A}$, is A together with all its accumulation points. The closure of a set A is closed, and $\overline{A} = A$ if and only if A is closed. In particular, $\overline{\overline{A}} = \overline{A}$. The set A is **dense** in X if $\overline{A} = X$. If A is dense in X then for each $x \in X$, there is an element $a \in A$ arbitrarily close to it. That is, for any $\epsilon > 0$ there is an element $a \in A$ such that $d(x, a) < \epsilon$.

The set $\mathbb{R}$ of real numbers is a metric space with $d(x, y) = | x - y |$ and for $a < b$, the interval $[a, b]$ is a closed subset of it. A classical result is that with respect to this metric d, if the union of any collection of open subsets of $\mathbb{R}$ contains $[a, b]$, then the union of finitely many of them also contains $[a, b]$. This is phrased by saying that any open cover contains a finite subcover. A subset A of a metric space is **compact** if any open cover of it contains a finite subcover. The subset A is **bounded** if $\sup\{d(x, y) : x, y \in A\} < \infty$. Compact subsets of the Euclidean spaces $\mathbb{R}^n$ are the closed and bounded ones.

If (X, d) and (Y, e) are metric spaces and f is a continuous map from X to Y, then the image of any compact subset of X is a compact subset of Y. If X itself is compact, then such a continuous map is uniformly continuous. That is, a continuous map on a compact space is uniformly continuous.

Let (X, d) be compact and $C(X)$ the set of all continuous real-valued functions on X with the metric $\sup_{x \in X} | f(x) - g(x) |$. If K is a compact subset of $C(X)$, then for each $\epsilon > 0$ there is a $\delta > 0$ such that if $d(x, y) < \delta$, then $| f(x) - f(y) | < \epsilon$ for *all* $f \in K$. Such a family K of functions is called an **equicontinuous** family.

A classical result known as the **Weierstrass theorem** is that continuous functions on a compact set $[a, b]$ can be approximated uniformly by polynomials. There is a generalization to metric spaces that we need. The **Stone-Weierstrass theorem** is this. Let (X, d) be a compact metric space. Let $H \subseteq C(X)$ satisfy the following conditions.

- H is a subalgebra of $C(X)$. That is, for $a \in \mathbb{R}$ and $f, g \in H$, we have af, $f + g$, and $fg \in H$.

- H vanishes at no point of X. That is, for $x \in X$, there is an $h \in H$ such that $h(x) \neq 0$.

- H separates points. That is, if $x, y \in X$ then there is an $h \in H$ such that $h(x) \neq h(y)$.

Then H is dense in $C(X)$. That is, $\overline{H} = C(X)$. This means that any real continuous function on X can be approximated arbitrarily closely by a function from H. Polynomials on an interval $[a, b]$ satisfy these conditions, and so continuous functions on $[a, b]$ can be approximated by polynomials. This is the classical Weierstrass theorem. For a proof, see [89].

8.4   Approximation capability

The design of a fuzzy system is aimed at approximating some idealistic input-output maps. The problem is well-known in various branches of science, for example, in system identification and in statistical regression. Fuzzy systems are viewed as a tool for approximating functions. In applications, the approximation techniques used depend on available data.

Typically, a fuzzy controller, or more generally, a fuzzy system, is a map from $\mathbb{R}^n$ to $\mathbb{R}$ constructed in some specific way. From a set of rules of the form "If for $i = 1, 2, ..., n$, $x_i \in A_{ij}$, then $y \in B_j$ for $j = 1, 2, ..., r$", where the variables x_i and y take values in $\mathbb{R}$ and the A_{ij} and B_j are fuzzy sets, one constructs a value $y^* = f^*(x_1, x_2, ..., x_n)$ by combining the rules above. For example, choosing a t-norm $\triangle$ for the logical connective "and" and a t-conorm $\triangledown$ for "or", one arrives at a membership function

$$\mu(y) = \triangledown_{1 \leq j \leq r} [A_{1j}(x_1) \triangle ... \triangle A_{nj}(x_n) \triangle B_j(y)]$$

and the quantity

$$y^* = \frac{\left(\int_{\mathbb{R}} y\mu(y)dy\right)}{\left(\int_{\mathbb{R}} \mu(y)dy\right)}$$

The map $x \rightarrow y^* = f^*(x)$ depends on

- the membership functions A_{ij} and B_j,

- the t-norm $\triangle$ and t-conorm $\triangledown$, and

- the "defuzzification procedure" $\mu(y) \rightarrow y^*$.

If we denote by $\mathcal{M}$ a class of membership functions of fuzzy concepts on $\mathbb{R}$, by $\mathcal{L}$ a class of fuzzy logical connectives, and by $\mathcal{D}$ a defuzzification procedure, then the triple $(\mathcal{M}, \mathcal{L}, \mathcal{D})$ is referred to as a **design methodology** and specifies the input-output map $y^* = f^*(x)$. The function f also depends on the number r of rules, and this dependency is indicated by the notation f_r. The approximation capability of fuzzy systems is that under suitable conditions on $(\mathcal{M}, \mathcal{L}, \mathcal{D})$, the class of functions $\{f_r : r \geq 1\}$ is dense in the space of continuous functions $C(K)$ from compact subsets K of $\mathbb{R}^n$ to $\mathbb{R}$ with respect to the sup-norm. To illustrate this, we look at two instances.

Let $\mathcal{F}$ be the class of functions $f : \mathbb{R}^n \rightarrow \mathbb{R}$ of the form

$$f(x) = \frac{\sum_{j=1}^{r} y_j \triangle (A_{1j}(x_1) \triangle A_{2j}(x_2) \triangle \cdots \triangle A_{nj}(x_n))}{\sum_{j=1}^{r} (A_{1j}(x_1) \triangle A_{2j}(x_2) \triangle \cdots \triangle A_{nj}(x_n))}$$

where $x = (x_1, x_2, ..., x_n)$, and $x_i, y_j \in \mathbb{R}$, $\triangle$ is a continuous t-norm, and A_{ij} are of **Gaussian type**. That is,

$$A_{ij}(x) = \alpha_{ij} e^{-(x-a_{ij})^2/k_{ij}}$$

If K is a compact subset of $\mathbb{R}^n$, then $\mathcal{F}_K$ will denote the restriction of elements of $\mathcal{F}$ to K.

Theorem 8.4.1 *Let* $a \triangle b = ab$, *or* $a \triangle b = a \wedge b$. *For any compact subset* K *of* $\mathbb{R}^n$, $\mathcal{F}_K$ *is dense in* $C(K)$ *with respect to the sup-norm.*

Proof. We prove the theorem for the case $a \wedge b$. It suffices to verify the hypotheses of the Stone-Weierstrass theorem. First, one needs that $\mathcal{F}_K$ is a subalgebra, that is, if $f, g \in \mathcal{F}_K$ and $\alpha \in \mathbb{R}$, then $f + g$, fg, and αf are in $\mathcal{F}_K$. This follows readily using the facts that $(\wedge\{a_i\})(\wedge\{b_j\}) = \wedge_i \wedge_j \{a_i b_j\}$ whenever the a_i and b_j are positive, and that products of Gaussian are Gaussian.

To show that $\mathcal{F}_K$ vanishes at no point, simply choose $y_j > 0$ for all j. Then $f(x) > 0$ since each $\mu_{ij}(x_i) > 0$.

We now show that $\mathcal{F}_K$ separates points in K. If $u, v \in K$ with $u \neq v$, then for f defined by

$$f(x) = \frac{\wedge_i \{e^{-\frac{1}{2}(x_i - u_i)^2}\}}{\wedge_i \{e^{-\frac{1}{2}(x_i - u_i)^2}\} + \wedge_i \{e^{-\frac{1}{2}(x_i - v_i)^2}\}}$$

$f(u) \neq f(v)$. Thus if $g \in C(K)$ and $\varepsilon > 0$, there is an $f \in \mathcal{F}_K$ such that $\|f - g\| < \varepsilon$, where $\|g\| = \vee_{x \in K} |g(x)|$. ∎

The class $\mathcal{F}$ of continuous functions appears in the design situations discussed in Section 8.1. We consider now a general class $\mathcal{F}(\mathcal{M}, \mathcal{L}, \mathcal{D})$ of designs as follows. $\mathcal{M}$ consists of those membership functions μ such that $\mu(x) = \mu_0(ax + b)$ for some $a, b \in \mathbb{R}$ and $a \neq 0$, and $\mu_0(x)$ is continuous, positive on some interval of $\mathbb{R}$, and 0 outside that interval. $\mathcal{L}$ consists of continuous t-norms and t-conorms. $\mathcal{D}$ is a defuzzification procedure transforming each membership function μ into a real number in such a way that if $\mu(x) = 0$ outside an interval (α, β), then $\mathcal{D}(\mu) \in [\alpha, \beta]$. For example,

$$\mathcal{D}(\mu) = \left(\int_{\mathbb{R}} x\mu(x)dx \right) / \left(\int_{\mathbb{R}} \mu(x)dx \right)$$

is such a procedure.

Theorem 8.4.2 *For any design methodology* $(\mathcal{M}, \mathcal{L}, \mathcal{D})$ *and any compact subset* K *of* $\mathbb{R}^n$, $\mathcal{F}(\mathcal{M}, \mathcal{L}, \mathcal{D})|_K$ *is dense in* $C(K)$ *with respect to the sup-norm.*

Proof. We need to show that for $f \in C(K)$ and for any $\varepsilon > 0$, there exists $g \in \mathcal{F}(\mathcal{M}, \mathcal{L}, \mathcal{D})|_K$ such that $\|f - g\| \leq \varepsilon$. Since f is uniformly continuous on the compact set K, there exists $\delta(\varepsilon)$ such that whenever

$$\bigvee \{|x_i - y_i| : i = 1, 2, ..., n\} \leq \delta(\varepsilon)$$

we have

$$|f(x) - f(y)| \leq \varepsilon/2$$

Let $r \geq 1$. Since K is compact, there is a covering by r open balls with the j-th centered at $z^{(j)}$ and with each of radius $\delta(\varepsilon)/2$. Consider the collection of r rules of the form "If x_1 is A_{1j} and x_2 is A_{2j} and ... and x_n is A_{nj} then y is B_j", where the membership functions are chosen as follows. Let μ_0 be a continuous function positive on an interval (α, β) and 0 outside. The function

$$\hat{\mu}_0(t) = \mu_0 \left(\frac{\beta - \alpha}{2} t + \frac{\beta + \alpha}{2} \right)$$

is in $\mathcal{M}$, is positive on $(-1, 1)$ and 0 outside. Take

$$A_{ij}(t) = \hat{\mu}_0 \left(\frac{t - z_i^{(j)}}{\delta} \right),$$

$$B_j(t) = \hat{\mu}_0 \left(\frac{t - f(z_i^{(j)})}{\varepsilon/2} \right)$$

Then for $\triangle, \triangledown \in \mathcal{L}$, let

$$\mu_x(y) = \bigvee_{1 \leq j \leq r} [A_{1j}(x_i) \triangle A_{2j}(x_i) \triangle \cdots \triangle A_{nj}(x_i) \triangle B_j(y)]$$

We will show that $g(x) = \mathcal{D}(\mu_x)$ approximates f to the desired accuracy.

In view of the properties of $\mathcal{D}$, it suffices to verify that μ_x is not identically 0 and $\mu_x(y) = 0$ when $y \notin (f(x) - \varepsilon, f(x) + \varepsilon)$. Now for $x \in K$, there is $z^{(j)}$ such that

$$\vee \left\{ \left| x_i - z_i^{(j)} \right| : i = 1, 2, ..., n \right\} \leq \delta(\varepsilon)$$

Thus

$$A_{ij}(x_i) = \hat{\mu}_0 \left(\frac{x_i - z_i^{(j)}}{\delta} \right) > 0$$

for all i since

$$\frac{x_i - z_i^{(j)}}{\delta} \in (-1, 1)$$

Taking $y = f(z^{(j)})$, we have $B_j(y) = \hat{\mu}_0(0) > 0$. By properties of t-norms and t-conorms, it follows that $\mu_x(y) > 0$.

Next, let $y \notin (f(x) - \varepsilon, f(x) + \varepsilon)$. By t-conorm properties, to show that $\mu_x(y) = 0$ it suffices to show that for $j = 1, 2, ..., r$,

$$p_j = [A_{1j}(x_i) \triangle A_{2j}(x_i) \triangle \cdots \triangle A_{nj}(x_i) \triangle B_j(y)] = 0$$

Since $\triangle$ is a t-norm, $p_j = 0$ if one of the numbers $A_{ij}(x_i), B_j(y)$ is 0. If all are positive, then $|f(x) - f(z^{(j)})| \leq \varepsilon/2$ by uniform continuity of f on K. On the other hand, by hypothesis, $|y - f(x)| \geq \varepsilon$, and so

$$\frac{y - f(z^{(j)})}{\varepsilon/2} \notin (-1, 1)$$

Thus

$$B_j(y) = \hat{\mu}_0 \left(\frac{y - f(z^{(j)})}{\varepsilon/2} \right) = 0$$

∎

In summary, there exist various classes of fuzzy systems which can approximate arbitrarily closely continuous functions defined on compact subsets of finite dimensional Euclidean spaces. Fuzzy systems themselves are finite-dimensional in the sense that the number of input variables is finite. However, as we will see, they can be used to approximate continuous maps defined on infinite-dimensional spaces. This is particularly useful for approximating control laws of distributed parameter processes. In view of results such as the last two theorems, to handle the infinite-dimensional case, it suffices to reduce it to the finite one.

Theorem 8.4.3 *Let F be a compact subset of $C(U)$, where U is a compact metric space. Let $J : F \to \mathbb{R}$ be continuous. Then for each $\varepsilon > 0$, there exists a continuous function π from F to the finite-dimensional Euclidean space $\mathbb{R}^q$ for some q, and a continuous function J_ε defined on the compact subset $\pi(F)$ such that for every $f \in F$,*

$$|J(f) - J_\varepsilon(\pi(f))| \le \varepsilon$$

Before giving the proof, we remark that the problem of approximating J by a fuzzy system is reduced to that of approximating J_ε, which is a continuous function of a finite number of variables. If a fuzzy system g approximates J_ε to within ε, then g also approximates J. In fact,

$$
\begin{aligned}
|J(f) - g(\pi(f))| &\le |J_\varepsilon(\pi(f)) - g(\pi(f))| + |J(f) - J_\varepsilon(\pi(f))| \\
&\le 2\varepsilon
\end{aligned}
$$

This reduction is necessary for practical implementation purposes. This procedure is parallel to computational techniques in H_∞-control in which the H_∞-optimization problem for distributed parameter systems is reduced to a finite dimensional problem of finding weighting filters [25].

Proof. We use the notation $\|\cdot\|$ to denote the various norms which appear. Since F is compact, J is uniformly continuous on F, so there exists $\delta(\varepsilon)$ such that whenever $\|f - g\| \le \delta(\varepsilon)$, we have $\|J(f) - J(g)\| \le \varepsilon$. Let G be a finite set of points in F such that for every $f \in F$, there exists a $g \in G$ with $\|f - g\| \le \delta(\varepsilon)/3$. Since F is a compact subset of $C(U)$, F forms a family of equicontinuous functions, so that there exists $\beta(\varepsilon) > 0$ such that whenever $\|u - v\| \le \beta(\varepsilon)$, we have $\|f(u) - f(v)\| \le \delta(\varepsilon)/4$ for every $f \in F$. Choose a finite set $\{v_1, v_2, ..., v_q\} = V \subseteq U$ such that for every $u \in U$, there exists a $v \in V$ such that $\|u - v\| \le \beta(\varepsilon)$.

Define $\pi : F \to \mathbb{R}^q$ by $\pi(f) = (f(v_1), f(v_2), ..., f(v_q))$. Obviously

$$
\begin{aligned}
\|\pi(f) - \pi(g)\| &= \bigvee_{1 \le i \le q} \left\{ \left| f(v_i) - \tilde{f}(v_i) \right| \right\} \\
&\le \|f - g\|
\end{aligned}
$$

so π is continuous and hence $\pi(F)$ is compact.

Define $J_\varepsilon : \pi(F) \to \mathbb{R}$ by

$$J_\varepsilon(\pi(f)) = \frac{\sum_G \alpha_g(f) J(g)}{\sum_G \alpha_g(f)}$$

where for every $g \in G$,

$$\alpha_g(f) = \vee \{0, \delta(\varepsilon)/2 - \|\pi(f) - \pi(g)\|\}$$

is a continuous function of $\pi(f)$. For every $f \in F$, there exists a $g \in G$ such that

$$
\begin{aligned}
\|\pi(f) - \pi(g)\| &\leq \|f - g\| \\
&\leq \delta(\varepsilon)/3 \\
&< \delta(\varepsilon)/2
\end{aligned}
$$

so $\sum_G \alpha_g(f) > 0$. Thus J_ε is well defined and continuous on $\pi(F)$.

Now

$$|J(f) - J_\varepsilon(\pi(f))| \leq \vee |J(f) - J(g))|$$

with the $\vee$ over all $g \in G$ such that $\|\pi(f) - \pi(g)\| \leq \delta(\varepsilon)/2$. For every $u \in U$, there exists a $v \in V$ such that $\|u - v\| \leq \beta(\varepsilon)$, so

$$
\begin{aligned}
|f(u) - g(u)|\, e &\leq |f(u) - f(v)| + |f(v) - g(v)| + |g(v) - g(u)| \\
&\leq \delta(\varepsilon)/4 + |f(v) - g(v)| + \delta(\varepsilon)/4
\end{aligned}
$$

Hence

$$
\begin{aligned}
\|f - g\| &= \bigvee_G \{|f(u) - g(u)|\} \\
&\leq \delta(\varepsilon)/2 + \|\pi(f) - \pi(g)\|
\end{aligned}
$$

Thus, when $\|\pi(f) - \pi(g)\| \leq \delta(\varepsilon)/2$, we have $\|f - g\| \leq \delta(\varepsilon)$. Hence $|J(f) - J(g)| \leq \varepsilon$, implying that for every $f \in F$, we have

$$|J(f) - J_\varepsilon(\pi(f))| \leq \varepsilon$$

The theorem follows. ∎

8.5   Exercises

1. Show that the examples following Definition 8.3.1 are indeed metric spaces.

2. Let $\mathcal{F}$ denote the class of all (real) polynomials in one variable x. Verify that $\mathcal{F}$ is a subalgebra of $\mathcal{C}([a,b])$, separating points in $[a,b]$ and vanishing at no point of $[a,b]$.

3. Let $\mathcal{F}$ be the class of all even polynomials in one variable, that is, those such that $f(x) = f(-x)$. Show that $\mathcal{F}$ does not separate points in, say, $K = [-2,+2]$.

4. Let $\mathcal{F} \subseteq \mathcal{C}(K)$ satisfying the hypotheses of the Stone-Weierstrass theorem. Let $x \neq y$.

   (a) Let $g, h \in \mathcal{F}$ such that $g(x) \neq g(y)$, and $h(x) \neq 0$. Consider $\varphi = g + \lambda h$ where λ is chosen as follows.
   If $g(x) \neq 0$,   then take $\lambda = 0$.
   If $g(x) = 0$,   then take $\lambda \neq 0$ such that

   $$\lambda[h(x) - h(y)] \neq g(y) \neq 0$$

   Verify that $\varphi \in \mathcal{F}$ and $\varphi(x) \neq 0$, $\varphi(x) \neq \varphi(y)$.
   (b) Let $\alpha = u^2(x) - u(x)u(y)$. Verify that $\alpha \neq 0$. Define

   $$f_1(\cdot) = \frac{1}{\alpha}[\varphi^2(\cdot) - \varphi(\cdot)\varphi(y)]$$

   Verify that $f_1 \in \mathcal{F}$ and $f_1(x) = 1$, $f_1(y) = 0$.
   (c) By symmetry, show that there exists $f_2 \in \mathcal{F}$ such that $f_2(x) = 0$ and $f_2(y) = 1$. Define $f = f_1 + f_2$. Verify that $f \in \mathcal{F}$ and $f(x) = f(y) = 1$.

5. Let $K = [0,1]$, and let G be a family of continuous functions on K. Let $\mathcal{F}$ be the class of polynomials with real coefficients in the elements of G, that is, polynomials P of the form

   $$P(x) = \sum_{i=1}^{k} \alpha_i [g_{i1}(x)]^{n_1} \cdots [g_{im}(x)]^{n_m}$$

   where α_i's are real, $g_{ij} \in G$, and n_j's are integers ≥ 0.

   (a) Use the Stone-Weierstrass theorem to show that if G separates points in K, then elements of $\mathcal{C}([0,1])$ are approximable by those of $\mathcal{F}$.

(b) Show that if G does not separate points in K, then there exist continuous functions in K which cannot be approximable by elements of $\mathcal{F}$. (Hint: There exist $x \neq y$ such that for all $g \in G$, $g(x) = g(y)$, implying that for all $p \in \mathcal{F}$, $p(x) = p(y)$. If f is a continuous function such that $f(x) \neq f(y)$, then f cannot be approximable by $\mathcal{F}$.)

6. Two metrics d and e on a metric space are **equivalent** if there are positive real numbers r and s such that $rd \leq e$ and $se \leq d$. Let $x = (x_1, x_2, ..., x_n)$ and $y = (y_1, y_2, ..., y_n) \in \mathbb{R}^n$. Show that

   (a) $d_1(x, y) = \sum_{k=1}^{n} |x_k - y_k|$,

   (b) $d_2(x, y) = \bigvee_{1 \leq i \leq k} \{|x_k - y_k|\}$, and

   (c) $d_3(x, y) = \sqrt{\sum_{k=1}^{n} (x_k - y_k)^2}$

   are equivalent metrics on $\mathbb{R}^n$.

7. Let d be the discrete metric on $\mathbb{R}^n$. That is, $d(x, y) = 1$ if $x \neq y$ and 0 if $x = y$. Show that the metric d is not equivalent to any of the three in the previous exercise.

8. Let X be a compact metric space and $C(X)$ the set of all continuous functions from X into $\mathbb{R}$. Operations on $C(X)$ are defined for $f, g \in C(X)$, $a \in \mathbb{R}$, and $x \in X$ by

$$
\begin{aligned}
(f + g)(x) &= f(x) + g(x) \\
(fg)(x) &= f(x)g(x) \\
(af)(x) &= af(x)
\end{aligned}
$$

   (a) Verify that $C(X)$ is an algebra of functions.

   (b) Let $[a, b]$ be an interval in $\mathbb{R}$ and 1 and x the maps $[a, b] \to \mathbb{R}$ such that $1(c) = 1$ and $x(c) = c$ for all $c \in [a,]$. Show that the smallest subalgebra containing 1 and x is the set $\mathbb{P}$ of all polynomials in x with coefficients in $\mathbb{R}$.

   (c) Show that $\mathbb{P}$ is not a closed set of the metric space $C([a, b])$.

9. Let X be a compact metric space. Then

$$
d(f, g) = \sup_{x \in X} |f(x) - g(x)|
$$

is a metric on $C(X)$.

(a) Let $\mathbb{H}$ be a subalgebra of $C(X)$. Use the Stone-Weierstrass theorem as stated in Section 8.3 to show that $\mathbb{H}$ is dense in $C(X)$ if for $x \neq y \in X$ and $a, b \in \mathbb{R}$, there is an $f \in \mathbb{H}$ such that $f(x) = a$ and $f(y) = b$.

(b) Verify that if the subalgebra $\mathbb{H}$ contains the constant function 1 and separates points, then it satisfies the condition of the previous part.

10. Let X be a compact metric space. Verify that $C(X)$ is a lattice under the order $f \leq g$ if $f(x) \leq g(x)$ for all $x \in X$.

Chapter 9

POSSIBILITY THEORY

In this chapter, we are concerned with uncertainty measures in the context of fuzziness. We start out by making a distinction between "possible" and "probable" and present a mathematical theory of possibility for analyzing, modeling, and processing natural language information.

9.1   Additive and nonadditive set functions

First we review the basics of the theory of probability. This review has two purposes. It spells out a way to propose a mathematical model to capture a notion of uncertainty. Secondly, viewing probability measures as special cases of additive measures, we give the basics of measure theory which are necessary for reviewing integration theory.

Let Ω be a set. A collection $\mathcal{A}$ of subsets of Ω is a σ-**field** if

1. $\Omega \in \mathcal{A}$,

2. if $A \in \mathcal{A}$ then $A' \in \mathcal{A}$, and

3. if $A_n \in \mathcal{A}$ then $\bigcup_{n=1}^{\infty} A_n \in \mathcal{A}$.

If the last condition is required only for a finite number of the A_n's, then $\mathcal{A}$ is called a **field** or an **algebra** of sets. The pair $(\Omega, \mathcal{A})$ is called a **measurable space**. An interpretation of probability theory is that Ω denotes the sample space, that is, the collection of all possible outcomes of an experiment, and $\mathcal{A}$ is the set of events—**the measurable sets**.

The power set 2^{Ω} of Ω is a σ-field. It is the largest σ-field of subsets of Ω, but in particular there is always one such and any collection $\mathcal{C}$ of subsets is contained in a σ-field. The intersection of σ-fields is a σ-field, so there is a smallest σ-field containing $\mathcal{C}$, called the σ-**field generated** by C. When $\Omega = \mathbb{R}$, the σ-field generated by the collection of all open subsets of $\mathbb{R}$ is called the **Borel** σ-field of $\mathbb{R}$ and denoted $\mathcal{B}$.

Definition 9.1.1 *Let* $(\Omega, \mathcal{A})$ *be a measurable space. A function* $\mu : \mathcal{A} \to [0, \infty]$ *is a* **measure** *if*

1. $\mu(\varnothing) = 0$;

2. *if* A_n, $n = 1, 2, 3, \ldots$ *is a sequence of pairwise disjoint elements of* $\mathcal{A}$, *then*

$$\mu\left(\bigcup_{n=1}^{\infty} A_n\right) = \sum_{n=1}^{\infty} \mu(A_n)$$

The triple $(\Omega, \mathcal{A}, \mu)$ *is a* **measure space**.

The second property is called σ-**additivity**. If it is required for only finitely many A_n's, it is called **additivity**. A set A in $\mathcal{A}$ is an **atom** if $\mu(A) > 0$ and for every $B \in \mathcal{A}$ with $B \subset A$, $\mu(B) = 0$. The measure μ is **nonatomic** if there are no atoms in $\mathcal{A}$.

If $(E, \mathcal{E})$ is another measurable space and $X : \Omega \to E$ with

$$X^{-1}(A) = \{\omega : X(\omega) \in A\} \in \mathcal{A}$$

for all $A \in \mathcal{E}$, then $\mu X^{-1} : \mathcal{E} \to [0, \infty)$ is a measure on $(E, \mathcal{E})$, referred to as the **image measure** of μ by X. On $(\mathbb{R}, \mathcal{B})$ the measure μ such that $\mu((a, b)) = b - a$ is called the **Lebesgue measure** on $\mathbb{R}$. The measure space $(\Omega, \mathcal{A}, \mu)$ is a **probability space** if $\mu(\Omega) = 1$ in which case X is called a **random variable** and μX^{-1} is the **probability law** of X on E. Probability measures on $(\mathbb{R}, \mathcal{B})$ can be defined in terms of **probability density functions**. These are functions $f : \mathbb{R} \to [0, \infty)$ such that $\int_{-\infty}^{\infty} f(x)dx = 1$. The associated probability measure is given by $\mu_f(A) = \int_A f(x)dx$ for $A \in \mathcal{B}$. The measure μ for a probability space is usually denoted by P. Probability spaces serve as mathematical models to describe random experiments.

Uncertainty due to randomness occupies a large place in natural phenomena as well as in scientific problems. The additivity property is essential in modeling uncertainty. Indeed, even from a subjective viewpoint, where probabilities are assigned to events subjectively, these numbers should satisfy the additivity property. This is referred to as the *coherence principle*. As we will see, various nonadditive set functions will be proposed to model other types of uncertainty. In an introductory course in fuzzy logic in which the uncertainty studied is fuzziness rather than randomness, it seems useful to discuss briefly the possibility of using non-additive set functions for general uncertainty modeling. We would like to know to what extent nonadditive set functions are compatible with the coherence principle.

A fairly general formulation of the so-called **coherence principle** is as follows. A **score function** is a real valued function f defined on $[0,1] \times \{0,1\}$ satisfying the following:

- $f(x,0)$ and $f(x,1)$ are differentiable functions of x on $(0,1)$ with continuous derivatives. (We denote these derivatives by $f'(x,0)$ and $f'(x,1)$, respectively.)

- There is an interval $(x_{0,f}, x_{1,f}) \subseteq [0,1]$ such that $f'(x_{0,f},0) = f'(x_{1,f},1) = 0$.

- On $(x_{0,f}, x_{1,f})$, $f'(x,0) > 0$ and $f'(x,1) < 0$.

Usually, one can take $x_{0,f}$ to be 0 and $x_{1,f}$ to be 1. The two functions $f(x,0)$ and $f(x,1)$ can be interpreted as follows. If x is the uncertainty measure of some event E, then $f(x,1)$ represents the "penalty" if E occurs, and $f(x,0)$ measures the "penalty" if E does not occur.

Let $\mathcal{A}$ be a Boolean algebra of subsets of a set Ω. By an **uncertainty measure** μ we mean a set function $\mu : \mathcal{A} \to [0,1]$ such that $\mu(\varnothing) = 0$ and $\mu(\Omega) = 1$. For $E \in \mathcal{A}$, we write $E = 1$ if E occurs, and $E = 0$ if it does not. Thus, the "score" is $f(\mu(E), E)$.

Given a score function f and n events E_i, $i = 1, 2, ..., n$, we define a **game** as a triple (Γ, χ, L) where $\Gamma = \{0,1\}^n$ is the space of realizations of the E_i, $\chi = [0,1]^\Gamma$ is the space of uncertainty measures, and L is the loss function $\chi \times \Gamma \to \mathbb{R}$ given by $L(\mu, E) = \sum_{i=1}^n f(\mu(E_i), E_i)$, where $E = (E_1, E_2, ..., E_n)$, and with the convention that $E_i = 1$ or 0 according to whether or not it occurs. The function μ is **inadmissible** with respect to L if there is an uncertainty measure ν such that $L(\nu, E) \leq L(\mu, E)$ for all E, with strict inequality for some E. Otherwise, μ is **admissible** with respect to L.

For each score function f define the transform

$$P_f : [x_{0,f}, x_{1,f}] \to [0,1] : x \to \frac{f'(x,0)}{f'(x,0) - f'(x,1)}$$

By the regularity conditions on f, the function P_f is continuous. If, in addition, the $f'(x,i)$ are strictly increasing, then P_f is strictly increasing, and thus P_f^{-1} exists.

A necessary condition for μ to be admissible with respect to f and E, E', where E' is the complement of E, is that $P_f(\mu(E)) + P_f(\mu(E')) = 1$. It is easy to construct a score function f such that even if μ is a probability measure, μ is not f-admissible. Reasonable uncertainty measures should be ones which are admissible with respect to some score function. For example, any probability measure is admissible since it suffices to consider proper score functions f, that is, those f such that $P_f(x) = x$. We refer to this type of admissibility as "general admissibility". Thus an uncertainty measure is not general admissible if there is no score function f such that it is f-admissible. Of course, if one can find a score function f for which an uncertainty measure μ is f-admissible, then μ is general admissible.

Theorem 9.1.2 *Let A and B be two disjoint events, and let f be a score function. If the uncertainty measure μ is admissible with respect to $f, A, B, A \cup B$, then*

$$(P_f \circ \mu)(A \cup B) = (P_f \circ \mu)(A) + (P_f \circ \mu)(B)$$

That is, $P_f \circ \mu$ is additive.

Proof. Since A and B are disjoint, the only configurations of the events A, B, and $A \cup B$ are $(1,0,1)$, $(0,1,1)$, and $(0,0,0)$. Setting $x = \mu(A)$, $y = \mu(B)$, and $z = \mu(A \cup B)$, the possible total scores are

$$
\begin{array}{ccccc}
f(x,1) & + & f(y,0) & + & f(z,1) \\
f(x,0) & + & f(y,1) & + & f(z,1) \\
f(x,0) & + & f(y,0) & + & f(z,0)
\end{array}
$$

The admissibility of (x, y, z) implies that

$$\det \begin{pmatrix} f'(x,1) & f'(y,0) & f'(z,1) \\ f'(x,0) & f'(y,1) & f'(z,1) \\ f'(x,0) & f'(y,0) & f'(z,0) \end{pmatrix} = 0$$

Expanding this determinant across row three gives the equality

$$(P_f \circ \mu)(A \cup B) = (P_f \circ \mu)(A) + (P_f \circ \mu)(B)$$

■

Another necessary condition for admissibility is this: given three events A, B, and C and an uncertainty measure μ, let $\mu(B|C) = x$, $\mu(A|BC) = y$, and $\mu(AB|C) = z$. If (x, y, z) is admissible with respect to a score function f, then $P_f(z) = P_f(x)P_f(y)$. This follows from the fact that the total score function is $f(x, B)C + f(y, A)BC + f(z, AB)C$ and that there are three possible realizations of the sequence, namely $(1, 1, 1), (0, 1, 0)$, and $(1, 0, 0)$.

Theorem 9.1.3 *Let f be a twice differentiable score function. Then $P_f = 1$ for any nonatomic probability measure admissible with respect to f.*

Proof. If the nonatomic probability measure μ is admissible with respect to a twice differentiable score function f, then

$$P_f(t) + P_f(1 - t) = 1 \text{ for all } t \in [0, 1],$$
$$P_f(xy) = P_f(x)P_f(y) \text{ for all } x, y \in [0, 1]$$

Differentiating the second equation with respect to x and then with respect to y, we get $x(\log P_f(x))' = y(\log P_f(y))$, so that $x(\log P_f(x))'$ is a constant c. Hence $P_f(x) = x^c$, and the first equation forces $c = 1$. ∎

An uncertainty measure μ is **general admissible** if there exists a score function f such that $P_f \circ \mu$ is additive. If there is a function $\triangledown :$ $[0, 1] \times [0, 1] \rightarrow [0, 1]$ such that $\mu(A \cup B) = \mu(A) \triangledown \mu(B)$ whenever $A \cap B = \varnothing$, then the uncertainty measure μ is $\triangledown$-**decomposable**. By **admissibility** of μ we simply mean that $P_f \circ \mu$ is additive for some f such that P_f^{-1} exists.

In the following we consider score functions f such that P_f^{-1} exists on $[0, 1]$. All uncertainty measures that are admissible with respect to such score functions are $\triangledown$-decomposable with $\triangledown$ a continuous, Archimedean t-conorm with additive generator h satisfying $h(1) = 1$. For example, if f is a proper score function (so that $P_f(x) = x$), and P is a probability measure, then P is $\triangledown$-decomposable with $x \triangledown y = (x + y) \wedge 1$, which is a continuous, Archimedean t-conorm with additive generator $h(x) = x$.

Theorem 9.1.4 *An uncertainty measure μ is admissible if and only if*

1. *there exists a continuous, increasing $h : [0, 1] \rightarrow [0, 1]$ with $h(0) = 0$, $h(1) = 1$ and*

$$h \circ \mu(A) + h \circ \mu(B) \leq 1 \text{ if } A \cap B = \varnothing \tag{9.1}$$

2. *μ is $\triangledown_h$-decomposable where $\triangledown_h$ is a continuous, Archimedean t-conorm with additive generator h.*

Proof. Suppose that μ is admissible. Take $h(x) = P_f(x)$. Since $P_f \circ \mu(A \cup B) \leq 1$, inequality 9.1 is satisfied. Now 9.1 implies that

$$\mu(A \cup B) = P_f^{-1}[P_f \circ \mu(A) + P_f \circ \mu(B)]$$

Set

$$x \nabla_h y = P_f^{-1}[P_f \circ \mu(A) + P_f \circ \mu(B)]$$

where $x = \mu(A)$ and $y = \mu(B)$.

For the converse we have that for $A \cap B = \varnothing$,

$$\begin{aligned}
\mu(A \cup B) &= \mu(A) \nabla_h \mu(B) \\
&= h^*(h \circ \mu(A) + h \circ \mu(B)) \\
&= h^{-1}(h \circ \mu(A) + h \circ \mu(B))
\end{aligned}$$

using equation 9.1. Thus

$$h \circ \mu(A \cup B) = h \circ \mu(A) + h \circ \mu(B)$$

Now set $P_f(x) = h(x)$ and solve for f.   ∎

Corollary 9.1.5 *Let Ω be finite and $\pi : \Omega \to [0,1]$ with $\sum_{\omega \in \Omega} \pi(\omega) \leq 1$. Define $\mu_p : \mathcal{P}(\Omega) \to [0,1]$ by $\mu_p(A) = \nabla_p(\pi(A))$, where*

$$x \nabla_p y = [(x^p + y^p) \wedge 1]^{\frac{1}{p}}$$

Then for $p \geq 1$, μ_p is admissible.

Proof. It is easy to check that if $\{x_1, x_2, ..., x_n\} \cap \{y_1, y_2, ..., y_n\} = \varnothing$, then

$$\begin{aligned}
&x_1 \nabla x_2 \nabla \cdots \nabla x_n \nabla y_1 \nabla y_2 \nabla \cdots \nabla y_n \\
&= [x_1 \nabla x_2 \nabla \cdots \nabla x_n] \nabla [\nabla y_1 \nabla y_2 \nabla \cdots \nabla y_n]
\end{aligned}$$

so that μ_p is ∇_p-decomposable. Note also that $x \nabla_p y) = h_p^*(h_p(x) + h_p(y))$, where $h_p(x) = x^p$. For $A \cap B = \varnothing$,

$$h_p \circ \mu_p(A) + h_p(B) = \left[h_p^* \left(\sum_{\omega \in A} \pi^p(\omega) \right)^p + h_p^* \left(\sum_{\omega \in B} \pi^p(\omega) \right)^p \right]^{\frac{1}{p}}$$

But since $\sum_{\omega \in \Omega} \pi(\omega) \leq 1$, we have $h_p \circ \mu_p(A) + h_p \circ \mu_p(B) \leq 1$. The result follows then from the last theorem.   ∎

If μ is such that $\mu(A \cup B) = \mu(A) \vee \mu(B)$ whenever $A \cap B \neq \varnothing$, then μ is inadmissible. This follows from the observation that since P_f is nondecreasing,

$$\begin{aligned}
P_f \circ \mu(A \cup B) &= P_f(\mu(A) \vee \mu(B)) \\
&= (P_f \circ \mu(A)) \vee (P_f \circ \mu(B)) \\
&< P_f \circ \mu(A) + P_f \circ \mu(B).
\end{aligned}$$

This is also explained by the fact that $\vee$ is not an Archimedean t-conorm. However, such μ are uniform limits of **admissible measures.** Indeed, we consider the proof of the previous corollary. Since $p \geq 1$,

$$
\begin{aligned}
\vee\{x_1, x_2, ..., x_n\} &\leq \nabla_p(x_1, x_2, ..., x_n) \\
&\leq \vee\{x_1, x_2, ..., x_n\} + n^{\frac{1}{p}} - 1
\end{aligned}
$$

It follows that

$$
\vee\{x_1, x_2, ..., x_n\} = \lim_{p \to \infty} \nabla_p(x_1, x_2, ..., x_n)
$$

and hence for $A \subseteq \Omega$, $\mu_p(A) = \nabla_p(\pi(A))$ converges to $\vee\{\pi(\omega), \omega \in A\} = \mu(A)$ uniformly in A.

As another example of admissible uncertainty measures, consider a finite set Θ and a probability measure P defined on its power set. Then for any positive integer n, the set function P^n is a belief function which is general admissible. (Belief functions are discussed in Chapter 10.) Indeed, for $n = 2$, let $\Theta = \{\theta_1, \theta_2, ..., \theta_k\}$ and define $m : 2^\Theta \to [0, 1]$ by $m(\{\theta\}) = P^2(\{\theta\})$ for all $\theta \in \Theta$, $m(\{\theta, \eta\}) = 2P(\{\theta\})P(\{\eta\})$ for $\theta \neq \eta$, and $m = 0$ otherwise. Then

$$
\begin{aligned}
\sum_{A \subseteq B} m(A) &= \sum_{\theta, \eta \in B} P(\{\theta\}) P(\{\eta\}) \\
&= \left(\sum_{\theta \in B} P(\{\theta\}) \right)^2 \\
&= P^2(B)
\end{aligned}
$$

Let $F(B) = \sum_{A \subseteq B} m(A)$. Then $F(\Theta) = 1$, F is a belief function on $(\Theta, 2^\Theta)$, and for any subset B of Θ, $F(B) = P^2(B)$.

In summary, if functions of additive set functions are considered as "admissible" for modeling uncertainty, then there is space for nonadditive set functions as candidates for uncertainty measures. Of course, the question remains: what types of uncertainty need such mathematical models?

9.2 Possibility

This section discusses the following main points:

- possibility as a quantitative representation of knowledge

- a distinction of possibility from subjective prior information in the Bayesian approach to decision analysis

- the mathematical concept of possibility and its calculus based on fuzzy sets

In ordinary everyday language, we often hear statements such as "It is *possible* but not *probable* that you will win the next lottery." Not only does such information carry two different types of uncertainty, but also it indicates a relationship between the two, namely that one is stronger than the other. In a statistical model, sets of values of a random variable X are quantified by probabilities. For example, if f is a discrete density, then $f(x)$ is the probability that X will take the value x. In such a case, the concept of "possibility" is unnecessary.

A common situation is that X is a random variable with density $f(x, \theta)$, where f is known up to a parameter θ, but the true value θ_0 of that parameter is not known. However, θ_0 is in some known set Θ, the **parameter space**. There is nothing random about it—θ_0 is not the outcome of an experiment. The information that $\theta_0 \in \Theta$ is deterministic, simply serving to specify the model. Every element of Θ is a possible candidate. They have the same degree of "possibility". One viewpoint is that this is a special case of a Bayesian model in which a probability distribution is put on the parameter space. For example, suppose that Θ is the interval $[a, b] \subset (-\infty, \infty)$. One can put the uniform density $f(x) = \frac{1}{b-a}\chi_{[a \leq x \leq b]}$ on $[a, b]$ giving the flavor that the $x \in \Theta$ are all of equal possibility. But what happens if Θ is an unbounded subset of $\mathbb{R}$ and we wish to express that each $x \in \Theta$ have the same degree of possibility? In such a case, the **possibility representation** using the indicator function χ_Θ is natural and does not raise any technical problems. If $\Theta = [0, \infty)$, for example, there is no probability distribution on Θ which expresses equal possibility for all $x \in \Theta$.

The Bayesian viewpoint is that if information besides that gotten from sampling is available, it should be used. Such information is usually subjective, such as knowledge furnished by an expert. But since the analysis of all the information is to be based on probability calculus, this additional information has to be represented probabilistically. Whether this is always possible is another issue. In the case we are discussing, the situation is rather simple. The knowledge is about the location of an unknown parameter $\theta_0 \in \Theta$. When Θ is a bounded interval of $\mathbb{R}$, the indicator function χ_Θ, or the uniform density function on Θ, gives the same information. If the additional information points to the fact that not all elements are "equally possible", but some are more "plausible" than others, then the probabilistic knowledge required for a Bayesian model is a (prior) probability density on Θ. Thus the Bayesian viewpoint is that θ_0 is a random variable with a density f. But how is f obtained? This is always a question for debate. Eliciting such a density from experts may be difficult. Even so, just as in the uniform case, having such a density is nothing more than quantifying different degrees of possibility for elements of Θ as candidates for θ_0.

With the information that $\theta_0 \in \Theta$, one can proceed directly to quantify different degrees of beliefs by a generalized indicator function, that is, by a fuzzy subset of Θ. Such a function $\varphi : \Theta \to [0, 1]$ has a clear semantic and can be obtained from experts by asking questions and giving numerical scales for answers. The value $\varphi(\theta)$ is the degree of possibility that θ can be θ_0. For example, if $\Theta = [0, \infty)$ and it is known that "θ is small", we model the label "small" as a fuzzy subset $\varphi : [0, \infty) \to [0, 1]$ of $[0, \infty)$ and take $\varphi(\theta)$ as a "possibility" distribution on Θ. An important point is that we are focusing on one specific value θ_0, while in statistics, a population is targeted rather than individuals.

In summary, the quantification of degrees of possibility for representing location information is natural without any postulating on the nature of the unknown parameter. In particular, there is no need to view such a parameter as a random variable, and some technical problems of the Bayesian approach are avoided. But next comes the problem of the calculus of the entities used to represent information. In the Bayesian framework, given a subset A of Θ, one can calculate the probability that θ_0 is in A just by integrating the density on Θ over the set A. If a possibility distribution φ is used, then how does one compute the degree of possibility $\mathrm{Poss}(A)$ that A contains θ_0? It is done using a principle analogous to the maximum probability principle in decision analysis, namely

$$\mathrm{Poss}(A) = \vee \{\varphi(\theta) : \theta \in A\}$$

The function Poss is not additive, but does obviously satisfy

$$\mathrm{Poss}(A \cup B) = \mathrm{Poss}(A) \vee \mathrm{Poss}(B)$$

Set-functions having this property are referred to as *maxitive set-functions* in Chapter 11. In fact, possibility measures satisfy much stronger conditions, as we will see in the definitions below.

As we have seen, such functions are limits of admissible uncertainty measures. Thus they can be used as an approximation operation.

Definition 9.2.1 *A **possibility distribution** on a set Θ is a function $\varphi : \Theta \to [0, 1]$ such that $\vee \{\varphi(\theta) : \theta \in \Theta\} = 1$.*

Definition 9.2.2 *A **possibility measure** on Θ is a function $\pi : 2^\Theta \to [0, 1]$ satisfying*

    1. $\pi(\varnothing) = 0$, and $\pi(\Theta) = 1$;

2. *for any family* $\{A_i\}_{i \in I}$ *of subsets of* Θ, $\pi\left(\bigcup_{i \in I} A_i\right) = \vee\{\pi(A_i) : i \in I\}$.

The restriction of π *to the one element subsets of* Θ, *which gives a map* $\Theta \to [0, 1]$, *is the* **possibility distribution associated with** π.

Property 2 deserves some comments. For I finite, it is the maxitivity mentioned above. This property is an enforcement of a well-known property of σ-additive measures, namely, if μ is a measure, then $\mu(A \cup B) = \mu(A)$ for all B such that $\mu(B) = 0$. It is clear that a set-function μ is maxitive if and only if $\mu(A \cup B) = \mu(A)$ for all B such that $\mu(B) \leq \mu(A)$.

Property 2 really distinguishes between maxitive set-functions and ordinary measures. This is somewhat similar to the distinction between measures and *dimensions*. In dimension theory, (see, for example, [24]) the property of maxitivity is called *stability, or finite stability*. *Box-counting dimension* is stable. As it will be shown in Chapter 11, stable or maxitive set-functions are alternating of infinite order, and hence are related to special Choquet capacities characterizing distributions of random sets. When the index set I is countable, the property is called *countable stability*. This is a property shared by all concepts of dimension, for example, Hausdorff dimension (see Chapter 11) and packing dimensions.

Technically, possibility distributions are special kinds of fuzzy sets. From a practical viewpoint, they represent a weak form of knowledge and can be encoded easily. Typically, possibility distributions are used to represent knowledge expressed in natural language such as "The speed of this car is high." This induces a possibility distribution π on the universe of discourse, say $U = [0, 100]$ of the variable "speed", where π is taken to be the membership function of the fuzzy subset "high" of U. Inferences with this type of knowledge will be discussed in the next section. In summary, the theory of possibility is applicable to real-world problems via the theory of fuzzy sets as a mathematical model for vagueness.

Probabilistic interpretation of possibility distributions

As in the case of membership functions of fuzzy concepts, possibility distributions have a probabilistic interpretation. This should not be taken to mean that possibility theory is subsumed by probability theory.

Let $f : U \to [0, 1]$ be a possibility distribution. Let $(\Omega, \mathcal{A}, P)$ be a probability space and let $\alpha : \Omega \to [0, 1]$ be a random variable uniformly distributed on $[0, 1]$. Consider the random set $f_\alpha : \Omega \to 2^U$ defined by

$$f_\alpha(\omega) = \{u \in U : f(u) \geq \alpha(\omega)\}$$

That is, we randomly select the α-cuts of f. Then

$$f(u) = P\{\omega : u \in f_\alpha(\omega)\}$$

In this form, a possibility distribution appears as a **one-point coverage function** of a random set. It is well known that, similar to moments of random variables, one or **many-point coverage functions** of a random set do not determine the probability measure governing the random evolution of that random set.

Possibility measures and imprecise data

Possibility measures appear also in the analyzing of **imprecise data**. Such data arises, for example, when the outcomes of an experiment cannot be obtained precisely, but only are located in subsets of the sample space. Random sets are employed in this situation.

Let U be a finite set and S a random set defined on the probability space $(\Omega, \mathcal{A}, P)$ with values in 2^U. So $S : \Omega \to 2^U$. Suppose that $\varnothing \neq A_1 \subseteq A_2 \subseteq \ldots \subseteq A_n \subseteq U$, that $\alpha_i = P\{\omega : S(\omega) = A_i\}$, and that $\sum \alpha_i = 1$. Let $\pi : 2^U \to [0,1]$ be defined by

$$\pi(A) = 1 - \sum_{A_i \subseteq A'} \alpha_i$$

Then π is a possibility measure. That $\pi(\varnothing) = 0$ and $\pi(\Omega) = 1$ are clear. Let $\{B_j : j \in I\}$ be a family of subsets of U. Then

$$\pi\left(\bigcup_{j \in J} B_j\right) = 1 - \sum_{A_i \subseteq \left(\bigcup_{j \in J} B_j\right)'} \alpha_i$$

$$= 1 - \sum_{A_i \subseteq \bigcap B_j'} \alpha_i$$

Since the A_i form a chain, those A_i with $A_i \subseteq \bigcap B_j'$ are those contained in B_k' containing the smallest number of A_i's, that is, the B_k which maximizes $\pi(B_k)$. Thus

$$\pi\left(\bigcup_{j \in J} B_j\right) = \vee\{\pi(B_j) : j \in J\}$$

and π is a possibility measure.

Let $F(A) = \sum_{A_i \subseteq A} \alpha_i$. This is the measure **dual to** π and is called a **necessity measure**. (See Exercise 8 below.)

The localization character of possibility measures

Let $(\Omega, \mathcal{A}, P)$ be a probability space. The value $-c \log P(A)$ for $c > 0$ is an **information measure** about the precision of A in the sense that it is a decreasing function on the elements of $\mathcal{A}$. An information measure need not be defined in terms of probability measures. (See, for example [42].) If μ is an information measure such that

1. $\mu(\varnothing) = \infty$ and $\mu(\Omega) = 0$, and

2. $\mu\left(\bigcup_I A_i\right) = \inf_I(\mu(A_i))$ for all index sets I,

then $1/\mu$ is a possibility measure.

9.3   Possibility theory based on fuzzy sets

With the concept of fuzzy sets, the concept of possibility can be made practical for representing and processing vague knowledge. Consider information given by a proposition of the form "X is A", where X is the name of a subject and A is a linguistic label in natural language viewed as a fuzzy subset of U. For example, consider the proposition "Brigitte is young". Let $a(x)$ denote the implicit attribute $a(\text{Brigitte}) = age$. Here, $A = \text{young}$ is a fuzzy subset of $U = [0, 100]$. What can be said about Brigitte's age given only the proposition "Brigitte is young". The fuzzy set A serves as an elastic constraint on the possible values of $a(x)$. The fuzzy set A plays the role of a possibility distribution for the variable $a(x)$. Thus, the proposition $p = $ "X is A" induces a possibility distribution $\pi_p(x) = A(x)$ on U if $\vee_x\{A(x)\} = 1$. Every fuzzy set is a possibility distribution if the sup of its values is 1. The interpretation is this. If the available information about Brigitte's age is the vague label $A = $ "young", then for each $x \in U$, $A(x)$ gives the degree of possibility that Brigitte's actual age is x. Of course, A is a *mathematical model* of the fuzzy concept "young".

The procedure above of extracting possibility distributions can also be viewed as a **meaning representation** in natural language. When a proposition $p = $ "X is A" contains implied attributes $a_i(x), i = 1, 2, ..., n$, the fuzzy subset A of the Cartesian product $U_1 \times U_2 \times ... \times U_n$ is a fuzzy relation. The meaning representation of p is expressed as the **joint possibility distribution**

$$\pi(u_1, u_2, ..., u_n) = A(u_1, u_2, ..., u_n)$$

In particular, if $A = A_1 \times A_2 \times ... \times A_n$, then

$$\pi(u_1, u_2, ..., u_n) = \vee \{A_i(u_i)\}$$

In general, from the joint possibility distribution π, one obtains various **marginal possibility distributions** by projections on suitable spaces. The possibility distribution of $(a_i(x), i \in I)$, where $I \subseteq \{1, 2, ..., n\}$ is taken to be

$$\pi_I(u_i, i \in I) = \vee \pi(u_1, u_2, ..., u_n)$$

where $\vee$ is taken over all u_j's for $j \notin I$. The meaning representation of fuzzy propositions via possibility distributions is used in approximate reasoning, the topic of the next section. For "If...then" statements, a concept of **conditional possibility distributions** might be needed. Unlike conditional probability densities, the issue of conditions in possibility theory is somewhat unsettled. For example, the conditional possibility distribution of a variable X given $Y = y$ should be of the form

$$\pi(x|y) = \pi(x, y) * \pi_y(x)$$

where $\pi(x, y)$ and π_y are the joint and the marginal possibility distributions, respectively, and $*$ is an operation such as $\wedge$ or product. For **conditional possibility measures**, the situation is the same.

9.4 Approximate reasoning

Approximate reasoning refers to processes by which imprecise conclusions are inferred from imprecise premises. When imprecision is fuzzy in nature, the term "fuzzy reasoning" is also used. We restrict ourselves to the formulation of some deductive procedures such as modus ponens and modus tollens. Some further general aspects of approximate reasoning will be discussed briefly.

Deduction processes are used in everyday human reasoning as well as in mathematics and other sciences. The most common formal deduction process has the following structural pattern: if we know that a implies b and we know a, then we conclude b. This pattern of reasoning is called **modus ponens**. In propositional calculus, as discussed in Chapter 4, if a and b are formulas and $a \Rightarrow b$ is true and a is true, then b is true. Modus ponens is written as follows:

$$\begin{array}{ccc} a & \Rightarrow & b \\ a & & \\ \hline b & & \end{array}$$

This argument is valid if for any assignment of truth values to a and b that make both $a \Rightarrow b$ and a true, then b is true. Now $a \Rightarrow b$ is material implication, which is $a' \cup b$ and its truth table is

a	b	$a \Rightarrow b$
T	T	T
T	F	F
F	T	T
F	F	T

From this we do see that whenever $a \Rightarrow b$ and a are true then so is b.

We wish to formulate the modus ponens pattern in a way suitable for generalization to the fuzzy case. The strategy is standard: identify sets with their indicator functions and generalize to fuzzy sets.

Consider $p = $ "$x \in a$", $q = $ "$y \in b$", where a and b are subsets of U and V, respectively. We write $a\,(x)$ and $b(x)$ for their indicator functions. Material implication $p \Rightarrow q$ is the relation $R = (a \times b) \cup (a' \times V) \subseteq U \times V$. Since b is the image of the projection of $a \times b$ into V, then $b(y) = \bigvee_{x \in U}(a \times b)(x, y)$. But

$$a \times b = [(a \times b) \cup (a' \times V)] \cap (a \times V)$$

so that

$$
\begin{aligned}
(a \times V)(x, y) &= \{a(x) \wedge V(y)\} \\
&= \{a(x) \wedge 1\} \\
&= a(x)
\end{aligned}
$$

we have

$$
\begin{aligned}
b(y) &= \bigvee_x \{R(x, y) \wedge (a \times V)(x, y)\} \\
&= \bigvee_x \{R(x, y) \wedge a(x)\}
\end{aligned}
$$

Thus, if we view a as a unary relation on U, then b is the $\max - \min$ composition $b = R \circ a$ of R with a. We also write $b = R\,(a)$.

Consider next a modified version of the modus ponens. Suppose $p \Rightarrow q$, but instead of having $p = $ "$x \in a$", we have $p^* = $ "$x \in a^*$", where a^* is a subset of some set X which might be different from a. The question is "what is the conclusion b^*" in

$$
\begin{array}{c}
p \Rightarrow q \\
\underline{p^*} \\
?
\end{array}
$$

By analogy, we take

$$R(a^*)(y) = \bigvee_x \{R(x,y) \wedge a^*(x)\}$$

for b^*. It can be checked that

$$R(a^*) = \begin{cases} b \text{ if } a^* \subseteq a \\ V \text{ otherwise} \end{cases}$$

From this, we have $R(a^*) = b$ when $a^* = a$.

This situation is similar to another form of deduction known as **modus tollens**. The structural pattern of this rule of deduction is

$$\frac{\begin{array}{c} p \Rightarrow q \\ q' \end{array}}{p'}$$

That is, if $p \Rightarrow q$ is true and q is false, then p is false. In terms of sets, let for $p =$ "$x \in a$" and $q =$ "$y \in b$". Then $(x,y) \in R = (a \times b) \cup (a' \times V)$ and $y \notin b$ imply that $(x,y) \in (a' \times V)$, and hence $x \in a'$. Thus p is false.

The conclusion a' from

$$\begin{cases} \text{"If } x \in a \text{ then } y \in b\text{"} \\ \text{"}y \in b\text{"} \end{cases}$$

is obtained as follows. The projection in V of

$$[(a \times b) \cup (a' \times V)] \cap (V \times b') = R \times (V \times b')$$

is a', so that

$$a'(x) = \bigvee_{y \in V} \{R(x,y) \vee b'(y)\}$$

For convenience, we denote this expression as $R^{-1}(b')$ so that by definition

$$R^{-1}(b')(x) = \bigvee_{y \in V} \{R(x,y) \vee b'(y)\}$$

Now if the fact turns out to be "$y \in b^*$" with $b^* \neq b'$ then the new conclusion is $a^* = R^{-1}(b^*)$

9.5 Approximate reasoning in expert systems

We saw above that it is possible to use control experts' knowledge to arrive at control laws. This is a special case of a much broader and ambitious perspective in this era of artificial intelligence—construction of machines capable of exhibiting human behavior in reasoning. It is well known that humans or experts in specific domains use very general forms of knowledge and different schemes of reasoning to solve problems. The attempt to mimic experts' behavior leads to expert systems. From a practical point of view, by an **expert system** we simply mean a computer system which can act like a human expert in some domain to reach "intelligent" decisions. Interested readers might consult Giarratono and Ripley [37] for a tutorial introduction to the field of expert systems. However, no prior knowledge about expert systems is needed to read this section and the next, since we focus our discussion only on the motivation for using fuzzy logic in knowledge representation and reasoning in a general setting.

Solving problems in human activities such as perception, decision making, language understanding, and so on requires knowledge and reasoning. These two main ingredients are obviously related, as we reason from our knowledge to reach conclusions. However, knowledge is the first ingredient to have, and we first address the problem of representing knowledge obtained, propose as associated reasoning procedure capturing, to a certain extent, the way humans reason. When knowledge is expressed by propositions that can be either true or false, classical binary logic is appropriate. In this case, the associated reasoning procedure will be based upon inference rules such as modus ponens and modus tollens. When this knowledge is uncertain in the sense that truth values of propositions are not known, other knowledge representation schemes should be called upon. For example, in the context of rule-based systems where knowledge is organized as rules of the form "If ... then ..." and propositions, both rules and facts might be uncertain. This is the case, for example, in the rule "If the patient coughs, then bronchitis is the cause" and the fact "The patient coughs". In this case, knowledge can be represented in the framework of probability theory, and the associated reasoning is termed **probabilistic reasoning** in which the rules of the calculus of probabilities are applied to derive "confidence" in possible decisions. More generally, when the concept of human belief is emphasized, the associated reasoning can be formulated using the theory of evidence

Observe that in typical decision problems, such as in a medical diagnosis problem, domain knowledge contains uncertainties coming from various different sources. To name a few, uncertain knowledge is due to imprecision, incomplete information, ambiguity, vagueness or fuzziness, and randomness. Thus *reasoning under uncertainty* is the main modeling task for building expert systems. Also, general knowledge is qualitative in nature, that is, is expressed in natural language in which concepts, facts, and rules are intrinsically fuzzy. For example, "high temperatures tend to produce high pressure" is such a general bit of knowledge. But it is well known that this type of qualitative, imprecise information is useful in everyday decision processes. And humans do reason with this common type of knowledge. Of course, from imprecise knowledge, we can expect only imprecise conclusions. The reasoning based on this principle is called *approximate reasoning.* In the next section we will consider the case where domain knowledge contains fuzzy concepts in a natural language, so that an appropriate candidate for modeling is the theory of fuzzy logic.

The context in which a theory of approximate reasoning seems indispensable is manifested by domain knowledge containing uncertainty and imprecision. This is the case when linguistic descriptions of knowledge contain

1. *fuzzy predicates* such as small, tall, high, young, etc.,

2. *predicate quantifiers* such as most, several, many, few, etc.,

3. *predicate modifiers* such as very, more or less, extremely, etc.,

4. *fuzzy relations* such as approximately equal, much greater than, etc.,

5. *fuzzy truth values* such as quite true, very true, mostly false, etc.,

6. *fuzzy probabilities* such as likely, unlikely, etc., and

7. *fuzzy possibilities* such as impossible, almost impossible, etc.

As first order logical systems extend propositional calculus to include the quantifiers $\forall$ and $\exists$, and predicates, fuzzy logic also deals with vague predicates, for example fuzzy propositions of the form "John is young", where the vague predicate "young" is modeled as a fuzzy subset of an appropriate space, as well as nonextreme quantifiers such as "most", "few", and linguistic hedges such as "much", "highly", "very", and "more or less". Also, fuzzy relations are modeled by fuzzy sets and are parts of the language of fuzzy logic.

As a generalization of exact predicates, a fuzzy predicate, such as "young" can be identified with the fuzzy subset "young" on a set S of people. Similarly, a binary fuzzy predicate such as "looks like" can be modeled as a fuzzy subset of S^2, that is, as a **fuzzy relation** on S. Thus, fuzzy predicates are fuzzy relations.

The quantifiers like "most", "few", and "some" are related to "fuzzy proportions". For example, consider "most birds fly". This fuzzy proposition is of the form "most a's are b's with a and b being the crisp sets "birds" and "flying animals", respectively. The common sense is that although a is not quite contained in b, it is almost so. So, formally, "most"

is a measure of the degree to which a is contained in b. The proportion of
elements of a which are in b is $\#(a \cap b)/\#(a)$, say in the finite case, where
$\#$ denotes cardinality. Thus, when $a \subseteq b$, we get "all" or 100%. When
we do not know exactly $\#(a \cap b)$, a subjective estimate of $\#(a \cap b)/\#(a)$
can be numerical. If the estimate is, for example, "most", then, as we
already mentioned, we have a choice between specifying the meaning of
the fuzzy quantifier "most" by a statistical survey or by fuzzy sets of the
unit interval $[0, 1]$. More generally, when a and b are fuzzy concepts such
as "young student" and "people without a job", and the proposition is
"most young students do not have a job", we are talking about "inclusion of one fuzzy set in another" with degrees. Of course, if $\mu_a \leq \mu_b$ then
we get the quantifier "for all". As far as cardinality and propositions
are concerned, one can extend these to the fuzzy case, using membership
functions. For example, let S be a finite set. For a fuzzy subset a of S,
one can generalize the concept of cardinality of crisp sets as follows:

$$\#(a) = \sum_{x \in S} \mu_A(x)$$

$$\#(b/a) = \frac{\#(a \cap b)}{\#(a)}$$

$$= \frac{\sum \mu_a(x) \wedge \sum \mu_b(x)}{\sum_{x \in S} \mu_a(x)}$$

Then "most a are b" is translated into "$\#(b/a)$ is most". Perhaps the
notion of degrees of inclusion of a fuzzy set a in a fuzzy set b provides
a clear semantics for fuzzy quantifiers. "Fuzzy inclusion" should not be
confused with inclusion of fuzzy sets.

We emphasize that the modeling of the concepts above by fuzzy sets
reflects a sort of gradual quantification, closely related to the meaning of
the linguistic labels, in the spirit that in natural language, meaning is a
matter of degree. Fuzzy logic, as a logic for vague propositions, is quite
different from probability logic.

Below are some examples of fuzzy modeling of quantifiers and hedges.

$$m(x) = \begin{cases} 0 & \text{if } x < 0 \\ 4x^2 & \text{if } 0 \leq x \leq 0.5 = \text{"most"} \\ 1 & \text{if } 0.5 < x \end{cases}$$

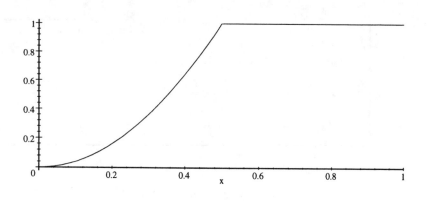

A fuzzy modeling of "most"

$$F(x) = \begin{cases} 1 & \text{if } x < 0 \\ 1 - 12x^2 + 16x^3 & \text{if } 0 \le x \le 0.5 \quad = \text{"few"} \\ 0 & \text{if } 0.5 < x \end{cases}$$

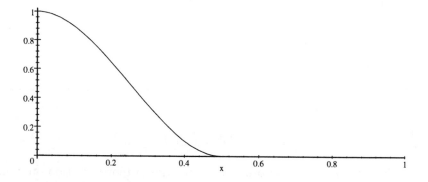

A modeling of "few"

$$t(x) = \begin{cases} 0 & \text{if } x < 5.8 \\ (1/0.7)\,x - 5.8/0.7 & \text{if } 5.8 \le x \le 6.5 \quad = \text{"tall"} \\ 1 & \text{if } 6.5 < x \end{cases}$$

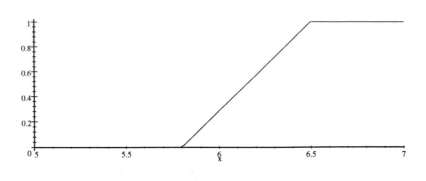

A fuzzy modeling of "tall"

From a membership functions $t(x)$ for "tall", a function commonly used to model "very tall" is $(t(x))^2$.

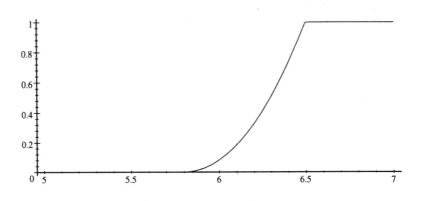

A fuzzy modeling of "very tall"

We describe now the problem of knowledge representation on which inference rules are based. Consider the simplest form, called the **canonical form**, of a fuzzy proposition, such as $P =$ "John is tall". A meaning representation of P is obtained by translating P into a mathematical equation. The *implicit variable* in P is $X =$ height of John. This variable is in fact a *linguistic variable* since its values could be tall, very, tall, short, and so on. The proposition P specifies a value of X, namely "tall". Under the "constraint" $A =$ "tall", X is viewed as a numerical variable with values in a set $S = [0, 10]$, say, and A acts as an elastic restriction on possible values of X. Now, suppose we model A by a fuzzy subset of U by specifying a membership function, still denoted by A. Then for each

$x \in U$, the assigned value $X = x$ is compatible with P to the degree $A(x)$. In other words, the possibility that $X = x$ is $A(x)$. Thus P induces a *possibility distribution* on the variable X, given by

$$\Pi_X : S \to [0, 1] : x \to A(x)$$

The translation of "X is A" into this possibility assignment equation is symbolized by "X is A" $\to \Pi_X = A$.

In the example above, the fuzzy predicate "tall" is somewhat simple, so that the implicit variable X is clearly related only to height. In more general cases, several implicit variables can be attributed to a fuzzy predicate. For example, when the linguistic variable "appearance" takes on the value "beautiful", several factors are responsible for the concept "beautiful", so that X is multidimensional.

Next, consider $P =$ "X is mA" where m is a modifier such as "not", "very", or "more or less". Then P is translated into $\Pi_X = A^+$, where A^+ is a fuzzy modification of A induced by m. For example, if $m =$ "not", then $A^+(x) = 1 - A(x)$, if $m =$ "very", then $A^+(x) = [A(x)]^2$, and if $m =$ "more or less", then $A^+(x) = \sqrt{A(x)}$.

Consider next compound propositions such as $P =$ "X is A" and "Y is B" where A and B are fuzzy subsets of $\mathcal{X}$ and $\mathcal{Y}$, respectively. P is translated into

$$\Pi_{(X,Y)} = A \times B$$

That is, the joint possibility distribution of (X, Y) is identified with the membership function of the Cartesian product of fuzzy sets by

$$\Pi_{(X,Y)}(x, y) = A(x) \bigwedge B(y)$$

When the logical connective is "or" instead of "and", the possibility assignment equation becomes

$$\Pi_{(X,Y)}(x, y) = A(x) \bigvee B(y)$$

A conditional proposition of the form "If X is A then Y is B" induces a *conditional possibility distribution* $\Pi_{(Y|X)}$ which is identified with a fuzzy implication $A \Rightarrow B$ given by

$$\Pi_{(Y|X)}(y|x) = A(x) \Rightarrow B(y)$$

where J is a fuzzy implication operator. Since there are various interpretations of fuzzy conditionals, in a given problem some choice of J should be made.

As an example of propositions involving fuzzy relations, consider $P =$ "X and Y are approximately equal". When we specify the binary fuzzy relation R = "approximately equal" by a membership function, the proposition P induces a joint possibility distribution for (X, Y) as

$$\Pi_{(X,Y)}(x, y) = R(x, y)$$

We turn now to *quantified propositions*.

9.5.1   Fuzzy syllogisms

Consider P = "QA's are B's". "Most tall men are fat" is an example. Q stands for a fuzzy quantifier such as "most". The concept of fuzzy quantifier is related to that of cardinality of fuzzy sets. Specifically, Q is a fuzzy characterization of the relative cardinality of B in A. When A and B are finite crisp sets, then the relative cardinality of B in A is the proportion of elements of B which are in A, that is, the ratio $\#(A \cap B)/\#(B)$. In this case, the associated implicit variable is $X = \#(A \cap B)/\#(A)$, so that "$Q\,A's$ are $B's$" is translated into $\Pi_X = Q$, where Q is specified by its membership function. For example, the membership function for Q = "most" might look like

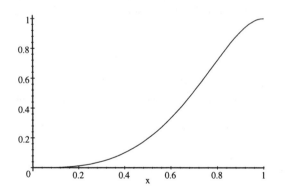

The notion of cardinality can be extended to fuzzy sets by defining $\#(A) = \sum_{x \in U} A(x)$ where A is a fuzzy subset of a finite set U. Thus the relative cardinality of the fuzzy set B is the fuzzy set A can be taken as the ratio $\#(A \cap B)/\#(A)$ where $\cap$ is modeled by a t-norm such as $T(x, y) = x \wedge y$. So, in general, "$Q\,A$'s are B's" is translated into

$$\Pi_{\#(A \cap B)/\#(A)}(x) = Q(x)$$

9.5.2   Truth qualification

Consider P ="X is A is τ", for example "John is young is very true", where τ is a linguistic truth value such as "very true", and is modeled as a fuzzy subset of $[0,1]$. This type of proposition is translated into $\Pi_X(x) = \tau(A(x))$.

9.5.3 Probability qualification

Consider $P =$"Tony is young is very likely", which is of the form "X is A is Λ" where Λ is a linguistic or fuzzy probability value which is modeled as a fuzzy subset of $[0,1]$. Note that, for simplicity, we denote the implicit variable by X, which is in fact the implied attribute of Tony, namely $X =$ age of Tony. The concept "young" is modeled as a fuzzy subset of $\mathcal{X} = [0,100]$. P can be rephrased as "The probability of (X is A) is Λ". Now A is a fuzzy event, that is, a fuzzy set whose membership function is measurable, so that within the basic framework of probability theory with X considered as a random variable with density f, we have

$$P(A) = \int_{\mathcal{X}} A(x)f(x)dx$$

This probability of the fuzzy event plays the role of the variable of interest and has the possibility distribution given by

$$\Pi_{P(A)}(u) = \Lambda(u)$$

for $u \in [0,1]$.

9.5.4 Possibility qualification

Consider $P =$ "X is A is possible". This is translated into $\Pi_X(x) = [A(x), 1]$. Note that the possibility distribution of X is interval-valued.

From the knowledge representation above, reasoning with possibility distributions can be carried through using various rules of inference as in 9.4.

9.6 A simple form of generalized modus ponens

In everyday life, we usually employ the following type of reasoning:

- It is known that "If a tomato is red, then it is ripe". Then if we see a red tomato, we will classify it as ripe.

- From "If x is large, then y is small" and "x is large" we deduce that "y is small".

The only difference from classical modus ponens is that the predicates involved are linguistic labels in natural language, and thus are modeled as fuzzy subsets of the appropriate sets. From

$$p = \text{``}x \text{ is } a\text{''}$$
$$q = \text{``}y \text{ is } b\text{''}$$

where a and b are fuzzy subsets of U and V, and x and y are variables taking values in U and V, respectively, we have the schema

$$
\begin{array}{ll}
p \Rightarrow q & : \text{``if } x \text{ is } a \text{ then } y \text{ is } b\text{''} \\
\underline{p \qquad\qquad} & \underline{: \text{``}x \text{ is } a\text{''} \qquad\qquad\qquad\qquad} \\
q & : \text{``}y \text{ is } b\text{''}
\end{array}
$$

Now $p \Rightarrow q$ is a fuzzy conditional R, and can be modeled by a fuzzy implication operator $\Rightarrow$. R is a fuzzy subset of $U \times V$ with membership function

$$R(x, y) = a(x) \Rightarrow b(y)$$

Note that in the crisp case,

$$
\begin{array}{rl}
\Rightarrow & : \{0, 1\} \times \{0, 1\} \rightarrow \{0, 1\} \\
u \Rightarrow v & = (u \wedge v) \vee (1 - u) \\
b(y) & = \bigvee_{x \in U} \{R(x, y) \wedge a(x)\}
\end{array}
$$

Now in the fuzzy case, we can look at

$$R(a)(y) = \bigvee_{x \in U} [(a(x) \Rightarrow b(y)) \triangle a(x)]$$

where $\triangle$ is some t-norm. The appearance of the t-norm $\triangle$ is due to the fact that $p \Rightarrow q$ and p being fuzzy propositions, we need to model the connective "and" by some t-norm. In the classical case, we have $R(a) = b$, but this is not true in the fuzzy case for arbitrary t-norms and arbitrary fuzzy implications $\Rightarrow$. (See Exercise 15.) But there are choices where $R(a) = b$ does hold. Here are a couple of examples.

Example 9.6.1 For

$$
\begin{array}{rl}
u \triangle v & = uv \\
(u \Rightarrow v) & = \begin{cases} 1 \text{ if } u = 0 \\ 1 \wedge v/u \text{ if } u \neq 0 \end{cases}
\end{array}
$$

we have

$$(a(x) \Rightarrow b(y))a(x) = a(x) \wedge b(y)$$

Thus

$$\bigvee_{x}(a(x) \Rightarrow b(y))a(x) = \bigvee_{x}\{a(x) \wedge b(y)\}$$

$$= b(y)$$

Example 9.6.2 For

$$u \triangle v = (u + v + 1) \vee 0$$
$$(u \Rightarrow v) = 1 \wedge (1 - u + v)$$

we have

$$(a(x) \Rightarrow b(y)) \triangle a(x) = \left\{ \begin{array}{l} a(x) \text{ if } a(x) \leq b(y) \\ b(y) \text{ if } a(x) > b(y) \end{array} \right.$$

and thus

$$\bigvee_{x}\{(a(x) \Rightarrow b(y)) \triangle a(x)\} = b(y)$$

9.7 The compositional rule of inference

Consider the rule "If a tomato is red then it is ripe", and the fact "The tomato is almost red". What can be said about the status of this tomato? This is a fairly general pattern of modus ponens. The so-called **compositional rule of inference** that we are going to formulate mathematically in this section is a simple proposed solution to this type of question. It constitutes an inference rule in approximate reasoning in which it is possible to draw vague conclusions from vague premises.

The mathematical pattern of the generalized modus ponens is this. Let X and Y be variables taking values in U and V, respectively. Let A, A^*, and B be fuzzy subsets of appropriate spaces. From "If X is A then Y is B", and "X is A^*", find a reasonable fuzzy subset B^* of U so that "Y is B^*" can be taken as a logical conclusion. As in Chapter 6, we can view the conditional statement above as a binary fuzzy relation R, that is, a fuzzy subset of $U \times V$, and A^* as a unary fuzzy relation on U. As such, the generalized modus ponens can be examined within a general framework of relations. First, if $f : U \to V$ is a function, then the value $b = f(a)$ may be viewed as the image of the projection of $\{a\}$ into V, that is as the {one-element) set $\{b \in V : (a, b) \in f\}$. When f is replaced by a relation R, and A is a subset of U, then the image of the projection of A into V is the set

$$B = \{v \in V : (u, v) \in R \text{ for some } u \in A\}$$

In terms of indicator functions,

$$
\begin{aligned}
B(v) &= \bigvee_{x \in U} \{(A \times V)(u, v) \wedge R(u, v)\} \\
&= \bigvee_{x \in U} \{A(u) \wedge R(u, v)\}
\end{aligned}
$$

This can be written as $B = R \circ A$. Note that the notation $A \circ R$ is also used to retain the analogy with matrix operations in the finite case, where addition and multiplication are replaced by maximum and minimum.

When R and A^* are fuzzy subsets of $U \times V$ and U, respectively, the same composition $R \circ A^*$ yields a fuzzy subset B of V. When applying this procedure to the generalized modus ponens schema

$$\text{If } X \text{ is } A \text{ then } Y \text{ is } B$$

$$X \text{ is } A^*$$

we get the conclusion $B^* = R \circ A^*$, where R is a fuzzy relation on $U \times V$ representing the conditional "If X is A then Y is B". Thus $R(u, v) = (A(u) \Rightarrow B(v))$ where $\Rightarrow$ is a fuzzy implication operator. More generally, the special t-norm $\wedge$ can be replaced by an arbitrary t-norm $\triangle$ in the composition operation among relations, leading to

$$B^*(v) = \bigvee_{u \in U} \{(A(u) \Rightarrow B(v)) \triangle A^*(u)\}$$

Note that if we insist on the coincidence with the classical pattern of modus ponens, that is, when $A^* = A$, we get $B^* = B$, then appropriate choices of $\triangle$ and $\Rightarrow$ are required. This choice problem can be broken down into two pieces:

- For each fixed fuzzy implication operator $\Rightarrow$, determine those t-norms $\triangle$ such that $B = R \circ A$.

- For each fixed t-norm $\triangle$, determine those $\Rightarrow$ so that $B = R \circ A$.

These problems belong to the area of functional equations and will not be discussed further.

9.8 Exercises

1. Show that the intersection of an arbitrary family of σ-fields on a set Ω is a σ-field.

2. Verify that the Borel σ-field of $\mathbb{R}$ is generated by open intervals.

3. Let $(\Omega, \mathcal{A}, \mu)$ be a measure space. Verify the following:

   (a) μ is monotone increasing.
   (b) $\mu(A \cup B) + \mu(A \cap B) = \mu(A) + \mu(B)$.
   (c) If A_n is an increasing sequence of elements of $\mathcal{A}$, then

   $$\mu\left(\bigcup_{n=1}^{\infty} A_n\right) = \lim_{n \to \infty} \mu(A_n)$$

   (d) Let $B \in \mathcal{A}$. Then the function $\nu(A) = \mu(A \cap B)$ is a measure on $(\Omega, \mathcal{A})$.

4. *Let Ω be a set and $\mathcal{F}$ the set of all finite subsets of Ω. For I and J in $\mathcal{F}$, let

   $$M(I, J) = \{A : A \subseteq \Omega, I \subseteq A, A \cap J = \varnothing\}$$

   Let $\mathcal{M}$ be the σ-field generated by the $M(I, J)$'s. Let π be a possibility measure on 2^{Ω}. Show that there exists a unique probability measure P on $(\Omega, \mathcal{M})$ such that for $I \in \mathcal{F}$

   $$\pi(I) = P\{A : A \cap I \neq \varnothing\}$$

   Hint: Use Mathéron's theorem in [64].

5. *Here is another probabilistic interpretation of possibility measures. Let f be a possibility distribution on $\mathbb{R}^n$. Suppose that f is upper semicontinuous, that is, that for any real number α, $\{x \in \mathbb{R}^n : f(x) \geq \alpha\}$ is a closed subset of $\mathbb{R}^n$. Let $\mathcal{F}$ be the set of all closed subsets of $\mathbb{R}^n$ and $\mathcal{C}$ the Borel σ-field of $\mathcal{F}$ viewed as a topological space. Let π be the possibility measure associated with f. Show that there exists a unique probability measure P on $(\mathcal{F}, \mathcal{C})$ such that for any compact set K of $\mathbb{R}^n$

   $$\pi(K) = P\{F : F \in \mathcal{F}, F \cap K \neq \varnothing\}$$

   Hint: Use Choquet's theorem in [64].

6. *Let S be a random set, defined on a probability space $(\Omega, \mathcal{A}, P)$ and taking values in the Borel σ-field of $\mathbb{R}$. Let $\mu(S)$ denote the Lebesgue measure of S. Under suitable measurability conditions, show that for $k \geq 1$ and $\pi(x_1, x_2, ..., x_k) = P\{\omega : \{x_1, x_2, ..., x_k\} \subseteq S(\omega)\}$,

$$E(\mu(S))^k = \int_{\mathbb{R}^k} \pi(x_1, x_2, ..., x_k) d\mu(x_1, x_2, ..., x_k)$$

Hint: See Robbins [88].

7. Let π be a possibility measure on U. Show that for $A \subseteq U$,

$$\pi(A) \vee \pi(A') = 1$$

8. Define $T : 2^U \to [0,1]$ by

$$T(A) = 1 - \pi(A')$$

(T is called a **necessity measure**.) Verify that

   (a) $T(\varnothing) = 0$, $T(U) = 1$.
   (b) For any index set I, $T(\bigcap_I A_i) = \inf_I T(A_i)$.

9. Let A_i, $i = 1, 2, \ldots, n$ be subsets of U such that $A_1 \subseteq A_2 \cdots \subseteq A_n$. For arbitrary subsets A and B of U, let i_1 (respectively i_2) be the largest integer such that $A_{i_1} \subseteq A$ (respectively $A_{i_2} \subseteq B$). Show that $A_i \subseteq A \cap B$ if and only if $i \leq i_1 \wedge i_2$.

10. Let π be a possibility measure on U. For $A, B \subseteq U$, show that

   (a) if $A \subseteq B$ then $\pi(A) \leq \pi(B)$;
   (b) if $\alpha(A) = 1 - \pi(A')$, then $\alpha(A) \leq \pi(A)$.

11. Let $\pi : U \times V \to [0,1]$. Show that for $(u_0, v_0) \in U \times V$,

$$\pi(u_0, v_0) \leq \left(\bigvee_v \pi(u_0, v) \right) \wedge \left(\bigvee_u \pi(u, v_0) \right)$$

12. Let $\pi : U \times V \to [0,1]$ be a joint possibility distribution of (X, Y).

   (a) Suppose $\pi(u, v) = \min\{\pi_Y(v), \pi(u|v)\}$. Show that if X and Y are independent in the possibilistic sense, that is,

$$\pi(u|v) = \pi_X(u) \quad \text{and} \quad \pi(v|u) = \pi_Y(v)$$

   then X and Y are noninteractive. That is, they satisfy

$$\pi(u, v) = \pi_X(u) \wedge \pi_Y(v)$$

(b) Suppose $\pi(u, v) = \pi_X(u) \wedge \pi_Y(v)$ and $\pi(u, v) = \pi_Y(v) \wedge \pi(u|v)$. Show that $\pi(u|v) = \pi_X(u)$ if $\pi_X(u) < \pi_Y(v)$ and $\pi(u|v) \in [\pi_X(u), 1]$ if $\pi_X(u) \geq \pi_Y(v)$.

(c) Define

$$\pi(u|v) = \begin{cases} \pi(u, v) \text{ if } \pi_X(u) \leq \pi_Y(v) \\ \pi(u, v) \dfrac{\pi_X(u)}{\pi_Y(v)} \text{ if } \pi_X(u) > \pi_Y(v) \end{cases}$$

Show that if $\pi(u, v) = \pi_X(u) \wedge \pi_Y(v)$, then $\pi(u|v) = \pi_X(u)$ for all u and v.

13. Let $(\Omega, \mathcal{A}, P)$ be a probability space. For $c > 0$, let $I_c(A) = -c \log P(A)$. Let

$$F_c : [0, \infty] \times [0, \infty] \to [0, \infty]$$

be given by

$$F_c(x, y) = 0 \vee \log(e^{-x/c} + e^{-y/c})$$

(a) Show that if $A \cap B = \emptyset$, then

$$I_c(A \cup B) = F_c(I_c(A), I_c(B))$$

(b) Show that

$$\lim_{c \searrow 0} F_c(x, y) = x \wedge y$$

(c) Let $I : \mathcal{A} \to [0, \infty]$. Show that if $I(A \cup B) = I(A) \wedge I(B)$ for $A \cap B = \emptyset$, then the same holds for arbitrary A and B.

14. Show that $b \neq R(a)$ for

$$u \,\triangle\, v = u \wedge b, \quad (u \Rightarrow v) = (u \wedge v) \vee (1 - v)$$

15. Is $b = R(a)$ in the following cases?

$$
\begin{array}{lll}
\text{(i)} & u \,\triangle\, v & = & u + v - 1 \vee 0 \\
& u \Rightarrow v & = & \begin{cases} 1 & \text{if } u \leq v \\ 0 & \text{if } u > v \end{cases} \\[2ex]
\text{(ii)} & u \,\triangle\, v & = & u + v - 1 \vee 0 \\
& u \Rightarrow v & = & 1 - u + uv \\[1ex]
\text{(iii)} & u \,\triangle\, v & = & u + v - 1 \vee 0 \\
& u \Rightarrow v & = & 1 - u \vee v
\end{array}
$$

16. For $I(u, v) = uv$ or $I(u, v) = u \wedge v$ at the place of J, verify that $b = R(a)$ for $u \bigtriangleup v = u \wedge v$ or $u \bigtriangleup v = uv$.

17. Verify that generalized modus ponens and generalized modus tollens do generalize their classical counterparts.

18. Let $u \Rightarrow v = 1 - u \wedge v$. Find an operator $\bigtriangleup$ such that $b = R(a)$.

19. Consider the following classical modus tollens:

Rule	:	If $X \in A$ then $Y \in B$
Fact	:	$Y \in B^*$
Conclusion	:	$X \in A^*$

Verify that

$$A^* = \left\{ \begin{array}{ll} A' & \text{if } B^* \subseteq B' \\ U & \text{otherwise} \end{array} \right.$$

where $A, A^* \subseteq U$ and $B, B^* \subseteq V$.

Chapter 10

PARTIAL KNOWLEDGE

This chapter is about mathematical tools for handling partial knowledge. We focus on three topics: some special types of nonadditive set functions with a special emphasis on belief functions; rough sets, which we present in an algebraic spirit and at the most basic level; and conditional events, where a brief survey is given.

10.1 Incomplete probabilistic information

Information needed for making decisions in real-world problems can take different forms. Possibility measures in Chapter 9 are examples of nonadditive set functions. In this chapter, we will consider situations arising from incomplete probabilistic information, leading to some new modeling tools.

To illustrate complete information being available in an uncertain environment, we use the familiar framework of probability and statistics. A random experiment is described by a **probability space** $(\Omega, \mathcal{A}, P)$, where Ω is a set representing the sample space, that is, the set of all possible outcomes of the experiment, $\mathcal{A}$ is a collection of subsets of Ω representing events, and P is a probability measure. The definition of probability space was given in Section 9.1.

When $\Omega = \{\omega_1, \omega_2, ..., \omega_n\}$ is finite, then the σ-field $\mathcal{A}$ can be taken to be the set 2^Ω of all subsets of Ω. Then the measure P is determined by its values on singletons. That is, P is known if the $P(\{\omega_i\})$ are known. This follows easily from the additivity of P. Also note that $\sum_i P(\{\omega_i\}) = 1$. A function $f : \Omega \to [0, 1]$ is a **probability density function** if

237

$\sum_i f(\omega_i) = 1$. Thus we have that f defined by $f(\omega_i) = P(\{\omega_i\})$ is a probability density function and $P(A) = \sum_{\omega \in A} f(\omega)$.

When modeling a physical situation with a probability space $(\Omega, \mathcal{A}, P)$, in general P is unknown, or only partially known. If we can gather some more information, such as performing an experiment, then perhaps we can estimate P. In many cases it is realistic to specify P as being in some set $\mathcal{P}$ of probability measures on $(\Omega, \mathcal{A})$ rather than to specify P itself. This is the case of incomplete probabilistic information, since P is known only to be a member of $\mathcal{P}$.

For example, consider a box containing 30 red balls and 60 other balls, some of which are white and the rest are black. A ball is drawn from the box. Suppose the payoffs for getting a red, black, and white ball are \$10, \$20, and \$30, respectively. What is the expected payoff? Of course there is not enough information to answer this question in the classical way since we do not know the probability distribution of the red, white, and black balls. We do not know the probability of getting a white ball, for example. We do have a set $\mathcal{P}$ of probability densities to which the true density must belong. It is given in the following table.

Ω	red	black	white
f_k	30/90	$k/90$	$(60-k)/90$

There are 61 of these densities, and from the information we have, any are possible. In this example, **imprecision** arises since we can only specify partially the true density f. It is one of the 61 just described.

Given such an imprecise model, what can be learned about P from knowledge of $\mathcal{P}$? If we insist on choosing a specific P, we can call upon various **principles** of rational choice, such as the **maximum entropy** principle. Or we may approximate P by getting upper and lower bounds for it as follows: for $A \in \mathcal{A}$, let

$$F(A) = \bigwedge\{Q(A) : Q \in \mathcal{P}\}$$
$$G(A) = \bigvee\{Q(A) : Q \in \mathcal{P}\}$$

Then $F(A) \leq P(A) \leq G(A)$ for all $A \in \mathcal{A}$. The functions F and G are the **lower probability** and **upper probability** functions, respectively. Noting that $G(A) = 1 - F(A')$, we need only focus on one of them.

Except in the cases where $\mathcal{P}$ is a singleton, corresponding to the case of complete probabilistic information, the functions F and G are *not* probability measures, not being additive. But the following do hold.

- $F(\varnothing) = G(\varnothing) = 0$,

- $F(\Omega) = G(\Omega) = 1$, and

- for $A \subseteq B$, $F(A) \leq F(B)$ and $G(A) \leq G(B)$.

The general mathematical treatment of such functions will be given in Chapter 11 when we study fuzzy measures. Here we will consider situations in which F and G possess additional properties.

10.2   Belief functions

To motivate the definition of a belief function, consider the following situation. Let Ω be a finite set and $\{\Omega_1, \Omega_2, ..., \Omega_k\}$ be a partition of Ω. For nonnegative numbers α_i such that $\sum_{i=1}^{k} \alpha_i = 1$, consider the class $\mathcal{P}$ of probability measures on Ω such that $P(\Omega_i) = \alpha_i$. Let F be the lower probability generated by $\mathcal{P}$. There is an important combinatorial fact about the function F. It is a generalization of Poincare's formula which asserts an equality below for probability measures.

Theorem 10.2.1 *Let $A_1, A_2, ..., A_n$ be subsets of Ω, and let $I = \{1, 2, ..., n\}$. Then*

$$F\left(\bigcup_{i \in I} A_i\right) \geq \sum_{\varnothing \neq J \subseteq I} (-1)^{|J|+1} F\left(\bigcap_{j \in J} A_j\right) \qquad (10.1)$$

Proof. For $A \subseteq \Omega$, let A^* be the union of all the Ω_i that are contained in A. Since all $P \in \mathcal{P}$ agree on A^*, we can define the map

$$\varphi : 2^\Omega \to [0,1] : A \to P(A^*)$$

Obviously, $\varphi(\varnothing) = 0$ and $\varphi(\Omega) = 1$. Moreover, $\varphi \leq P$ for any $P \in \mathcal{P}$, so that $\varphi \leq F$. Since Ω is finite, there is a $P \in \mathcal{P}$ such that $P(A) = P(A^*)$ simply by putting zero mass on $A^{*\prime} \cap A$. Thus $\varphi(A) = P(A^*) = P(A) \geq F(A)$, so $\varphi = F$. So

$$
\begin{aligned}
F\left(\bigcup_{i \in I} A_i\right) &= P\left(\bigcup_{i \in I} A_i\right)^* \\
&\geq P\left(\bigcup_{i \in I} A_i^*\right) \\
&= \sum_{\varnothing \neq J \subseteq I} (-1)^{|J|+1} P\left(\bigcap_{j \in J} A_j^*\right) \\
&= \sum_{\varnothing \neq J \subseteq I} (-1)^{|J|+1} P\left(\bigcap_{j \in J} A_j\right)^* \\
&= \sum_{\varnothing \neq J \subseteq I} (-1)^{|J|+1} F\left(\bigcap_{j \in J} A_j\right)
\end{aligned}
$$

∎

In this situation, it turns out that the set $\mathcal{P}$ is the set of all probability measures dominating F. That is,

$$\mathcal{P} = \{P : F \leq P\}$$

To see this, suppose that $P \geq F$. Then for any $Q \in \mathcal{P}$, $F(\Omega_i) = Q(\Omega_i) = \alpha_i \leq P(\Omega_i)$. But since P is a probability measure, $\sum P(\Omega_i) = 1$, and so $P(\Omega_i) = \alpha_i$. Thus $P \in \mathcal{P}$. Clearly if a probability measure is in $\mathcal{P}$, then it dominates F.

It should be noted that the theorem implies that F is monotone increasing. That is, if $A \subseteq B$ then $F(A) \leq F(B)$. For an arbitrary class $\mathcal{P}$ of probability measures on a set Ω, the associated F might not satisfy 10.1. If it does, F is an **infinite monotone capacity**. Such functions are used to model general degrees of belief.

Definition 10.2.2 *A function $F : 2^{\Omega} \to [0, 1]$ is a **belief function** if*

  1. $F(\varnothing) = 0$, $F(\Omega) = 1$ and

  2. for all $n \geq 1$, and $A_i \subseteq \Omega$,

$$F\left(\bigcup_{i=1}^{n} A_i\right) \geq \sum_{\varnothing \neq J \subseteq \{1,2,\ldots,n\}} (-1)^{|J|+1} F\left(\bigcap_{j \in J} A_j\right)$$

The function $G(A) = 1 - F(A')$ is a **plausibility function**. For a subset A, we have

$$1 = F(A \cup A') \geq F(A) + F(A')$$

so that $G(A) = 1 - F(A') \geq F(A)$.

For each $A \subseteq \Omega$, the value $F(A)$ is interpreted as a lower bound on the likelihood of A, or more generally as the **degree of belief** in A. The upper probability $G(A)$ is an upper bound for the likelihood. Since $F(A) \leq G(A)$, the interval $[F(A), G(A)]$ represents the **imprecision** about beliefs.

Theorem 10.2.3 *If F is a belief function, then the function f defined on 2^{Ω} by*

$$f(A) = \sum_{B \subseteq A} (-1)^{|A-B|} F(B)$$

is nonnegative.

Proof. Let $A = \{\omega_1, \omega_2, ..., \omega_n\} \subseteq \Omega$, and let $A_i = A - \{\omega_i\}$. Then from 10.1, we have

$$F(A) \geq \sum_{i=1}^{n} F(A_i) - \sum_{i<j} F(A_i \cap A_j) + ... + (-1)^n \sum_{i=1}^{n} F\left(\bigcap_{j \neq i} A_j\right)$$

noting that $\bigcap_{i=1}^{n} A_i = \varnothing$. This means that $f(A) \geq 0$. ∎

The function f is the **Möbius inversion** of the function F. There is a general theorem.

Theorem 10.2.4 *Let f and F be functions $2^\Omega \to \mathbb{R}$. Then for $A \subseteq \Omega$,*

$$F(A) = \sum_{B \subseteq A} f(B)$$

if and only if

$$f(A) = \sum_{B \subseteq A} (-1)^{|A-B|} F(B)$$

Proof. Suppose that $F(A) = \sum_{B \subseteq A} f(B)$. Then

$$\sum_{B \subseteq A} (-1)^{|A-B|} F(B) = \sum_{B \subseteq A} (-1)^{|A-B|} \sum_{C \subseteq B} f(C)$$

$$= \sum_{C \subseteq B \subseteq A} (-1)^{|A-B|} f(C)$$

If $C = A$, this last expression is $f(A)$. If $C \neq A$ then $A - C$ has $2^{|A-C|}$ subsets, so there are an even number of subsets B with $C \subseteq B \subseteq A$, exactly half of which have an even number of elements. Thus half of the numbers $(-1)^{|A-B|}$ are 1 and half -1. Thus for each C with $C \neq A$, $\sum_{C \subseteq B \subseteq A} (-1)^{|A-B|} f(C) = 0$ with the summation taken over B. Therefore $\sum_{B \subseteq A} (-1)^{|A-B|} F(B) = f(A)$. The proof of the other half is exactly the same. ∎

The interpretation of the Möbius inversion f of a belief function F is this. For each $A \subseteq \Omega$, $f(A)$ is the **weight of evidence** in support of A which is in addition to the weights already given to proper subsets of A. From the last two theorems, we see that if F is a belief function, then its Möbius inversion f satisfies the following:

- f is nonnegative, $f(\varnothing) = 0$, and $\sum_{A \subseteq \Omega} f(A) = 1$. In particular, f is a density on 2^Ω.

These properties characterize the Möbius inversions of belief functions.

Corollary 10.2.5 *Suppose* $f : 2^\Omega \to [0,1]$ *satisfies the properties just above. Then*

$$F(A) = \sum_{B \subseteq A} f(B)$$

is a belief function.

Proof. Clearly $F(\varnothing) = 0$ and $F(\Omega) = 1$, so only infinite monotonicity remains to be checked. Let $I = \{1, 2, ..., n\}$, A_i be subsets of Ω. We need

$$F\left(\bigcup_{i \in I} A_i\right) \geq \sum_{\varnothing \neq J \subseteq I} (-1)^{|J|+1} F\left(\bigcap_{j \in J} A_j\right)$$

Let Γ be the set of those subsets of $\cup_{i \in I} A_i$ that are contained in at least one A_i. We have

$$F\left(\bigcup_{i \in I} A_i\right) = \sum_{B \subseteq \cup_i A_i} f(B)$$

$$\geq \sum_{B \in \Gamma} f(B)$$

and

$$\sum_{\varnothing \neq J \subseteq I} (-1)^{|J|+1} F\left(\bigcap_{j \in J} A_j\right) = \sum_{\varnothing \neq J \subseteq I} (-1)^{|J|+1} \sum_{B \subseteq \cap_{j \in J} A_j} f(B)$$

$$= \sum_{B \in \Gamma} \sum_{\varnothing \neq J \subseteq I} (-1)^{|J|+1} f(B)$$

$$= \sum_{B \in \Gamma} f(B)$$

since $\sum_{\varnothing \neq J \subseteq I} (-1)^{|J|+1} = 1$, as is easily seen. ∎

Noting that $1 = \sum_{A \subseteq \Omega} f(A)$, the plausibility function $G(A)$ can be expressed in terms of f.

$$G(A) = 1 - F(A')$$

$$= \sum_{B \subseteq \Omega} f(B) - \sum_{B \subseteq A'} f(B)$$

$$= \sum_{B \cap A \neq \varnothing} f(B)$$

Example 10.2.6 Let $\Omega = \{\omega_1, \omega_2, \omega_3, \omega_4\}$, and

$$\mathcal{P} = \{P : P\{\omega_1\} \geq 0.4, P\{\omega_2\} \geq 0.2, P\{\omega_3\} \geq 0.2, P\{\omega_4\} \geq 0.1\}$$

Define

$$
\begin{array}{rcl}
f\{\omega_1\} & = & 0.4 \\
f\{\omega_2\} & = & 0.2 \\
f\{\omega_3\} & = & 0.2 \\
f\{\omega_4\} & = & 0.1 \\
f(\Omega) & = & 0.1 \\
f(A) & = & 0 \text{ for all other subsets } A
\end{array}
$$

So f is a probability density, or a probability mass assignment, on 2^Ω, and $F(A) = \sum_{B \subseteq A} f(B)$ is a belief function. Now for any $P \in \mathcal{P}$,

$$
\begin{array}{rcl}
F(A) & = & \sum_{B \subseteq A} f(B) \\
& = & \sum_{\omega_i \in A} f\{\omega_i\} \\
& \leq & \sum_{\omega_i \in A} P\{\omega_i\} \\
& = & P(A)
\end{array}
$$

If for a particular $A \neq \Omega$, we set $P(A) = \sum_{\omega_i \in A} f\{\omega_i\}$, then P can be extended to a probability measure in $\mathcal{P}$ by assigning it appropriate values $\geq f\{\omega_i\}$ on the sets $\{\omega_i\}$ with $\omega_i \notin A$. Thus $F(A) = \inf\{P(A) : P \in \mathcal{P}\}$, and $\mathcal{P} \subseteq \{P : F \leq P\}$. But if $F \leq P$, then clearly $P\{\omega_i\} \geq f\{\omega_i\}$, and $\mathcal{P} = \{P : F \leq P\}$.

Thus from our results, including the last corollary, we see that the Möbius inversion of a belief function F on 2^Ω is a density f on 2^Ω. This density is called the **probability mass assignment** of F. This may be viewed as the probability density of a random variable X whose values are subsets in 2^Ω. Such a random variable is referred to as a **random set**. For each $A \in 2^\Omega$, $f(A) = P(X = A)$, and

$$F(A) = \sum_{B \subseteq A} f(B) = \sum_{B \subseteq A} P(X = B) = P(X \subseteq A)$$

Thus we see that F plays the role of the cumulative distribution function of a random variable. This situation is general. Let X be a random variable on some probability space $(\Theta, \mathcal{E}, P)$ with values in a σ-field $\mathcal{A}$. Then X and P induce a nonadditive function F on $\mathcal{A}$ by

$$F(A) = P\{\theta : X(\theta) \subseteq A\}$$

We of course assume the appropriate measurability condition on X, namely that for all $A \in \mathcal{A}$, $\{\theta : X(\theta) \subseteq A\} \in \mathcal{E}$. This is referred to as **strong measurability** of X.

It can be shown that the function $F : \mathcal{A} \to [0, 1]$ is a belief function.

10.3   Reasoning with belief functions

We will restrict ourselves to some basic mathematical tools for use in
reasoning with belief functions, such as in combination of evidence and
in updating procedures. We begin with the following theorem.

Theorem 10.3.1 *Let Ω be finite and let $F : 2^{\Omega} \to [0,1]$ be a belief
function. Let $\mathcal{P}$ be the set of probability measures P on 2^{Ω} such that
$F \leq P$. Then for each $A \in 2^{\Omega}$,*

$$F(A) = \bigwedge\{P(A) : P \in \mathcal{P}\}$$

Proof. Let f be the Möbius inversion of F. Then f is a probability
density on 2^{Ω}, and we are going to build from f and each nonempty
subset A of Ω a probability density g_A on Ω. For $\varnothing \neq A \subseteq \Omega$, define

$$g_A(\omega) = \begin{cases} \sum_{\omega \in B \subseteq A} \dfrac{f(B)}{|B|} \text{ for } \omega \in A \\[2em] \sum_{\omega \in B} \dfrac{f(B)}{|B - A \cap B|} \text{ for } \omega \notin A \end{cases}$$

Then $g_A(\omega) \geq 0$, and

$$\sum_{\omega \in \Omega} g_A(\omega) = \sum_{\omega \in A} g_A(\omega) + \sum_{\omega \notin A} g_A(\omega)$$

But

$$\begin{aligned} \sum_{\omega \in A} g_A(\omega) &= \sum_{\omega \in A} \sum_{\omega \in B \subseteq A} \frac{f(B)}{|B|} \\ &= \sum_{B \subseteq A} \frac{f(B)}{|B|} \sum_{\omega \in B} 1 \\ &= \sum_{B \subseteq A} f(B) \end{aligned}$$

and

$$\begin{aligned} \sum_{\omega \notin A} g_A(\omega) &= \sum_{B \not\subseteq A} \frac{f(B)}{|B - A \cap B|} \sum_{\omega \in B - A \cap B} 1 \\ &= \sum_{B \not\subseteq A} f(B) \end{aligned}$$

Thus

$$\sum_{\omega \in \Omega} g_A(\omega) \;=\; \sum_{B \subseteq \Omega} f(B)$$
$$=\; 1$$

and g_A is a probability density on Ω. Let P_A be the corresponding probability measure. Then for $B \subseteq \Omega$,

$$P_A(B) \;=\; \sum_{\omega \in B} g_A(\omega)$$

$$=\; \sum_{\omega \in A \cap B} g_A(\omega) + \sum_{\omega \in A' \cap B} g_A(\omega)$$

$$=\; \left(\sum_{\omega \in A \cap B} \sum_{\omega \in X \subseteq A} \frac{f(X)}{|X|} \right)$$

$$+ \left(\sum_{\omega \in X \cap A'} \sum_{\omega \in X \cap A' \neq \varnothing} \frac{f(X)}{|X \cap A'|} \right)$$

$$\geq\; \sum_{X \subseteq A \cap B} f(X) + \sum_{X \cap A' \neq \varnothing} f(X)$$

$$=\; \sum_{X \subseteq B} f(X)$$

$$=\; F(B)$$

Thus

$$F(B) = \sum_{C \subseteq B} f(C) \leq P_A(B)$$

so that $P_A \in \mathcal{P}$. In particular, $P_A(A) = F(A)$, so that for each $A \subseteq \Omega$,

$$F(A) = \bigwedge \{P(A) : P \in \mathcal{P}\}$$

∎

The set of probability measures $\mathcal{P} = \{P : F \leq P\}$ is the set of measures **compatible** with F. Thus a belief function F is always the lower envelope of the class $\mathcal{P}$ of probability measures compatible with F, and this set is of course nonempty for any belief function. If F is a lower probability of some given set Q of probability measures and F is a belief function, then Q might be properly contained in the set $\mathcal{P}$ of compatibles ones. If $Q = \mathcal{P}$, we say that Q can be **modeled**, or **described by** the belief function F.

The set $\mathcal{P}$ of compatible probability measures with a given F can be described in terms of the Möbius inversion f of F. For $\varnothing \neq A \subseteq \Omega$ and $\omega \in A$, let the $\alpha(\omega, A)$ be nonnegative numbers such that $\sum_{\omega \in A} \alpha(\omega, A) = f(A)$. For example, one may take each $\alpha(\omega, A)$ to be $f(A)/|A|$, but there are many other choices. Such an α is called an **allocation**. Then $g_\alpha(\omega) = \sum_{\omega \in A} \alpha(\omega, A)$ is a density. It can be shown that the probability measure $P_\alpha(A) = \sum_{\omega \in A} g_\alpha(\omega)$ dominates F, and if $P \geq F$, then $P = P_\alpha$ for some allocation α.

Since the Möbius inversion f of a belief function F is a density on 2^Ω, F can be viewed as the distribution of some random set S. Thus an **evidence** is represented by a random set. Two bodies of evidence are **independent** if their corresponding random sets S and T are stochastically independent. That is, for $A,\ B \subseteq \Omega$,

$$P(S = A, T = B) = P(S = A)P(T = B)$$

The random set $S \cap T$ represents a form of **combination of evidence**, or **fusion of data**. The associated belief function is

$$F(A) = P(S \cap T \subseteq A)$$

When S and T are independent, then the Möbius inversion f of F is expressed in terms of the Möbius inversions F and G of the random sets S and T by

$$F(A) = \sum f(X)g(Y)$$

where the summation is over pairs (X, Y) such that $X \cap Y = A$. In fact,

$$
\begin{aligned}
F(A) &= P(S \cap T \subseteq A)\\
&= \sum_{B \subseteq A} P(S \cap T = B)\\
&= \sum_{B \subseteq A} \sum_{X \cap Y = B} P(S = X, T = Y)\\
&= \sum_{B \subseteq A} \sum_{X \cap Y = B} P(S = X)P(T = Y)\\
&= \sum_{B \subseteq A} f(B)
\end{aligned}
$$

Although S and T are such that $P(S = \varnothing) = P(T = \varnothing) = 0$, the combined random set $S \cap T$ might have positive mass at $\varnothing$. That is, it is possible that $P(S \cap T = \varnothing) > 0$.

To remove this situation so that $V = S \cap T$ corresponds to F with $F(\varnothing) = 0$ proceed as follows: let the range of V be $\{A_1, A_2, ..., A_k, \varnothing\}$. Then $\sum_{i=1}^{k} f(A_i) = 1 - f(\varnothing)$. Let U have the range $\{A_1, A_2, ..., A_k\}$ with density $g(A_i) = \alpha_i f(A_i)$, where the α_i are chosen so that $g(A_i) > 0$ and $\sum_{i=1}^{k} \alpha_i f(A_i) = 1$, and hence $g(\varnothing) = 0$. A canonical choice is $\alpha_i = 1/(1 - f(\varnothing))$, in which case $g(A_i) = f(A_i)/(1 - f(\varnothing))$. Then we have

$$
\begin{aligned}
G(A) &= P(S \subseteq A) : S \neq \varnothing) \\
&= \begin{cases} 0 \text{ if } A = \varnothing \\ P(\varnothing \neq S \subseteq A)/P(S \neq \varnothing) \text{ for } A \neq \varnothing) \end{cases}
\end{aligned}
$$

We turn now to the formulation of **updating** with belief functions. Given a belief function F on a finite set Ω such that $F(B) > 0$, one would like to define a belief function F_B representing the modification of F based on the evidence B, in other words, a **conditional belief function** in the spirit of probability theory. One way to achieve this is to call upon the existence of the nonempty class $\mathcal{P} = \{P : F \leq P\}$ of compatible probability measures.

Now

$$
F(A) = \inf\{P(A) : P \in \mathcal{P}\}
$$

When $F(A) > 0$, we have $P(A) > 0$ since $F \leq P$. The converse holds: if $P(A) > 0$ for all $P \in \mathcal{P}$, then $F(A) > 0$. This can be seen by looking at the density g_A constructed above.

Given B, it is natural to update all P in $\mathcal{P}$ and thus to change $\mathcal{P}$ to

$$
\mathcal{P}_B = \{P(\cdot \mid B) : P \in \mathcal{P}\}
$$

where $P(A \mid B) = P(A \cap B)/P(B)$, and to consider

$$
F_B = \inf\{P : P \in \mathcal{P}_B\}
$$

as a candidate for a conditional belief function. Fortunately, it can be shown that F_B is indeed a belief function and that it can be expressed in terms of F and B in a simple form. For $A \subseteq B$,

$$
F_B(A) = \frac{F(A)}{F(A) + 1 - F(A \cup B')}
$$

For other forms of conditional beliefs, see the exercises at the end of this chapter.

10.4   Decision making using belief functions

In the framework of incomplete probabilistic information, we will investigate the following simple decision problem. A decision problem consists of choosing an action among a collection $\mathbb{A}$ of relevant actions in such a way that utility is maximized. Specifically, if Θ denotes the possible "states of nature", the true state being unknown, then a utility function

$$u : \mathbb{A} \times \Theta \to \mathbb{R}$$

is specified, where $u(a, \theta)$ is the payoff when action a is taken and nature presents θ. In the Bayesian framework, the knowledge about Θ is described by a probability measure Q on it. Then the expected value $E_Q u(a, \cdot)$ is used to make a choice as to which action a to choose. When Q is specified, the "optimal" action is the one that maximizes $E_Q u(a, \cdot)$ over $a \in \mathbb{A}$. But suppose that Q is specified to lie in a set $\mathcal{P}$ of probability measures on Θ, and that the lower envelope F given by

$$F(A) = \inf\{P(A) : P \in \mathcal{P}\}$$

is a belief function and $\mathcal{P} = \{P : F \leq P\}$. This is the same as assuming that the knowledge about Θ is given by an "evidence" whose representation is a belief function F over Θ. Decision procedures based on F will be derived by viewing F as equivalent to its class of compatible probability measures $\mathcal{P}$. Some choices of actions that can be made follow.

10.4.1   A minimax viewpoint

Choose action to maximize

$$\bigwedge\{E_P u(a, \cdot) : P \in \mathcal{P}\}$$

We will show that in the case where Θ is finite, the inf is attained and is equal to a **generalized integral** of $u(a, \cdot)$ with respect to the belief function F.

Generalizing the ordinary concept of integrals, we consider the integral of a function $u : \Theta \to \mathbb{R}$ defined by

$$E_F(u) = \int_0^\infty F(u > t)dt + \int_{-\infty}^0 [F(u > t) - 1]dt$$

Since Θ is assumed to be finite, each $P \in \mathcal{P}$ is characterized by a probability density function $g : \Theta \to [0, 1]$ where $g(\omega) = P(\{\omega\})$. When g is specified, we write the corresponding P as P_g and the corresponding set of densities for $\mathcal{P}$ is denoted $\mathcal{D}$.

Theorem 10.4.1 *Let Θ be finite, $u : \Theta \to \mathbb{R}$ and F be a belief function on Θ. Then there exists an $h \in \mathcal{D}$ such that*

$$E_F(u) = E_{P_h}(u) = \inf\{E_{P_g}(u) : g \in \mathcal{D}\}$$

Proof. Let $\Theta = \{\theta_1, \theta_2, ..., \theta_n\}$, with $u(\theta_1) \le u(\theta_2) \le ... \le u(\theta_n)$. Then

$$E_F(u) = \sum_{i=1}^{n} u(\theta_i)\left[F(\{\theta_i, \theta_{i+1},...,\theta_n\}) - F(\{\theta_{i+1}, \theta_{i++2}...,\theta_n\})\right]$$

Let

$$h(\theta_i) = F(\{\theta_i, \theta_{i+1},...,\theta_n\}) - F(\{\theta_{i+1}, \theta_{i++2}...,\theta_n\})$$

If $A_i = \{\theta_i, \theta_{i+1},...,\theta_n\}$, then

$$
\begin{aligned}
h(\theta_i) &= F(A_i) - F(A_i - \{\theta_i\}) \\
&= \sum_{B \subseteq A_i} f(B) - \sum_{B \subseteq (A_i - \{\theta_i\})} f(B) \\
&= \sum_{\theta_i \in B \subseteq A_i} f(B)
\end{aligned}
$$

where f is the Möbius inversion of F. Thus $h \in \mathcal{D}$. Next, for each $t \in \mathbb{R}$ and $g \in \mathcal{D}$, it can be checked that $P_h(u > t) \le P_g(u > t)$ since $(u > t)$ is of the form $\{\theta_i, \theta_{i+1},...,\theta_n\}$. Thus for all $g \in \mathcal{D}$, $E_{P_h}(u) \le E_{P_g}(u)$. ∎

The special density above depends on u since the ordering of Θ is defined in terms of u.

Example 10.4.2 Let $\Theta = \{\theta_1, \theta_2, \theta_3, \theta_4\}$ and let F be the Möbius inversion of f given by

$$
\begin{aligned}
f(\{\theta_1\}) &= 0.4 \\
f(\{\theta_2\}) &= 0.2 \\
f(\{\theta_3\}) &= 0.2 \\
f(\{\theta_4\}) &= 0.1 \\
f(\Omega) &= 0.1
\end{aligned}
$$

We have

h	θ_1	θ_2	θ_3	θ_4
	0.5	0.2	0.2	0.1

If

$$
\begin{array}{c|cccc}
u & \theta_1 & \theta_2 & \theta_3 & \theta_4 \\
\hline
 & 1 & 5 & 10 & 20
\end{array}
$$

then

$$
E_F(u) = \sum_{i=1}^{4} u(\theta_i) h(\theta_i) = 5.5
$$

10.4.2 An expected value approach

Since the Möbius inversion f of the belief function F is a probability density on 2^Ω, the "true" expected value $E_Q(u)$ is replaced by the expectation of a suitable function of S, namely

$$
E_f \varphi(S) = \sum_{A \subseteq \Theta} \varphi(A) f(A)
$$

where $\varphi : 2^\Theta \to \mathbb{R}$. This is possible since for each $g \in \mathcal{D}$, one can find many φ such that $E_{P_g}(u) = E_f(\varphi)$. For example, define φ arbitrarily on A for which $f(A) \neq 0$ by

$$
\varphi(A) = \left[E_{P_g}(u) - \sum_{B \neq A} \varphi(B) f(B) \right] / f(A)
$$

As an example for φ, for $\rho \in [0, 1]$, take

$$
\varphi_\rho(A) = \rho \bigvee \{ u(\theta) : \theta \in A \} + (1 - \rho) \bigwedge \{ u(\theta) : \theta \in A \}
$$

Then $E_f(\varphi_\rho) = E_{P_g}(u)$ where g is constructed as follows. Order the θ_i so that $u(\theta_1) \leq u(\theta_2) \leq ... \leq u(\theta_n)$ and define

$$
g(\theta_i) = \rho \sum_A f(A) + (1 - \rho) \sum_B f(B)
$$

where the summations are over A and B of the form

$$
\begin{aligned}
\theta_i \in A &\subseteq \{\theta_1, \theta_2, ..., \theta_i\} \\
\theta_i \in B &\subseteq \{\theta_i, \theta_{i+1}, ..., \theta_n\}
\end{aligned}
$$

10.4.3 Maximum entropy principle

If the choice of a canonical g in $\mathcal{D}$ is desired, then the inference principle known as the "maximum entropy principle" can be used. For a probability density function g, its **entropy** is

$$H(g) = -\sum_{\theta \in \Theta} g(\theta) \log g(\theta)$$

If g is the element of $\mathcal{D}$ maximizing H, then decisions can be based on $E_{P_g}(u(a, \cdot))$. This means choosing an action $b \in \mathbb{A}$ such that

$$E_{P_g}(u(b, \cdot)) = \bigvee \{E_{P_g}(u(a, \cdot))\}$$

There recently have been two algorithms developed for finding the g that maximizes H [66]. We will present the easier of the two (called Algorithm 2 in [66]) and give an example of its use.

Here, then, is the situation. Let Θ be a finite set and B be a belief function on it. Corresponding to this belief function is the set $\mathcal{P}$ of all probability measures P on Θ such that $P \geq B$. Each P corresponds to a density f on Θ via $P(A) = \sum_{a \in A} f(a)$, yielding a set $\mathcal{D}$ of densities. The Möbius inversion of B is a probability density m on 2^{Θ}. The elements of $\mathcal{D}$ arise from allocations of m, as described earlier. The following algorithm calculates the density g in $\mathcal{D}$ with maximum entropy directly from the belief function B.

Algorithm 2 Define a density g on Θ as follows. Inductively define a decreasing sequence of subsets Θ_i of Θ, and numbers b_i, as follows, quitting when Θ_i is empty.

- $\Theta_0 = \Theta$

- $b_i = \max_{\varnothing \neq K \subset \Theta_i} \dfrac{B(K \cup \Theta_i^c) - B(\Theta_i^c)}{|K|}$

- K_i is the largest subset of Θ_i such that $B(K_i \cup \Theta_i^c) - B(\Theta_i^c) = b_i |K_i|$ (There is a unique such K_i.)

- $\Theta_{i+1} = \Theta_i \setminus K_i$

If $x \in K_i$, then set $g(x) = b_i$. Then $g \in \mathcal{D}$.

Example 10.4.3 Let $\Theta = \{a, b, c, d\}$ and let B and m be given as in the following table:

2^Θ	m	B
$\varnothing$	0	0
$\{a\}$	0.12	0.12
$\{b\}$	0.01	0.01
$\{c\}$	0.00	0.00
$\{d\}$	0.18	0.18
$\{a,b\}$	0.12	0.25
$\{a,c\}$	0.10	0.22
$\{a,d\}$	0.22	0.52
$\{b,c\}$	0.08	0.09
$\{b,d\}$	0.04	0.23
$\{c,d\}$	0.02	0.20
$\{a,b,c\}$	0.03	0.46
$\{a,c,d\}$	0.01	0.65
$\{a,b,d\}$	0.00	0.69
$\{b,c,d\}$	0.03	0.36
Θ	0.04	1

First we must find

$$\max_{K \subset \Theta_i} \frac{B(K \cup \Theta_i^c) - B(\Theta_i^c)}{|K|}$$

for $i = 0$. That is, we need $\max_{K \subset \Theta} B(K)/|K|$, which is .26, corresponding to the set $K_0 = \{a,d\}$. Next we consider the set $\Theta_1 = \Theta \backslash K_0 = \{b,c\}$, and find

$$\max_{\varnothing \neq K \subset \{b,c\}} \frac{B(K \cup \{a,d\}) - B(\{a,d\})}{|K|}$$

The sets K to be considered are $\{b\}, \{c\}$, and $\{b,c\}$ corresponding to the numbers

$$\frac{B(\{a,b,d\}) - B(\{a,d\})}{|\{b\}|} \;=\; 0.17$$

$$\frac{B(\{a,c,d\}) - B(\{a,d\})}{|\{c\}|} \;=\; 0.13$$

$$\frac{B(\{a,b,c,d\}) - B(\{a,d\})}{|\{b,c\}|} \;=\; 0.24$$

So the appropriate K is $\{b,c\}$ and Θ_2 is empty and we are done. The probability assignments are thus

$$g(a) = 0.26 \quad g(b) = 0.24 \quad g(c) = 0.26 \quad g(d) = 0.24$$

This is the density on Θ that maximizes $H(g) = -\sum_{\theta \in \Theta} g(\theta) \log g(\theta)$ among all those densities compatible with m, or equivalently, among all those corresponding to probability distributions on Θ dominating B.

The density on any finite set that maximizes entropy is the one assigning the same probability to each element. So an algorithm that maximizes entropy will make the probability assignments as "equal as possible". In particular, Algorithm 2 above could end in one step—if $B(\Theta)/|\Theta|$ is at least as large as $B(X)/|X|$ for all the nonempty subsets X of Θ. It is clear that the algorithm is easily programmed. It simply must evaluate and compare various $B(X)/|X|$ at each step. The algorithm does produce the density g in at most $|\Theta|$ steps.

There is exactly one density in the family $\mathcal{D}$ that maximizes entropy. Algorithm 2 gives that density. An allocation can be gotten from this density using a linear programming procedure. Algorithm 1 in [66] actually produces an allocation, and from that allocation the entropy maximizing density is readily produced. The interested reader is encouraged to consult [66] for proofs.

10.5 Rough sets

Rough sets were introduced by Pawlak [82] and since have been the subject of many papers. We refer to the references for the many mathematical approaches and variants of the theory [51, 85, 74, 94, 16, 52, 31]. The material we present here is in the algebraic spirit and is at the most basic level. It follows that presented by Pawlak himself. For numerous applications and examples, we refer to Pawlak's book [83].

The starting point is this. Let U be a set, and let $\mathcal{E}$ be the set of equivalence classes of an equivalence relation on U. This set $\mathcal{E}$ of equivalence classes is, of course, a partition of U. For a subset X of U, let $\underline{X} = \cup\{E \in \mathcal{E} : E \subseteq X\}$ and $\overline{X} = \cup\{E \in \mathcal{E} : E \cap X \neq \varnothing\}$. The elements $\underline{X}$ and $\overline{X}$ are viewed as lower and upper approximations, respectively, of X. It is clear that $\underline{X} \subseteq X \subseteq \overline{X}$. The set $\underline{X}$ is the union of all the members of the partition contained in it, and the set $\overline{X}$ is the union of all the members of the partition that have nonempty intersection with it. Calling X and Y equivalent if $\underline{X} = \underline{Y}$ is clearly an equivalence relation, and similarly for $\overline{X} = \overline{Y}$. So meeting both these conditions is an equivalence relation, being the intersection of the two. We have the following.

Proposition 10.5.1 *Let $\mathcal{E}$ be the equivalence classes of an equivalence relation on a set U. For subsets X and Y of U, let $X \sim Y$ if $\underline{X} = \underline{Y}$ and $\overline{X} = \overline{Y}$. Then $\sim$ is an equivalence relation on the set of subsets 2^U.*

Definition 10.5.2 *The equivalence classes of the equivalence relation* $\sim$ *in the previous proposition are* **rough sets** *[82], and the set of rough sets will be denoted by* $\mathcal{R}$.

Of course, the set of rough sets $\mathcal{R}$ depends on the set U and the partition $\mathcal{E}$. By its very definition, a rough set R is uniquely determined by the pair $(\underline{X}, \overline{X})$, where X is any member of R. *The pair* $(\underline{X}, \overline{X})$ *is independent of the representative* X *chosen, and we identify rough sets with the pairs* $(\underline{X}, \overline{X})$. There is a natural partial order relation on $\mathcal{R}$, namely that given by $(\underline{X}, \overline{X}) \leq (\underline{Y}, \overline{Y})$ if $\underline{X} \subseteq \underline{Y}$ and $\overline{X} \subseteq \overline{Y}$. Our primary interest will be in this mathematical structure $(\mathcal{R}, \leq)$, and the first job is to show that this is a lattice. There is a technical difficulty: $\underline{X} \cup \underline{Y} \neq \underline{X \cup Y}$ for example. This is because an element of $\mathcal{E}$ can be contained in $X \cup Y$ and not be contained in either. Thus it can be in $X \cup Y$, hence in $\underline{X \cup Y}$, and not in $\underline{X} \cup \underline{Y}$. The following lemma solves most of the technical problems.

Lemma 10.5.3 *For each rough set* $R \in \mathcal{R}$, *there is an element* $X_R \in R$ *such that for every pair of rough sets* R *and* S *in* $\mathcal{R}$, *the following equations hold.*

$$\underline{X_R} \cup \underline{X_S} = \underline{X_R \cup X_S},$$

$$\underline{X_R} \cap \underline{X_S} = \underline{X_R \cap X_S},$$

$$\overline{X_R} \cup \overline{X_S} = \overline{X_R \cup X_S},$$

$$\overline{X_R} \cap \overline{X_S} = \overline{X_R \cap X_S},$$

Proof. For each $E \in \mathcal{E}$ containing more than one element of U, pick a subset A_E such that $\varnothing \subsetneqq A_E \subsetneqq E$, and let $A = \cup\{A_E : E \in \mathcal{E}\}$. Pick a representative X of each rough set R, and let $X_R = \underline{X} \cup (\overline{X} \cap A)$. Observe that X_R is a representative of the rough set R, and is independent of which representative X of R was chosen. We have picked out exactly one representative X_R of each rough set R. Now, consider the equation $\underline{X_R} \cup \underline{X_S} = \underline{X_R \cup X_S}$. If an element $E \in \mathcal{E}$ is contained in $\underline{X_R \cup X_S}$, then it is contained in either $\underline{X_R}$ or $\underline{X_S}$ by their very construction. So $\underline{X_R} \cup \underline{X_S} \supseteq \underline{X_R \cup X_S}$, and the other inclusion is clear. The other equalities follow similarly. ∎

In the proof above, we chose one representative from each rough set, and we chose the sets A_E. This requires the *axiom of choice*, a set theoretic axiom whose use here should be noted. This axiom says that if $\mathcal{S}$ is a nonempty set of nonempty sets, then there is a set T consisting of an element from each of the members of $\mathcal{S}$. None of this would be an issue were U finite, but in any case we choose to use the axiom of choice.

Theorem 10.5.4 $(\mathcal{R}, \leq)$ *is a bounded lattice.*

Proof. We write rough sets as the upper and lower approximations of the sets X_R picked out in the Lemma. So let $(\underline{X_R}, \overline{X_R})$ and $(\underline{X_S}, \overline{X_S})$ be rough sets. Now, $(\underline{X_R}, \overline{X_R}) \leq (\underline{X_S}, \overline{X_S})$ means that $\underline{X_R} \subseteq \underline{X_S}$ and $\overline{X_R} \subseteq \overline{X_S}$. This is clearly a partial order, and is the one induced by the lattice $\mathcal{P}(U) \times \mathcal{P}(U)$. In this lattice, $(\underline{X_R}, \overline{X_R}) \vee (\underline{X_S}, \overline{X_S}) = (\underline{X_R} \cup \underline{X_S}, \overline{X_R} \cup \overline{X_S}) = (\underline{X_R \cup X_S}, \overline{X_R \cup X_S})$, so this sup is the sup in this induced partial order. The existence of inf is analogous. Clearly $(\varnothing, \varnothing)$ and (U, U) are in $\mathcal{R}$, and so are the 0 and 1 of the lattice $(\mathcal{R}, \leq)$, The Theorem follows. ∎

The following is an easy consequence.

Corollary 10.5.5 $(\mathcal{R}, \vee, \wedge, 0, 1)$ *is a bounded distributive lattice.*

It is perhaps useful to summarize at this point. We start with a finite set U and a partition $\mathcal{E}$ of it, or equivalently, an equivalence relation. This partition $\mathcal{E}$ of U gives an equivalence relation on the set $\mathcal{P}(U)$ of all subsets of U, namely, two elements X and Y of $\mathcal{P}(U)$ are equivalent if their lower and upper approximations are equal, that is, if $\underline{X} = \underline{Y}$ and $\overline{X} = \overline{Y}$. The equivalence classes of this latter equivalence relation are rough sets, and the set of them is denoted $\mathcal{R}$. The set $\mathcal{R}$ is thus in one-to-one correspondence with the set $\{(\underline{X}, \overline{X}) : X \in \mathcal{P}(U)\}$ of pairs. Of course, lots of X give the same pair. We identify these two sets: $\mathcal{R} = \{(\underline{X}, \overline{X}) : X \in \mathcal{P}(U)\}$. This set of pairs is a sublattice of the bounded distributive lattice $\mathcal{P}(U) \times \mathcal{P}(U)$. This last fact comes from the lemma: there is a uniform set of representatives from the equivalence classes of rough sets such that the equations in the lemma hold.

Now $\mathcal{R}$ is actually a sublattice of the lattice $\mathcal{P}(U)^{[2]} = \{(X, Y) : X, Y \in \mathcal{P}(U), X \subseteq Y\}$. We know that for any Boolean algebra B, $B^{[2]}$ is a Stone algebra with pseudocomplement $(a.b)^* = (b', b')$, where $'$ is the complement in B. So $\mathcal{P}(U)^{[2]}$ is a Stone algebra.

Theorem 10.5.6 $\mathcal{R}$ *is a subalgebra of the Stone algebra* $\mathcal{P}(U)^{[2]}$, *and in particular is a Stone algebra.*

Proof. All we need is that if $(\underline{X}, \overline{X}) \in \mathcal{R}$, then so is $(\overline{X}', \overline{X}')$. But this is very easy to see. ∎

The algebra $\mathcal{P}(U)^{[2]}$ has another operation on it, namely $(X, Y)' = (Y', X')$.) This operation $'$ on this Stone algebra $\mathcal{P}(U)^{[2]}$ is a **duality**. It reverses inclusions and has order two. Clearly $(X, Y)'' = (X, Y)$. A Stone algebra with a duality is called **symmetric**. Now on the subalgebra $\mathcal{R}$, $(\underline{X}, \overline{X})' = (\overline{X}', \underline{X}')$, which is still in $\mathcal{R}$.

Corollary 10.5.7 $\mathcal{R}$ *is a symmetric Stone algebra.*

There is another algebraic object around that will give more insight into rough sets. Let $\mathcal{B}$ be the complete Boolean subalgebra of $\mathcal{P}(U)$ generated by the elements of the partition of U. Now $\mathcal{B}^{[2]}$ is a symmetric Stone algebra, indeed a subalgebra of $\mathcal{P}(U)^{[2]}$, and it should be clear that $\mathcal{R}$ is a symmetric Stone subalgebra of $\mathcal{B}^{[2]}$. So given the partition of U, all the action is taking place in the symmetric Stone algebra $\mathcal{B}^{[2]}$.

One final pertinent comment. We got a uniform set of representatives of the set of rough sets, namely the X_R above. Now these actually form a sublattice of the bounded lattice $\mathcal{P}(U)$, and this sublattice is isomorphic to the bounded lattice $\mathcal{R}$. Of course this sublattice can be made into a symmetric Stone algebra, but not using in a natural way the operations of the Boolean algebra $\mathcal{P}(U)$.

Stone algebras have two fundamental building blocks, their centers and their dense sets. The center of a Stone algebra is the image of its pseudocomplement, and its dense set consists of those elements that the pseudocomplement takes to 0, or in our case, to $\varnothing$. It is easy to verify the following.

Corollary 10.5.8 *The center of* $\mathcal{R} = \{(\underline{X}, \overline{X}) : \underline{X} = X = \overline{X}\}$, *and the dense set is* $\{(\underline{X}, \overline{X}) : \overline{X} = U\}$.

In rough set theory, the sets X such that $\underline{X} = X = \overline{X}$ are the *definable* rough sets. Those X such that $\overline{X} = U$ are called *externally undefinable*.

10.5.1   An example

We give now the simplest nontrivial example of all this. We give a list of the various entities and then draw some pictures.

- $U = \{a, b, c\}$. To save on notation, we write the set $\{a\}$ as a, $\{a, b\}$ as ab, and so on.

- $\mathcal{E} = \{a, bc\}$

- $\mathcal{P}(U) = \{\varnothing, a, b, c, ab, ac, bc, U\}$

- $\mathcal{R} = \{\{\varnothing\}, \{a\}, \{bc\}, \{b, c\}, \{ab, ac\}, \{U\}\}$.

- $A = b$. (The only choices for A are b and c.)

- $\{X_R\} = \{\varnothing, a, bc, b, ab, U\}$, these corresponding in order to the list of elements of $\mathcal{R}$ above.

- $\mathcal{B} = \{\varnothing, a, bc, U\}$

- $\mathcal{B}^{[2]} = \{(\varnothing, \varnothing), (\varnothing, a), (\varnothing, bc), (\varnothing, U), (a, a),$

  $(a, U), (bc, bc), (bc, U), (U, U)\}.$

- $\mathcal{R} = \{(\varnothing, \varnothing), (a, a), (\varnothing, bc), (bc, bc), (a, U), (U, U)\}$, again these corresponding in order to the lists of elements of $\mathcal{R}$ above.

In the lattice $\mathcal{P}(U)$ pictured just below, the sublattice gotten by leaving out c and ac is the lattice $\{X_R\}$, and hence is isomorphic to $\mathcal{R}$.

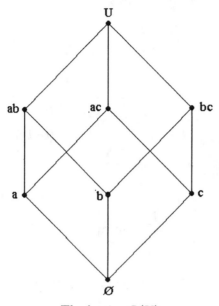

The lattice $\mathcal{P}(U)$

The lattice pictured below is the Boolean subalgebra of $\mathcal{P}(U)$ generated by the elements of the partition $\mathcal{E}$.

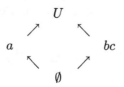

The lattice B

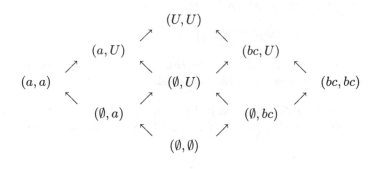

The lattice $B^{[2]}$

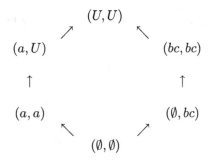

The lattice of rough sets R

10.5.2   The structure of R

To get at the structure of R, we will now construct the pairs $[\underline{X}, \overline{X}]$ in a different way. In the partition $\mathcal{E}$ of U, let $\mathcal{E}_1$ be the elements of $\mathcal{E}$ that have one element, and $\mathcal{E}_2$ be the elements of $\mathcal{E}$ that have more than one element. In the example above, $\mathcal{E}_1 = a$ and $\mathcal{E}_2 = bc$. We identify **2** with the two element Boolean algebra $\{0, 1\}$, and **3** with the three element chain $\{0, u, 1\}$, with $0 < u < 1$. For an element $(f, g) \in 2^{\mathcal{E}_1} \times 3^{\mathcal{E}_2}$, consider the pair

$$\underline{(f, g)} = (\cup\{E \in \mathcal{E}_1 : f(E) = 1\}) \cup (\cup\{E \in \mathcal{E}_2 : g(E) = 1\}),$$

and

$$\overline{(f, g)} = \underline{(f, g)} \cup (\cup\{E \in \mathcal{E}_2 : g(E) = u\}).$$

For each element $E \in \mathcal{E}_2$, let $A_E \subset U$ with $\varnothing \subsetneqq A_E \subsetneqq E$. (Here, the axiom of choice is used explicitly.) Now it should be clear that if

$$X = \underline{(f,g)} \cup (\cup \{A_E : g(E) = u\}),$$

then

$$(\underline{X}, \overline{X}) = (\underline{(f,g)}, \overline{(f,g)}).$$

Thus each pair $(\underline{(f,g)}, \overline{(f,g)})$ is a rough set. For a rough set $(\underline{X}, \overline{X})$, let $f \in 2^{\mathcal{E}_1}$ be defined by $f(E) = 1$ if $E \subseteq \underline{X}$ and $f(E) = 0$ otherwise, and let $g \in 3^{\mathcal{E}_2}$ be defined by $g(E) = 1$ if $E \subseteq \underline{X}$, $g(E) = u$ if $E \subseteq \overline{X} \cap \underline{X}'$, and $g(E) = 0$ otherwise. Since

$$
\begin{aligned}
\overline{X} \;\; &= X \cap (\cup \{E : E \in \mathcal{E}_1 \cup \mathcal{E}_2\} \\
&= \cup \{X \cap E : E \in \mathcal{E}_1 \cup \mathcal{E}_2, E \leq \underline{X}\} \cup (\cup \{X \cap E : E \in \mathcal{E}_2\} \\
&= \underline{X} \cup (\cup \{X \cap E : E \in \mathcal{E}_2, E \leq \overline{X} \cap \underline{X}'\},
\end{aligned}
$$

it follows that $[\underline{X}, \overline{X}] = [\underline{(f,g)}, \overline{(f,g)}]$. Since different pairs (f,g) clearly give different pairs $[\underline{(f,g)}, \overline{(f,g)}]$,

$$2^{\mathcal{E}_1} \times 3^{\mathcal{E}_2} \to \mathcal{R} : (f,g) \to (\underline{(f,g)}, \overline{(f,g)})$$

is a one-to-one correspondence. Identifying the rough sets $\mathcal{R}$ with pairs as we are doing, $\mathcal{R}$ has an order structure given coordinate-wise. Now $2^{\mathcal{E}_1} \times 3^{\mathcal{E}_2}$ is a Stone algebra with ordering given by $(f,g) \leq (h,k)$ if $f \leq h$ and $g \leq k$ pointwise. The mapping above is order preserving by its very definition. Thus under componentwise operations, $\mathcal{R} \cong 2^{\mathcal{E}_1} \times 3^{\mathcal{E}_2}$ as Stone algebras. Further, this tells us exactly what $\mathcal{R}$ is in terms of the partition of U. On one extreme, if $\mathcal{E}_2 = \varnothing$, then for every $X \subseteq U$, $\underline{X} = X = \overline{X}$. In this case, all approximations are exact, and rough sets are just elements of U. On the other hand, if $\mathcal{E}_1 = \varnothing$, then $\mathcal{R} \cong 3^{\mathcal{E}_2} \cong (2^{[2]})^{\mathcal{E}_2} \cong (2^{\mathcal{E}_2})^{[2]} = B^{[2]}$ [31]. This last object is the space of conditional events of the Boolean algebra B, which we will see in the next section. We have the following theorem.

Theorem 10.5.9 *If $\mathcal{E}$ is a partition of the set U, $\mathcal{E}_1$ is the set of elements of $\mathcal{E}$ that have one element, and $\mathcal{E}_2$ is the set of elements of $\mathcal{E}$ that have more than one element, then $\mathcal{R} \cong 2^{\mathcal{E}_1} \times 3^{\mathcal{E}_2}$ as symmetric Stone algebras.*

One consequence of this theorem is that $\mathcal{R}$ is determined solely by the number of elements in $\mathcal{E}_1$ and in $\mathcal{E}_2$, not by the size of U. In the example given above, $\mathcal{E}$ has two elements, the singleton a and the two element set bc. Thus in that example $\mathcal{R}$ is the six element Stone algebra $\mathbf{2} \times \mathbf{3}$.

The results just given can be generalized in various ways. That setup is a particular instance of the following situation. Let $\mathcal{A}$ be a completely distributive lattice and $\mathcal{B}$ be a complete sublattice of $\mathcal{A}$ that is atomic, that is, every element of $\mathcal{B}$ is the supremum of the atoms below it. Here $\mathcal{A}$ plays the role of the power set of U and $\mathcal{B}$ plays the role of the complete Boolean subalgebra of $\mathcal{A}$ generated by the partition $\mathcal{E}$ of U. The scheme is to approximate elements of $\mathcal{A}$ by elements of $\mathcal{B}$, again getting upper and lower approximations. In this situation, rough sets are defined analogously, as follows.

Definition 10.5.10 *Let $\mathcal{A}$ be a completely distributive lattice, and let $\mathcal{B}$ be a subalgebra of $\mathcal{A}$ as a complete lattice. Further, assume that $\mathcal{B}$ is atomic (and hence Boolean). For an element $a \in \mathcal{A}$, let*

$$\underline{a} = \vee\{b \in \mathcal{B} : b \leq a\},$$

$$\overline{a} = \wedge\{b \in \mathcal{B} : a \leq b\}.$$

The elements $\underline{a}$ and $\overline{a}$ are the lower *and* upper *approximations, respectively, of a. Two elements x and y of $\mathcal{A}$ are* equivalent *if they have the same upper and the same lower approximations. The resulting equivalence classes are called* **rough** *sets.*

The theory above goes through in this more general case [31].

10.6   Conditional events

Another form of partial knowledge surfaces when we deal with uncertain rules. These are rules in knowledge-based systems of the form "If X is B, then Y is A", where X and Y are input and output variables, respectively, and A and B are subsets, possibly fuzzy, of appropriate spaces. These rules are uncertain in the sense that if the premise "$X \in B$" is true, the consequent "$Y \in A$" might not always be true, but is only true some percentage of the time. This is typical, for example, in rules formed by physicians in medical diagnosis, where each uncertain rule is accompanied by a degree of confidence or a strength of the rule which is taken to be the conditional probability $P(a|b)$. To be rigorous, we let $(\Omega, \mathcal{A}, P)$ be a probability space on which the random variables X and Y are defined, and $a = \{\omega : Y(\omega) \in A\}$, $b = \{\omega : X(\omega) \in B\}$ being events, that is, elements of $\mathcal{A}$, and with A and B being Borel sets of the appropriate Euclidean space.

The point is this. Each uncertain rule of the form $b \Longrightarrow a$, read "if b then a", is quantified as $P(a|b)$. Moreover, a rule base is in fact a collection of such rules, and it is necessary to combine these rules in some fashion to produce outputs from inputs. Thus, we need to model $b \Longrightarrow a$ mathematically as well. Now, $b \Longrightarrow a$ is a *conditional*, having the flavor of an implication. Can we model $b \Longrightarrow a$ as a material implication, that is, as $b' \vee a$? No, we cannot model it as such and preserve probabilities since $P(b' \vee a) = P(a|b) + P(b')P(a'|b) > P(a|b)$ in general. In fact, there is no binary operation $\diamond$ on $\mathcal{A}$ such that $P(a \diamond b) = P(a|b)$ for all a and b. This is known as **Lewis' Triviality Result** which is left as an exercise. Its implication is that one must look outside $\mathcal{A}$ for an appropriate model. That is, $\mathcal{A}$ needs to be extended, with this extension accommodating not only the elements of $\mathcal{A}$, that is, the events, but also the "conditional events" $a|b$. And then one must worry about an appropriate algebraic structure on this extension, extending the one already on $\mathcal{A}$, and respecting in some sense the probabilities of events and conditional events. There are essentially two solutions to this problem, and we present them now.

To set some notation, let $(\Omega, \mathcal{A})$ be a measurable space, and $\mathcal{C}(\mathcal{A}) = \mathcal{A}^{[2]} = \{(a,b) : a, b \in \mathcal{A}, a \leq b\}$. We view $\mathcal{A}$ sitting inside $\mathcal{C}(\mathcal{A})$ via $a \to (a, 1)$. An element (a, b) will be written $a|b$ to remind us that we are thinking of conditionals. We call $\mathcal{C}(\mathcal{A})$ the space of **conditional events**. The set $\mathcal{A}$ has been enlarged to the set $\mathcal{C}(\mathcal{A})$, and for any probability measure P on $\mathcal{A}$, we have P defined on $\mathcal{C}(\mathcal{A})$ by $P(a|b) = P(a)/P(b)$ as usual. (When $b = 0$, then $a = 0$, and we set $P(0/0) = 1$.) Now it is a matter of defining operations on $\mathcal{C}(\mathcal{A})$ that behave "properly" with respect to P. The usual coordinatewise operations on $\mathcal{C}(\mathcal{A})$ are not appropriate here because that would mean that for $a \leq c$, $P(a) < P(c)$, and $P(d) < 1$, $P(a|1 \vee c|d) = P((a \vee c)|1) = P(c) < P(c|d)$. But we should have $P(a|1 \vee c|d) \geq P(c|d)$ for joins to act properly with respect to P. So we must extend the Boolean algebra operations on $\mathcal{A}$ in another way. At this point, it is not clear that there is such an extension. But there is, and we refer the reader to [41] for the details. Here is the theorem.

Theorem 10.6.1 *Let $(\Omega, \mathcal{A})$ be a measurable space and let $\mathcal{C}(\mathcal{A})$ be the space of conditional events, with the imbedding of $\mathcal{A}$ into $\mathcal{C}(\mathcal{A})$ as indicated. Let join, meet, and negation on $\mathcal{A}$ be denoted by $\vee, \wedge$, and $'$, respectively. Let negation on $\mathcal{C}(\mathcal{A})$ be given by $(a|b)' = a'b|b$. Then there is exactly one way to extend $\vee$ and $\wedge$ to $\mathcal{C}(\mathcal{A})$ so that both these operations are commutative and associative, have identities, De Morgan's laws hold, and for all probability measures P on $\mathcal{A}$,*

1. $P(a|b \vee c|d) \geq P(a|b)$ and

2. $P(a|b \wedge c|d) \leq P(a|b)$.

This extension is given by the equations

$$a|b \vee c|d = (a \vee c)|(a \vee c \vee bd)$$
$$a|b \wedge c|d = (a \wedge c)|((a' \wedge b) \vee (c' \wedge d) \vee (b \wedge d))$$

With these operations, $\mathcal{C}(\mathcal{A})$ is a bounded distributive lattice. But it is more: it is a Stone algebra. The pseudocomplement * is given by $(a|b)^* = (a'b)|1$. These facts are left as exercises.

One unsatisfactory aspect of the construction just made is that it does not yield a measurable space in which the events contain $\mathcal{C}(\mathcal{A})$. Here is a construction that does so. Again, let $(\Omega, \mathcal{A})$ be a measurable space. Let $\Omega^* = \Omega \times \Omega \times \Omega \times ...$, the Cartesian product of infinitely many copies of Ω. Let $\mathcal{A}^*$ be the σ-field of subsets of Ω^* generated by sets of the form $a_1 \times a_2 \times ... \times a_n \times \Omega \times \Omega \times ...$, with each $a_i \in \mathcal{A}$, and $n \geq 0$. If P is a probability measure on $(\Omega, \mathcal{A})$, let P^* be the infinite product measure on $(\Omega^*, \mathcal{A}^*)$ determined by $P^*(a_1 \times a_2 \times ... \times a_n \times \Omega \times \Omega \times ...) = \prod_{i=1}^{n} P(a_i)$. Denote $a_1 \times a_2 \times ... \times a_n \times \Omega \times \Omega \times ...$ simply by $a_1 \times a_2 \times ... \times a_n$, and let $\alpha : \mathcal{C}(\mathcal{A}) \to \mathcal{A}^*$ be defined by

$$\alpha(a, b) = (ab) \vee (b' \times ab) \vee (b' \times b' \times ab) \vee ...$$

This is a disjoint union in Ω^*, and hence

$$
\begin{aligned}
P^*(\alpha(a, b)) &= \sum_{n=0}^{\infty} P((b')^n \times ab \times \Omega \times \Omega...) \\
&= \sum_{n=0}^{\infty} P(ab)(P(b'))^n \\
&= P(ab) \sum_{n=0}^{\infty} (P(b'))^n \\
&= \frac{P(ab)}{1 - P(b')} \\
&= \frac{P(ab)}{P(b)} \\
&= P(a|b)
\end{aligned}
$$

Now $(\Omega^*, \mathcal{A}^*, P^*)$ is a probability space, and $\alpha(a, b)$ is an event in the σ-field $\mathcal{A}^*$, a much bigger σ-field $\mathcal{A}^*$ than $\mathcal{A}$.

10.7   Exercises

1. Let $\mathcal{P}$ be a set of probability measures on a measurable space $(\Omega, \mathcal{A})$. For $A \in \mathcal{A}$, let

$$
\begin{aligned}
F(A) &= \bigwedge \{P(A) : P \in \mathcal{P}\} \\
G(A) &= \bigvee \{P(A) : P \in \mathcal{P}\}
\end{aligned}
$$

Show that

   (a) $G(A) = 1 - F(A')$ and
   (b) if P is a probability measure, then $F \leq P$ if and only if $P \leq G$.

2. Let $\Omega = \{\omega_1, \omega_2, \omega_3, \omega_4\}$, and make the probability mass assignments

f	ω_1	ω_2	ω_3	ω_4	Ω
	0.4	0.2	0.2	0.1	0.1

   (a) Write down all the numerical values of $F(A) = \sum_{B \subseteq A} f(B)$.
   (b) Let $\mathcal{P}$ denote the set of probability measures P such that

$$
\begin{aligned}
P(\{\omega_1\}) &\geq 0.4 \\
P(\{\omega_2\}) &\geq 0.2 \\
P(\{\omega_3\}) &\geq 0.2 \\
P(\{\omega_4\}) &\geq 0.1
\end{aligned}
$$

   Verify that for $A \subseteq \Omega$,

$$
F(A) = \inf\{P(A) : P \in \mathcal{P}\}
$$

3. Let Q be a probability measure on a finite set Ω, and let $0 < \varepsilon < 1$. Let

$$
\mathcal{P} = \{\varepsilon P + (1 - \varepsilon)Q : P \text{ is a probability measure on } \Omega\}
$$

   Show that

$$
F(A) = \inf\{P(A) : P \in \mathcal{P}\}
$$

   is a belief function.

4. Let f be a probability mass assignment, that is, a probability density function on 2^Ω, with Ω finite. For $\varnothing \neq A \in 2^\Omega$ and for $\omega \in A$, let

$$
\begin{aligned}
\alpha(A, \omega) &= f(A)/ \mid A \mid \\
g(\omega) &= \sum_{\omega \in A} \alpha(A, \omega)
\end{aligned}
$$

(a) Verify that g is a probability density.

(b) Show that if $P_g(A) = \sum_{w \in A} g(w)$, then

$$F(A) = \sum_{B \subseteq A} f(B) \leq P_g(A)$$

5. Let S be a constant random set. That is, $P(S = B) = 1$ for some $B \subseteq \Omega$.

    (a) Determine the associated belief function $F(A) = P(S \subseteq A)$.

    (b) Let T be a random set with belief function G. Show that the belief function of $S \cap T$ is $G(A \cup B')$.

    (c) Suppose that G is additive. Show that in general, $G(A \cup B') \neq G(A|B)$.

6. Let $(\Omega, \mathcal{A}, P)$ be a probability space. Using induction, verify Poincare's inequality: for $n \geq 1$ and $A_1, A_2, ..., A_n \in \mathcal{A}$,

$$P\left(\bigcup_{i=1}^{n} A_i\right) = \sum_{\varnothing \neq J \subseteq \{1,2,...,n\}} (-1)^{|J|+1} P\left(\bigcap_{j \in J} A_j\right)$$

7. Let F be a belief function on a finite set Ω.

    (a) Show that F is monotone increasing.

    (b) Show that F is 2-monotone (or super modular, or convex). That is, show that

$$F(A \cup B) + F(A \cap B) \geq F(A) + F(B)$$

8. Let $\Omega = \{w_1, w_2, w_3, ..., w_k\}$. Let

$$T(w_j) = \{A : w_j \in A \subseteq \{w_j, w_{j+1}, ..., w_k\}\}$$

Let F be a belief function on Ω and f its Möbius inversion. Define $g : \Omega \to \mathbb{R}$ by

$$g(w_j) = \sum_{A \in T(w_j)} f(A)$$

    (a) Show that g is a probability density on Ω.

    (b) Let P_g denote the probability measure with density g. Show that $F \leq P_g$.

(c) Let $\mathcal{F} = \{A : F(A) > 0\}$. For $A \in \mathcal{F}$ and $\omega \in A$, let $\alpha(\omega, A) \geq 0$ satisfy

$$\sum_{\omega \in A} \alpha(\omega, A) = f(A)$$

and define

$$g_\alpha(\omega) = \sum_{\omega \in A \in \mathcal{F}} \alpha(\omega, A)$$

Verify that g_α is a probability density and that $F \leq P_{g_\alpha}$.

9. Let F be a belief function on the finite set Ω. Let $B \subseteq \Omega$ with $F(B) > 0$. For $A \subseteq B$ show that the formula

$$F_B(A) = \frac{F(A)}{F(A) + 1 - F(A \cup B')}$$

reduces to the definition of conditional probability if F is additive.

10. Verify that the following conditioning operators for belief functions generalize conditional probability measures. Assume that no denominator is 0.

(a) $F(A|B) = \dfrac{F(A \cup B') - F(B')}{1 - F(B')}$

(b) $F(A|B) = \dfrac{F(A \cap B)}{F(B)}$

(c) $F(A|B) = \dfrac{F(A) - F(A \cap B')}{1 - F(B')}$

11. Let P be a probability measure on the finite set Ω. Assume that no denominator is 0.

(a) For any positive integer n, show that $F(A) = [P(A)]^n$ is a belief function.

(b) Let $F(A) = [P(A)]^2$ and

$$\begin{aligned}
F_B(A) &= F(A|B) \\
&= \frac{F(A \cap B)}{F(B)} \\
&= \left[\frac{P(A \cap B)}{P(B)}\right]^2
\end{aligned}$$

Verify that for each B, the Möbius inversion of the belief function $F(\cdot, |B)$ is

$$
f_B(A) = \begin{cases} [P(A|B)]^2 & \text{when } A = \{\omega\} \\ 2P(\omega_1|B)P(\omega_2|B) & \text{when } A = \{\omega_1, \omega_2\} \\ 0 & \text{when } A = \varnothing \text{ or } |A| > 2 \end{cases}
$$

(c) Show that the $F(\cdot|B)$ in the previous part is commutative, that is, that $(F_B)_C = (F_C)_B$.

(d) Show that $F(\cdot|B)$ satisfies the "sandwich principle", namely that for $A \subseteq \Omega$,

$$
F(A) \geq \min\{F(A|B), F(A|B')\}
$$

12. Let $(\Omega, \mathcal{A}, P)$ be a probability space, and $X : \Omega \to \mathbb{R}^+$ be a non-negative random variable. Show the expected value $\int_\Omega X(\omega)dP(\omega)$ of X can also be written $\int_0^\infty P(X > t)dt$.

13. Let n be a positive integer. If P is a probability measure on the finite set Ω whose density f on Ω is the uniform one, then the density on Ω which is compatible with the belief function P^n and has maximum entropy is f.

14. Show that in Algorithm 2 there is a unique K such that

$$
\frac{B(K \cup \Theta_i^c) - B(\Theta_i^c)}{|K|}
$$

is maximum.

15. If m is uniform on the nonempty subsets of 2^Ω, how many steps does Algorithm 2 take?

16. Let U be a set, $\mathcal{E}$ be a partition of U, and $\sim$ the equivalence relation on $\mathcal{P}(U)$ giving rough sets. Show that $\sim$ is not a congruence on the partially ordered set $(\mathcal{P}(U), \subseteq)$.

17. (Lewis' Triviality Result) Let B be a Boolean algebra with more than four elements. Then there is no binary operation $\diamond$ on B such that for all probability measures P on B, and all $a, b \in B$ with $P(b) > 0$, we have $P(a \diamond b) = P(a|b)$.

18. Let $P : \mathcal{A} \to [0, 1]$ be a probability measure on a Boolean algebra $\mathcal{A}$. For $b \in \mathcal{A}$ with $P(b) > 0$, define $P(a|b) = (P(a \cap b))/P(b)$. Let $(b \Rightarrow a) = (b' \cup a)$. Show that

$$
P(b \Rightarrow a) = P(a|b) + P(a'|b)P(b).
$$

Thus $P(b \Rightarrow a) > P(a|b)$, in general. (See Section 9.1 for the definition of probability measure.)

19. Show that the space $\mathcal{C}(\mathcal{A})$ of conditional events as given in Theorem 10.6.1 is a Stone algebra.

20. Complete the proof of Lemma 10.5.3.

21. Write out in detail the proof of Theorem 10.5.6.

22. Show that the X_R in Lemma 10.5.3 form a sublattice of the bounded lattice $\mathcal{P}(U)$ that is isomorphic to R.

23. Prove Corollary 10.5.8.

24. (Coset representation of conditional events) A Boolean ring is a ring R with identity such that every element is idempotent, that is, satisfies $a^2 = a$. Let $a' = 1 + a$, $a + Rb = \{a + rb : r \in R\}$, $a \vee b = a + b + ab$, $a \leq b$ if $ab = a$, and $[a, b] = \{x : a \leq x \leq b\}$.

  (a) Show that $a + Rb' = c + Rd'$ if and only if $ab = cd$ and $b = d$.

  (b) Show that $a + Rb' = [ab, b' \vee a]$.

  (c) Show that $\{[a, b] : a \leq b\}\} = \{a + Rb' : a, b \in R\}$.

25. Let $(\Omega, \mathcal{A})$ be a measure space. For $a, b \in \mathcal{A}$, let $\phi_{a,b} : \Omega \to \{0, u, 1\}$ be defined by

$$\phi_{a,b}(\omega) = \begin{cases} 1 & \text{if} \quad \omega \in ab \\ 0 & \text{if} \quad \omega \in a'b \\ u & \text{if} \quad \omega \in b' \end{cases}$$

Show that $\{\phi_{a,b} : a, b \in \mathcal{A}\}$ is in one-to-one correspondence with $\mathcal{C}(\mathcal{A})$.

Chapter 11

FUZZY MEASURES

Possibility measures and belief functions are special nonadditive set functions. In this chapter, we consider a more general type of nonadditive set functions called fuzzy measures. They are used to model uncertainty in subjective evaluations.

11.1 Motivation and definitions

The simplest situation where subjective evaluations might be called upon is exemplified by the following: suppose Ω represents the states of nature, the unknown true state being ω_0. For a subset A of Ω, you are asked to guess whether A contains ω_0. You may answer "yes" but are not quite sure. This situation is similar to the one in the setting of belief functions, and is reminiscent of the statistical concept of confidence interval estimation. For the latter, in the process of constructing an interval estimate of, say, the mean μ of a population, we have a random sample $X_1, X_2, ..., X_n$ from that population and a random set $S(X_1, X_2, ..., X_n)$. Before collecting the data, S is a random set, and we can talk about the probability $P(\mu \in S)$ of coverage. Suppose that $P(\mu \in S) = \alpha$. We collect the random sample $x_1, x_2, ..., x_n$ and form the set $S(x_1, x_2, ..., x_n)$. Either $\mu \in S(x_1, x_2, ..., x_n)$ or it is not, but we say that we are confident to a degree that $\mu \in S(x_1, x_2, ..., x_n)$. Now it is agreed that even without a random mechanism as in statistics, humans still can express subjectively their degrees of trust or "grades of fuzziness", with values in $[0, 1]$. Thus, to each $A \subseteq \Omega$, a value $\nu(A)$ is assigned expressing a belief that $\omega_0 \in A$. Obviously, $\nu(\varnothing) = 0$ and $\nu(\Omega) = 1$ and ν is monotonic increasing. Since ν is assigned subjectively, one refers to $\nu(A)$ as a grade of fuzziness, where fuzziness is used in analogy with the subjectivity in assigning a membership function to a fuzzy concept. In this interpretation, the set function ν

is called a **fuzzy measure** and is used to model human subjective evaluations. One might argue that subjective evaluations could be subjective probabilities. The main difference is that, as set functions, subjective evaluations need not be additive. Thus, the general framework for modeling subjective evaluations is flexible for real-world applications. Note that one might also use the adjective "fuzzy" as opposed to "stochastic", as in fuzzy reasoning or fuzzy control. The concept of fuzzy measures, on the other hand, is also backed by considerations of imprecise probability models in which one is forced to deal with nonadditive set functions such as lower and upper probabilities. As we will see, except for special cases, these "imprecise" probabilities are only monotone set functions.

The terms "fuzzy measure" and "fuzzy integral" were introduced in Sugeno's thesis in 1974. In the context of Lebesgue abstract integration theory, fuzzy measures surface as a weakening of the σ-additivity of standard measures. If $(\Omega, \mathcal{A})$ is a measurable space, then the original Sugeno fuzzy measure concept is defined as follows: A set function $\mu : \mathcal{A} \to [0, 1]$ is called a fuzzy measure if $\mu(\varnothing) = 0$, $\mu(\Omega) = 1$, and for any monotone increasing sequence $A_n \in \mathcal{A}$, $\mu(\cup A_n) = \lim_{n \to \infty} \mu(A_n)$. This last condition is **monotone continuity**. The condition $\mu(\Omega) = 1$ is not crucial, and if it is dropped, then abstract measures are examples of fuzzy measures. The basic additivity property of measures is dropped. For an ordinary measure, the σ-additivity property implies monotone continuity, which is essential in establishing the Lebesgue theory. But, integration with respect to fuzzy measures, either using Sugeno's integral or Choquet's integral, does not rely on the monotone continuity property, but only on the fact that μ is monotone. Thus the axiom of monotone continuity is also dropped. *And* more generally, the domain $\mathcal{A}$ of μ can be taken as an arbitrary class of subsets of Ω.

We will elaborate on this point. In probability theory, or more generally, in measure theory, the domain of σ-additive measures is taken to be a σ-field. For example, consider the experiment of selecting at random a point in $[0, 1]$. For any subset A of $[0, 1]$ we can declare whether or not the "event" A occurred—it did if the point selected was in A and did not if it was not. However, what we are really interested in is the probability $P(A)$ of A before performing the experiment. Unlike the discrete case where $P(A)$ can be determined by specifying a probability on singletons, the continuous case is more complicated. For any $\omega \in [0, 1]$, we must specify that $P(\{\omega\}) = 0$. Were $P(\{\omega\}) > 0$, then since we are selecting a point at random, each $\omega \in [0, 1]$ would have the same probability, and so $\sum_{i=1}^{n} P(\{\omega_i\}) > 1$ for sufficiently large n. We can assign $P(A)$ to be the length of A for subintervals of $[0, 1]$ but cannot extend this assignment to the set of all subsets of $[0, 1]$ and keep σ-additivity. But it can be extended uniquely to the σ-field $\mathcal{B}$ generated by the subintervals, namely to the Borel subsets of $[0, 1]$.

Now the essential property of a fuzzy measure μ is its monotonicity. Suppose that a fuzzy measure μ is specified on an arbitrary class $\mathcal{C}$ of subsets of a set Ω. Then μ can be extended in many ways to all of 2^{Ω} keeping its monotonicity. For example, $\mu_*(A) = \sup\{\mu(B) : B \in \mathcal{C}, B \subseteq A\}$ is one such extension and $\mu^*(A) = \inf\{\mu(B) : B \in \mathcal{C}, B \supseteq A\}$ is another. Thus the domain of a fuzzy measure can be an arbitrary set of subsets of a set. The range of μ can be taken to be $[0, 1]$ or $[0, \infty)$ if need be.

Definition 11.1.1 *Let $\mathcal{A}$ be a family of subsets of a set Ω, with $\varnothing \in \mathcal{A}$. A mapping $\mu : \mathcal{A} \to [0, \infty)$ is called a **fuzzy measure** if*

1. $\mu(\varnothing) = 0$ and

2. if A, $B \in \mathcal{A}$ and $A \subseteq B$, then $\mu(A) \leq \mu(B)$.

*The triple $(\Omega, \mathcal{A}, \mu)$ is a **fuzzy measure space**.*

With this general definition, fuzzy measures are set functions which are monotone increasing with respect to set inclusion. They appear in various areas of mathematics. In the following two sections, we will illustrate some basic situations.

11.2 Fuzzy measures and lower probabilities

Consider again the case of incomplete probabilistic information. Let $\mathcal{P}$ be a class of probability measures on a measurable space $(\Omega, \mathcal{A})$. For $A \in \mathcal{A}$, let $\mu(A) = \inf\{P(A) : P \in \mathcal{P}\}$. Obviously, μ is a fuzzy measure with $\mu(\Omega) = 1$. In this case, μ is a **lower probability**. It is the **lower envelope** of $\mathcal{P}$. The monotonicity of a fuzzy measure μ is referred to as monotonicity of order 1. More generally, μ is **monotone of order** n if for $A_i \in \mathcal{A}$,

$$\mu(\cup_{i=1}^{n} A_i) \geq \sum_{\varnothing \neq I \subseteq \{1,2,\ldots,n\}} (-1)^{|I|+1} \mu(\cap_{k \in I} A_k)$$

Probability measures and belief functions on finite Ω are monotone of order n for all positive integers n. Monotone of order 2 means

$$\mu(A \cup B) \geq \mu(A) + \mu(B) - \mu(A \cap B)$$

For a fuzzy measure μ, the set of all probability measures P such that $P \geq \mu$ is denoted $\mathcal{P}(\mu)$. The elements of $\mathcal{P}(\mu)$ are those probability measures that **dominate** μ, or are **compatible** with μ.

Example 11.2.1 Let $\Omega = \{\omega_1, \omega_2\}$. Consider the fuzzy measure μ defined by

$$
\begin{aligned}
\mu(\varnothing) &= 0 \\
\mu(\omega_1) &= 0.7 \\
\mu(\omega_2) &= 0.7 \\
\mu(\Omega) &= 1
\end{aligned}
$$

No probability measures P on Ω can dominate μ in the sense that if $A \subseteq \Omega$ then $\mu(A) \leq P(A)$. That is, there is no probability measure compatible with μ. Indeed, for P to dominate μ, $P\{\omega_i\} \geq 0.7$, an impossibility. Thus the class of probability measures compatible with μ is empty. Unlike belief functions, fuzzy measures do not have a probabilistic interpretation in terms of classes of compatible probability measures. Thus fuzzy measures can be used to model different types of uncertainty.

The fuzzy measure in this example is not monotone of order 2 since for $A = \{\omega_1\}$ and $B = \{\omega_2\}$, we have

$$
\begin{aligned}
\mu(A \cup B) &= \mu(\Omega) = 1 \\
\mu(A \cap B) &= \mu(\varnothing) = 0
\end{aligned}
$$

while $\mu(A) + \mu(B) = 1.4$. However, if a fuzzy measure is monotone of order at least 2, then it is monotone of lower order. In this example, $\mathcal{P}(\mu) = \varnothing$ is due precisely to the fact that μ is *not* monotone of order 2.

Example 11.2.2 This example shows that, even when $\mathcal{P}(\mu) \neq \varnothing$, μ might not be the lower envelope of $\mathcal{P}(\mu)$. Let $\Omega = \{\omega_1, \omega_2, \omega_3\}$. Consider the fuzzy measure

$$
\begin{aligned}
\mu(\varnothing) &= 0 \\
\mu(\omega_1) &= 1/3 \\
\mu(\omega_2) &= 1/3 \\
\mu(\omega_3) &= 1/6 \\
\mu(\omega_1, \omega_2) &= 2/3 \\
\mu(\omega_1, \omega_3) &= 2/3 \\
\mu(\omega_2, \omega_3) &= 2/3 \\
\mu(\Omega) &= 1
\end{aligned}
$$

If $P \in \mathcal{P}(\mu)$, then since

$$\mu(\omega_1) = \tfrac{1}{3} \leq P\{\omega_1\}$$
$$\mu(\omega_2) = \tfrac{1}{3} \leq P\{\omega_2\}$$
$$1 = P\{\omega_1\} + P\{\omega_2\} + P\{\omega_3\}$$
$$\mu\{\omega_1, \omega_3\} = \tfrac{2}{3} \leq P\{\omega_1\} + P\{\omega_3\}$$

we get $P\{\omega_2\} = \tfrac{1}{3}$. Similarly, $P\{\omega_1\} = \tfrac{1}{3}$, and therefore so is $P\{\omega_3\}$. Thus P is the uniform distribution on Ω, and clearly is not μ.

Example 11.2.3 This example shows that if $\mu = \inf \mathcal{P}$, the lower envelope of a class of probability measures $\mathcal{P}$, its class $\mathcal{P}(\mu)$ of compatible measures might be different from $\mathcal{P}$. Let $\Omega = \{\omega_1, \omega_2, \omega_3, \omega_4\}$, and let probability measures P_i be given by the table below.

Ω	ω_1	ω_2	ω_3	ω_4
P_1	0.5	0.2	0.2	0.1
P_2	0.4	0.3	0.2	0.1
P_3	0.4	0.2	0.3	0.1
P_4	0.4	0.2	0.2	0.2

If $\mathcal{P} = \{P_1, P_2, P_3, P_4\}$ and $\mu(A) = \inf\{P_i(A)\}$, then $\mathcal{P} \subseteq \mathcal{P}(\mu)$, but the probability density P given by $P\{\omega_1\} = 0.425$, $P\{\omega_2\} = P\{\omega_3\} = 0.225$ and $P\{\omega_4\} = 0.125$ dominates μ.

Example 11.2.4 The discussion in this example emphasizes the fact that, unlike belief functions, general fuzzy measures do not have the formal interpretation in terms of random sets. The notion of **Möbius inversion** can be defined for arbitrary set functions on finite sets. Let Ω be finite and let $\mu : 2^\Omega \to \mathbb{R}$. Now define $f : 2^\Omega \to \mathbb{R}$ by

$$f(A) = \sum_{B \subseteq A} (-1)^{|A - B|} \mu(B)$$

This f is the Möbius inversion of μ, and

$$\mu(A) = \sum_{B \subseteq A} f(B)$$

In the case of a fuzzy measure μ with $\mu(\Omega) = 1$, we have

$$\sum_{A \subseteq \Omega} f(A) = \mu(\Omega) = 1$$

but unless f is nonnegative, we cannot view f as a density function for some random set. Here is an example. Let $\Omega = \{\omega_1, \omega_2, \omega_3\}$, let μ be 0 except for $\mu\{\omega_i, \omega_j\} = .5$ for distinct pairs, and $\mu(\Omega) = 1$. Then

$$f(\Omega) = \sum_{A \subseteq \Omega} (-1)^{|\Omega - A|} \mu(A) = -.5$$

The Möbius inversion gives a nonnegative function only in the case of belief functions, that is, fuzzy measures monotone of infinite order.

Example 11.2.5 A very general and useful class of fuzzy measures consists of those which are monotone of order 2. These fuzzy measures are sometimes referred to as **convex fuzzy measures**, a term in game theory where the monotonicity of order 2 is the analog of the positivity of the second derivative of convex functions. This type of fuzzy measure also appears in robust Bayesian statistics.

Let $\Omega = \{\omega_1, \omega_2, ..., \omega_n\}$, and let μ be a fuzzy measure monotone of order 2. Let f be the Möbius inversion of μ. Define $P : 2^\Omega \to \mathbb{R}$ by

$$P(A) = \sum_{\varnothing \neq B \subseteq \Omega} f(B) \frac{|A \cap B|}{|B|}$$

We are going to show that P is a probability measure on Ω. That $P(\varnothing) = 0$ and $P(\Omega) = 1$ are easy. Also it is easy to check that for $A \cap B = \varnothing$, $P(A \cup B) = P(A) + P(B)$. To finish the proof, it suffices to show that for $\omega \in \Omega$, $P(\{\omega\}) \geq 0$.

Now $\sum_{\omega \in D \subseteq \Omega} [\mu(D \cup \{\omega\}) - \mu(D)] \geq 0$ since μ is monotone. Thus it suffices to show that

$$\sum_{\{\omega\} \cup D \subseteq B} \frac{(-1)^{|B - D \cup \{\omega\}|}}{|B|} \geq 0$$

For $m = |\Omega| - |D| - 1, d = |D|$,

$$\sum_{D\cup\{\omega\}\subseteq B\subseteq\Omega} \frac{(-1)^{|B\backslash(D\cup\{\omega\})|}}{|B|} = \sum_{n=|D|+1}^{|\Omega|} \sum_{\substack{D\cup\{\omega\}\subseteq B\subseteq\Omega \\ |B|=n}} \frac{(-1)^{n-|D|-1}}{n}$$

$$= \sum_{n=|D|+1}^{|\Omega|} \binom{|\Omega|-|D|-1}{n-|D|-1} \frac{(-1)^{n-|D|-1}}{n}$$

$$= \sum_{k=0}^{|\Omega|-|D|-1} \binom{|\Omega|-|D|-1}{k} \frac{(-1)^k}{k+|D|+1}$$

$$= \sum_{k=0}^{m} \frac{(-1)^k}{k+d+1} \binom{m}{k}$$

$$= \frac{d!m!}{(m+d+1)!} > 0$$

All equalities are clear except the last, and that is proved in the lemma below.

Lemma 11.2.6 *For $m \geq 0, d \geq 0$,*

$$\sum_{k=0}^{m} \frac{(-1)^k}{k+d+1} \binom{m}{k} = \frac{d!m!}{(m+d+1)!}$$

Proof. Induct on m. For $m = 0$, the summation reduces to

$$\frac{(-1)^0}{0+d+1}\binom{0}{0} = \frac{1}{d+1}$$

Assume that for some $m = m_0 \geq 0$ the equality holds for all $d \geq 0$ and for all $m \leq m_0$ and consider

$$\sum_{k=0}^{m+1} \frac{(-1)^k}{k+d+1}\binom{m+1}{k}$$

The following relationship

$$\binom{m+1}{k} = \binom{m}{k} + \binom{m}{k-1}$$

holds between binomial coefficients, and we apply it in the induction step.

$$\sum_{k=0}^{m+1} \frac{(-1)^k}{k+d+1} \binom{m+1}{k}$$

$$= \frac{(-1)^0}{0+d+1} + \sum_{k=1}^{m} \frac{(-1)^k}{k+d+1} \left(\binom{m}{k} + \binom{m}{k-1} \right)$$

$$+ \frac{(-1)^{m+1}}{m+1+d+1} \binom{m+1}{m+1}$$

$$= \frac{1}{d+1} + \sum_{k=1}^{m} \frac{(-1)^k}{k+d+1} \binom{m}{k} + \sum_{k=1}^{m} \frac{(-1)^k}{k+d+1} \binom{m}{k-1} + \frac{(-1)^{m+1}}{m+d+2}$$

$$= \frac{d!\,m!}{(m+d+1)!} - \sum_{k=0}^{m-1} \frac{(-1)^k}{k+d+2} \binom{m}{k} + \frac{(-1)^{m+1}}{m+d+2}$$

$$= \frac{d!\,m!}{(m+d+1)!} - \frac{(d+1)!\,m!}{(m+d+2)!} + \frac{(-1)^m}{m+d+2} \binom{m}{m} + \frac{(-1)^{m+1}}{m+d+2}$$

$$= \frac{d!\,m!}{(m+d+1)!} \frac{(m+d+2)}{(m+d+2)} - \frac{(d+1)!\,m!}{(m+d+2)!}$$

$$= \frac{d!\,m!\,(m+d+2) - (d+1)!\,m!}{(m+d+2)!}$$

$$= \frac{d!}{(m+d+2)!} \left(m!\,(m+1) \right)$$

$$= \frac{d!\,(m+1)!}{(m+d+2)!}$$

This probability measure dominates μ, so that $\mathcal{P}(\mu) \neq \varnothing$. The proof of this fact is quite lengthy and is omitted. The interested reader can consult [14]. ■

11.3   Fuzzy measures in other areas

11.3.1   Capacities

As in measure theory, classical capacities on Euclidean spaces in potential theory are extended to an abstract setting as follows. Let Ω be a set. A **precapacity** on Ω is a map I from the power set 2^{Ω} to the extended real line $\overline{\mathbb{R}} = [-\infty, \infty]$ in such a way that

- I is monotone increasing and

- if A_n is an increasing sequence in 2^{Ω}, then $I(\cup A_n) = \sup I(A_n)$.

A simple precapacity is this: for any set Ω, let $I(A) = 1$ if A is an uncountable subset of Ω and 0 otherwise. Of course, letting $I(A) = 0$ for all A is a precapacity also.

Let $(\Omega, \mathcal{A}, P)$ be a probability space. Define I on 2^Ω by

$$I(A) = \inf\{P(B) : B \in \mathcal{A}, A \subseteq B\}$$

This precapacity has an important additional property. Since P is a probability measure, for any decreasing sequence A_n in $\mathcal{A}$,

$$P(\cap A_n) = I(\cap A_n) = \inf P(A_n) = \inf I(A_n)$$

Precapacities having this additional structure are called **capacities**. A collection $\mathcal{F}$ of subsets of Ω is called a **pavage** if it contains the empty set and is closed under finite unions and intersections. An $\mathcal{F}$-capacity on Ω is a precapacity I such that if F_n is a decreasing sequence in $\mathcal{F}$, then

$$I(\cap F_n) = \inf I(F_n)$$

Here are some examples.

Example 11.3.1 Let $\mathcal{K}$ and $\mathcal{O}$ denote, respectively, the set of compact and the set of open subsets of the Euclidean space $\mathbb{R}^d$. Classical capacities in potential theory have some special properties. Let $I : \mathcal{K} \to [0, \infty]$ satisfy the following:

1. I is increasing on $\mathcal{K}$.

2. I is **strongly subadditive**. That is, for $A, B \in \mathcal{K}$,

$$I(A \cup B) + I(A \cap B) \leq I(A) + I(B)$$

3. I is **continuous on the right**. That is, for $A \in \mathcal{K}$ and $\varepsilon > 0$, there is an open set V containing A such that for every $B \in \mathcal{K}$ satisfying $A \subseteq B \subseteq V$ we have $I(B) \leq I(A) + \varepsilon$.

The **inner capacity** associated with I is defined for any subset A of $\mathbb{R}^d$ by

$$I_*(A) = \sup\{I(K) : K \in \mathcal{K}, K \subseteq A\}$$

and the **outer capacity** I^* by

$$I^*(A) = \inf\{I_*(X) : X \in \mathcal{O}, A \subseteq X\}$$

It can be verified that I^* is a $\mathcal{K}$-capacity.

Example 11.3.2 Let $(\Omega, \mathcal{A}, P)$ be a probability space. Define $I : 2^\Omega \to [0,1]$ by

$$I(A) = \inf\{P(B) : B \in \mathcal{A}, A \subseteq B\}$$

Then I is an $\mathcal{A}$-capacity.

Example 11.3.3 When Ω is finite, every increasing set function on 2^Ω is a 2^Ω-capacity. For example, let $\varphi : 2^\Omega \to [0,1]$ satisfy $\sum_{A \subseteq \Omega} \varphi(A) = 1$. Then $I(A) = \sum_{B \subseteq A} \varphi(B)$ is a 2^Ω-capacity. Without referring to the pavage 2^Ω in the finite case, fuzzy measures on finite sets are often referred to as capacities.

11.3.2   Measures and dimensions

We give here some examples of measures and dimensions.

Example 11.3.4 An interval J in $\mathbb{R}^n$ is the Cartesian product $\prod_{i=1}^n [a_i, b_i]$ of intervals of $\mathbb{R}$. Let $v(J) = \prod_{i=1}^n (b_i - a_i)$. For $A \subseteq \mathbb{R}^n$, define $\mu(A) = \inf \sigma(\mathcal{J})$ where the inf is over all possible coverings $\mathcal{J} = \{J_j\}_{j=1}^\infty$ of A by countable intervals of $\mathbb{R}^n$, and $\sigma(\mathcal{J}) = \sum_{j=1}^\infty v(J_j)$. Clearly $\mu : 2^{\mathbb{R}^n} \to [0,\infty)$ is a fuzzy measure. Note that $\mu(\cup_{j=1}^\infty A_j) \leq \sum_{j=1}^\infty \mu(A_j)$, so μ is subadditive. Such a set function is called an **outer measure**.

Example 11.3.5 Let (Ω, d) be a metric space. The diameter of $A \subseteq \Omega$ is $\delta(A) = \sup\{d(x,y) : x, y \in A\}$. For each $\alpha > 0$ and each $\varepsilon > 0$, consider $\mu_\alpha^\varepsilon(A) = \inf \sum_k [\delta(A_k)]^\alpha$, where the inf is taken over all countable collections $\{A_k\}$ such that $A \subseteq \cup A_k$ and $\delta(A_k) < \varepsilon$ for each k. Let $\mu_\alpha(A) = \lim_{\varepsilon \to 0} \mu_\alpha^\varepsilon(A)$. Then μ_α is an outer measure, called the **Hausdorff outer measure of dimension** α. Now define $\mu : 2^\Omega \to [0,\infty]$ by

$$\mu(A) = \inf\{\alpha > 0 : \mu_\alpha(A) = 0\}$$

The quantity $\mu(A)$ is called the **Hausdorff dimension** of A. The function μ is not additive, but is monotone increasing.

Example 11.3.6 Let $(\Omega, \mathcal{A}, P)$ be a probability space. Define $\mu : \mathcal{A} \to [0,\infty]$ by $\mu(A) = [-c \log P(A)]^{-1}$, where $c > 0$. This fuzzy measure has the following special property.   If $A \cap B = \varnothing$, then

$$\begin{aligned}
\mu(A \cup B) &= [-c \log(P(A) + P(B))]^{-1} \\
&= \left[\max\{0, -c\log\{e^{\frac{-\mu(A)}{c}} + e^{\frac{1\mu(B)}{c}}\}\}\right]^{-1}
\end{aligned}$$

That is, when $A \cap B = \varnothing$, $\mu(A \cup B)$ is a function of $\mu(A)$ and $\mu(B)$. So $\mu(A \cup B) = \varphi(\mu(A), \mu(B))$. Such fuzzy measures are called **decomposable** with **decomposition operation** φ. If μ is additive, then it is decomposable with operation $\varphi(x, y) = x + y$.

Example 11.3.7 Let $(\Omega, \mathcal{A})$ be a measurable space. Let $\mu : \mathcal{A} \to [0, 1]$ satisfy

1. $\mu(\Omega) = 1$ and

2. if $A \cap B = \varnothing$, then

$$\mu(A \cup B) = \mu(A) + \mu(B) + \lambda\mu(A)\mu(B)$$

for some $\lambda > -1$. Then μ is called a λ-measure. Since

$$\begin{aligned} 1 &= \mu(\Omega \cup \varnothing) \\ &= \mu(\Omega) + \mu(\varnothing) + \lambda\mu(A)\mu(B) \\ &= 1 + \mu(\varnothing) + \lambda\mu(\varnothing) \end{aligned}$$

we have $\mu(\varnothing)(1 + \lambda) = 0$, implying that $\mu(\varnothing) = 0$. For $A \subseteq B$, we have

$$\mu(B) = \mu(A) + \mu(B - A) + \lambda\mu(A)\mu(B - A)$$

But

$$\mu(B - A) + \lambda\mu(A)\mu(B - A) \geq 0$$

since $\lambda > -1$ and $0 \leq \mu(A) \leq 1$. Thus $\mu(A) \leq \mu(B)$, so that μ is a fuzzy measure. Note that this λ-measure can be expressed in the form $\mu(A) = f(P(A))$ where P is a probability measure and f is a monotone increasing function. Indeed,

$$\mu(A) = \frac{1}{\lambda}[e^{\log(1+\lambda\mu(A))} - 1]$$

But observe that

$$\begin{aligned} e^{\log(1+\lambda\mu(A))} &= [(1 + \lambda)^{\frac{1}{\log(1+\lambda)}}]^{\log(1+\lambda\mu(A))} \\ &= (1 + \lambda)^{\frac{\log(1+\lambda\mu(A))}{\log(1+\lambda)}} \end{aligned}$$

Thus for $\lambda > -1$, we take

$$f(x) = \begin{cases} \frac{1}{\lambda}[(1 + \lambda)^x - 1] & \text{if } \lambda \neq 0 \\ x & \text{if } \lambda = 0 \end{cases}$$

Then $f(0) = 0$ and $f(1) = 1$ and f is increasing. Next take

$$P(A) = \frac{\log(1 + \lambda\mu(A))}{\log(1 + \lambda)}$$

It remains to verify the P is additive. For $A \cap B = \varnothing$, we have

$$P(A \cup B) = \frac{\log(1 + \lambda\mu(A \cup B))}{\log(1 + \lambda)}$$

But

$$
\begin{aligned}
\log(1 + \lambda\mu(A \cup B)) &= \log\{1 + \lambda(\mu(A) + \mu(B) + \lambda(\mu(A)\mu(B)) \\
&= \log[(1 + \lambda(\mu(A)))(1 + \lambda(\mu(B))] \\
&= \log(1 + \lambda(\mu(A))) + \log(1 + \lambda(\mu(B)) \\
&= [P(A) + P(B)\}\log(1 + \lambda)
\end{aligned}
$$

Note that Sugeno's λ-measures are decomposable with $\varphi_\lambda(x, y) = x + y + \lambda y$. If $\lambda = 0$, then μ_λ is additive.

11.3.3 Game theory

The mathematical framework for the coalition form of game theory is as follows [7]: let Ω be the set of players, and $\mathcal{A}$ be a σ-field of subsets of Ω, representing the collection of all possible coalitions. A **game** is a map $\nu : \mathcal{A} \to \mathbb{R}$ with the interpretation that for $A \in \mathcal{A}$, $\nu(A)$ is the worth of A, that is, the payoff that the coalition A would get. The function ν is very general, and not necessarily additive. For example, ν can be the square of an additive measure.

Set functions in game theory are sometimes superadditive, that is they satisfy $\nu(A \cup B) \geq \nu A) + \nu(B)$ for A and B disjoint. This is supposed to reflect the fact that disjoint coalitions do not lose by joining forces. A similar situation with belief functions is this: the **core** of a game ν is the set of all bounded, finitely additive signed measures μ on $\mathcal{A}$ dominating ν, that is for which $\nu(A) \leq \mu(A)$ for $A \in \mathcal{A}$ and $\nu(\Omega) \leq \mu(\Omega)$.

Another fundamental concept in game theory is that of a **value.** Roughly speaking, it is an operator that assigns to each player of a game a number representing the amount the player would be willing to pay in order to participate. Thus a value is a map from games to payoff distributions, which are bounded, finitely additive signed measures.

It is interesting to note that in the study of value and core in game theory [7], there is a need to generalize—or idealize—the notion of set by specifying for each point a weight between 0 and 1, which indicates the "degree" to which that point belongs to the set. This "ideal" kind of set is used to formalize the intuitive notion of "evenly spread" sets. Subsets of the set Ω of players are coalitions. Evenly spread subsets are viewed as "ideal" coalitions which are simply fuzzy subsets of Ω. The extension of a game from $\mathcal{A}$ to fuzzy events plays a key role in the study. (For details, see [7].) The study of nonadditive set functions and their associated linear operators is an interesting topic in game theory. Results in Chapter 10 concerning belief functions viewed as special games can be partially extended to arbitrary monotone games.

11.3.4 Distributions of random sets

In Chapter 1 we touched on sets obtained randomly. We will elaborate more here on these **random sets**. They play a special role in connection with fuzzy analysis. As spelled out in Chapter 1, membership functions of fuzzy sets are covering functions of random sets. On the other hand, distributions of random sets are characterized by a special class of fuzzy measures. This last point is the main thrust of this section.

Random sets are mathematical models for describing random patterns in fields such as stochastic geometry, spatial statistics, pattern recognition, stereology, mathematical morphology, and image analysis.

Here is a simple example of a random set. Let X be a nonnegative random variable, defined on some probability space $(\Omega, \mathcal{A}, P)$. Then $[0, X]$ is a **random interval** of the real line.

Distributions of random sets on a finite set U are easy to characterize. Let $S : \Omega \to \mathcal{P}(U)$ be a random set on U. Then, as in the case of random variables, its distribution, that is a probability measure on $\mathcal{P}\mathcal{P}(U)$, is characterized by its probability density function $f : \mathcal{P}(U) \to [0,1]$, with $\sum_{A \subseteq U} f(A) = 1$, and $f(A) = P\{\omega : S(\omega+ = A\}$. Equivalently, the distribution of S is characterized by the belief function $F(A) = \sum_{B \subseteq A} f(B)$. Note that F is a fuzzy measure that is **monotone of infinite order**, that is, for $A_1, A_2, ..., A_k$ in $\mathcal{P}(U)$,

$$F(\cup_{j=1}^k A_j) \geq \sum_{\varnothing \neq I \subseteq \{1,2,...,k\}} (-1)^{|I|+1} F(\cap_{i \in I} A_i)$$

It is convenient to use the dual of F, namely $T(A) = 1 - F(A') = P(S \cap A \neq \varnothing)$. The set function T is a fuzzy measure which is **alternating of infinite order**. That is, T is monotone increasing and for $k \geq 2$, and $A_1, A_2, ..., A_k$ in $\mathcal{P}(U)$,

$$T(\cap_{j=1}^k A_j) \leq \sum_{\varnothing \neq I \subseteq \{1,2,...,k\}} (-1)^{|I|+1} T(\cup_{i \in I} A_i)$$

Without going into technical details, we merely say that the simple situation above generalizes as follows.

By a **random closed set** we mean a map S from Ω to the set $\mathcal{F}$ of closed subsets of $\mathbb{R}^m$. Let $\mathcal{K}$ and $\mathcal{O}$ be respectively the set of compact and open sets of $\mathbb{R}^m$. Then the distribution of S is completely characterized by its capacity functional $T(K) = P(S \cap K) \neq \varnothing)$, which is alternating of infinite order on $\mathcal{K}$, and satisfies the condition that if $K_n \searrow K$, then $T(K_n) \searrow T(K)$. This characterization result is the **Choquet theorem** in random set theory.

Here is an example. Let $\pi : \mathbb{R} \to [0,1]$ be upper semicontinuous. That is, for each $a \in [0,1], \{x : \pi(x) \geq a\}$ is closed. Then $T(K) = \sup_{x \in K} \pi(x)$ is the capacity functional of the random closed set $S(\omega) = \{x : \pi(x) \geq \alpha(\omega)\}$, where α is a random variable uniformly distributed on $[0,1]$. It is easy to see that $T(K) = P(S \cap K \neq \varnothing)$ since $\{\omega : S\omega) \cap K \neq \varnothing\} = \{\omega : \alpha(\omega) \leq \sup_{x \in K} \pi(x)\}$. The fact that T is alternating of infinite order follows from the fact that $T(A \cup B) = \max(T(A), T(B))$, that is, that T is a **maxitive set function**. First, it is clear that $T(\cup_{i=1}^n (A) = \max\{T(A_i) : i = 1, 2, ..., n\}$. We can assume that

$$0 \leq s_n = T(A_n) \leq ... \leq s_2 = T(A_2) \leq s_1 = T(A_1)$$

For $k \in \{1, 2, ..., n\}$, let $\mathcal{I}(k) = \{I \subseteq \{1, 2, ..., n\} : |I| = k\}$, and for $I \in \mathcal{I}(k)$ and $i = 1, 2, ..., n - k + 1$, let $\mathcal{I}_i(k) = \{I \in \mathcal{I}(k) : \min\{I\} = i\}$. Then $T(\cup_{j \in I} A_j) = s_i$ for every $I \in \mathcal{I}_i(k)$, and $i = 1, 2, ..., n - k + 1$. Next

$$\sum_{I \in \mathcal{I}(k)} T(\cup_{i \in I} A_i)$$

$$= \sum_{i=1}^{n-k+1} \sum_{I \in \mathcal{I}(k_j)} T(\cup_{j \in I} A_j)$$

$$= \sum_{i=1}^{n-k+1} \sum_{I \in \mathcal{I}(k_j)} s_i$$

$$= \sum_{i=1}^{n-k+1} \binom{n-1}{k-1} s_i$$

Thus

$$\sum_{\{I:\Omega\neq I\subseteq\{1,2,...,k\}\}} (-1)^{|I|+1} T(\cup_{i\in I} A_i)$$

$$= \sum_{I=1}^{n} (-1)^{k+1} \sum_{I\in\mathcal{I}(k)} T(\cup_{i\in I} A_i)$$

$$= \sum_{i=1}^{n-k+1} (-1)^{k+1} \sum_{i=1}^{n-k+1} \binom{n-1}{k-1} s_i$$

$$= \sum_{i=1}^{n1} (-1)^{k+1} \left[\sum_{k=0}^{n-1} \binom{n-i}{k} (-1)^k \right] s_i$$

But $\sum_{k=0}^{n-1} \binom{n-i}{k}(-1)^k = 0$ for every $i = 1, 2, ..., n-1$, so we have

$$\sum_{\{I:\Omega\neq I\subseteq\{1,2,...,k\}\}} (-1)^{|I|+1} T(\cup_{i\in I} A_i) = s_n = T(A_n) \geq T(\cap_{k=1}^{n} A_k)$$

whence T is alternating of infinite order.

The interest in maxitive functions lies in the fact, as seen just above, that they are possible models of distributions of random sets. Possibility measures are of course maxitive. For a general construction of maxitive set functions, see the exercises.

Here is another example. Let (Ω, d) be a metric space. Let $B(x, r) = \{y \in \Omega : d(x, y) < r\}$. For $A \subseteq \Omega$ and $n \geq 1$, let

$$T_n(A) = \inf\{r > 0 : A \subseteq B(x_1, r) \cup B(x_1, r) \cup B(x_2, r) \cup ... \cup B(x_n, r)\}$$

Since the sequence $(T_n(A), n \geq 1)$ is decreasing, its limit exists, is denoted $T(A)$, and called the **Kuratowski measure of non-compactness**. Now observe that $T_{2n}(A \cup B) \leq \max(T_n(A), T_n(B))$ for $n \geq 1$. Thus $T(A \cup B) \leq \max(T(A), T(B))$ and hence $T(A \cup B) = \max(T(A), T(B))$ since T is increasing.

11.4 Conditional fuzzy measures

Let $(\Omega, \mathcal{A}, \mu)$ be a fuzzy measure space. We take $\mathcal{A}$ to be an algebra of subsets of Ω. The function μ is monotone increasing with $\mu(\varnothing) = 0$ and $\mu(\Omega) \in \mathbb{R}^+$. The **dual fuzzy measure** of μ is given by $\mu^*(A) = \mu(\Omega) - \mu(A')$, where A' is the set complement of A. By analogy with conditional probabilities, define $\mu(\cdot|B)$ by

$$\mu(A|B) = \frac{\mu(A \cap B)}{\mu(A \cap B) + \mu^*(A' \cap B)}$$

This is well-defined only when the denominator $\neq 0$. So the domain $\mathcal{D}(B)$ of $\mu(\cdot|B)$ is the set of those A such that $\mu(A \cap B) + \mu^*(A' \cap B) > 0$. In particular, $\varnothing \in \mathcal{D}(B)$ if $\mu^*(B) > 0$. Note that $\mathcal{D}(B) = \mathcal{A}$ if, for example, $\mu^*(B) > 0$ and μ is subadditive. Subadditive means that $\mu(A_1 \cup A_2) \leq \mu(A_1) \cup \mu(A_2)$ for disjoint A_1 and A_2.

We show that $\mu(\cdot|B)$ is monotone increasing on $\mathcal{D}(B)$. Let $A_1 \subseteq A_2$. Then since μ and μ^* are increasing,

$$\begin{aligned}
&\mu(A_1 \cap B)[\mu(A_2 \cap B) + \mu^*(A_2' \cap B)] \\
=\ &\mu(A_1 \cap B)\mu(A_2 \cap B) + \mu(A_1 \cap B)\mu^*(A_2' \cap B) \\
\leq\ &\mu(A_1 \cap B)\mu(A_2 \cap B) + \mu(A_2 \cap B)\mu^*(A_1' \cap B)
\end{aligned}$$

and so $\mu(A_1|B) \leq \mu(A_2|B)$. Thus $\mu(\cdot|B)$ is a fuzzy measure. Other forms of conditional fuzzy measures are possible. The following are examples.

- $\mu(A|B) = \dfrac{\mu(A \cap B)}{\mu(B)}$ for $\mu(B) > 0$.

- $\mu(A_1|B) = \dfrac{\mu((A \cap B) \cup B') - \mu(B)}{\mu(\Omega) - \mu(B')}$ for the denominator $\mu^*(B) > 0$.

We discuss now the case when $\mu(B) = 0$. Without exception, fuzzy inference is based essentially on conditional information. Probabilistic reasoning and the use of conditional probability measures rest on a firm mathematical foundation and are well justified. The situation for non-additive set functions seems far from satisfactory. This is an important open problem and we will elaborate on it in some detail. The material is essentially probabilistic.

Let X and Y be random variables defined on a probability space $(\Omega, \mathcal{A}, P)$. The product Borel σ-field on $\mathbb{R} \times \mathbb{R}$ is denoted $\mathcal{B} \times \mathcal{B}$ and the joint probability measure of (X, Y) on $\mathcal{B} \times \mathcal{B}$ is determined by

$$Q(A \times B) = P(X \in A, Y \in B)$$

The marginal probability measures are, respectively,

$$\begin{aligned}
Q_X(A) &= Q(A \times \mathbb{R}) \\
Q_X(B) &= Q(\mathbb{R} \times B)
\end{aligned}$$

We now review briefly the concept of conditional probability. Setting

$$a = (X \in A) = X^{-1}(A)$$
$$b = (Y \in B) = Y^{-1}(B)$$

and denoting $a \cap b$ by ab, we set

$$P(b|a) = \frac{P(ab)}{P(a)}$$

whenever $P(a) \neq 0$. For each such a, $P(b|a)$ is a probability measure on $(\Omega, \mathcal{A})$.

Now suppose X and Y are defined on $(\Omega, \mathcal{A}, \mathbb{R})$, and X is a continuous random variable. Then for $x \in \mathbb{R}$, $P(X = x) = 0$. How do we make sense of $P(Y \in B|X = x)$? Suppose X is uniformly distributed on $[0,1]$, and Y is a binomial random variable defined as follows. If $X = x$, then the value of Y is the number of heads gotten in n tosses of a coin where the probability of getting a head is the value x of X. The probability of getting at most j heads is given by the formula

$$\sum_{k=0}^{j} \binom{n}{k} x^k (1-x)^{n-k}$$

and this value is $P(Y \leq j|X = x)$. The density of X is $f(x) = 1_{[0,1]}(x)$ and the density of Y is

$$g(k|x) = \binom{n}{k} x^k (1-x)^{n-k}$$

under the condition that $X = x$. If we define g by this formula for $x \in [0,1]$ and 0 otherwise, then the variable Y, given $X = x$ should have g as a "conditional density", and then

$$\sum_{k=0}^{j} \binom{n}{k} x^k (1-x)^{n-k}$$

can be written as

$$P(Y \leq j|X = x) = \sum_{k=0}^{j} g(k|x)$$

Thus, for any $B \in \mathcal{B}$, one might take $P(Y \in B|X = x)$ as a probability measure, even when $P(X = x) = 0$. The set function $B \to P(Y \in B|X = x)$ is the conditional probability measure of Y when $X = x$. We discuss now the existence of such conditional probability measures in a general setting. Noting that

$$P(Y \in B) = E(1_{(Y \in B)}) = \int_{\Omega} 1_{(Y \in B)}(\omega)dP(\omega) < \infty$$

the general problem becomes the existence of the "conditional expectation" of an integrable random variable Y given $X = x$. This conditional expectation is denoted $E(Y|X = x)$. Since Y is integrable, the set function on $\mathcal{B}$ defined by

$$M(B) = \int_{(X \in B)} Y(\omega)dP(\omega)$$

is a signed measure, absolutely continuous with respect to Q_X, and hence by the Radon-Nikodym theorem from measure theory, there exists a $\mathcal{B}$-measurable function $f(x)$, unique to a set of Q_X measure zero, such that

$$\int_{(X \in B)} Y(\omega)dP(\omega) = \int_B f(x)dQ_X(x)$$

As a special case, when Y is of the form $1_{(Y \in B)}$, we write

$$Q(B|X) = E[1_{(Y \in B)}|X = x]$$

It is easy to check that the function

$$K : \mathcal{B} \times \mathbb{R} \to [0,1]$$

defined by $K(B, x) = Q(B|x)$ satisfies

- for each fixed $B \in \mathcal{B}$, $K(B, \cdot)$ is $\mathcal{B}$-measurable and

- for each $x \in \mathbb{R}$, $K(\cdot, x)$ is a probability measure.

Such a function is called a **Markov kernel.** By the first property, $\int_A K(B, x)dQ_X(x)$ is well defined for $A, B \in \mathcal{B}$, and

$$Q(A \times B) = \int_A K(B, x)dQ_X(x)$$

thus relating the joint measure Q on $\mathcal{B} \times \mathcal{B}$ with the marginal measure Q_X of X. When $A = \mathbb{R}$, we have

$$Q(\mathbb{R} \times B) = Q_Y(B) = \int_{\mathbb{R}} K(B, x)dQ_X(x) \qquad (11.1)$$

This is the basic relation between marginals Q_X, Q_Y and the Markov kernel. Thus we can state that the conditional probability measure $P(Y \in B|X = x)$ is a Markov kernel $K(B, x)$ such that 11.1 holds. This can be used as a guideline for defining conditional fuzzy measures.

By analogy with probability theory, one can define a conditional fuzzy measure, denoted $\mu(B|x)$, to be a "fuzzy kernel", that is, satisfying

- for each $B \in \mathcal{B}$, $\mu(B|\cdot)$ is a $\mathcal{B}$-measurable function;

- for each $x \in \mathbb{R}$ (say), $\mu(\cdot|x)$ is a fuzzy measure.

The basic problem is the existence of μ satisfying the counterpart of 11.1 in the fuzzy setting. "Integrals" with respect to fuzzy measures need to be defined in such a way that they generalize abstract Lebesgue integrals. We postpone this analysis until fuzzy integrals have been defined, but will indicate the formulation.

Let $(\Omega_i, \mathcal{A}_i)$ be two measurable spaces. A **fuzzy kernel** K from Ω_1 to Ω_2 is a map $K : \mathcal{A}_2 \times \Omega_1 \to \mathbb{R}^+$ such that

- for each $A_2 \in \mathcal{A}_2$, $K(A_2, \cdot)$ is $\mathcal{A}_1$-measurable and

- for each $\omega_1 \in \Omega_1$, $K(\cdot, \omega_1)$ is a normalized fuzzy measure on $\mathcal{A}_2$.

If P_1 is a probability measure on $(\Omega_1, \mathcal{A}_1)$, then

$$Q_2(A_2) = \int_{\Omega_1} K(A_2, \omega_1) dP_1(\omega_1)$$

is a fuzzy measure on $(\Omega_2, \mathcal{A}_2)$. The same is true when P_1 is replaced by a fuzzy measure Q_1 on $(\Omega_1, \mathcal{A}_1)$, where integration is in the sense of Choquet. That is,

$$\int_{\Omega_1} K(A_2, \omega_1) dP_1(\omega_1) = \int_0^1 Q_1\{\omega_1 : K(A_2, \omega_1) \geq t\} dt$$

In the following, the product σ-field on the Cartesian product space $\Omega_1 \times \Omega_2$ is denoted $\mathcal{A}_1 \otimes \mathcal{A}_2$. If a normalized fuzzy measure Q on $(\Omega_1 \times \Omega_2, \mathcal{A}_1 \otimes \mathcal{A}_2)$ is such that

$$
\begin{aligned}
Q(A_1 \times A_2) &= \int_{A_1} K(A_2, \omega_1) dQ_1(\omega_1) \\
&= \int_0^1 Q_1\{\omega_1 : 1_{A_1}(\omega_1) K(A_2, \omega_1) \geq t\} dt
\end{aligned}
$$

then Q has Q_1 and Q_2 as marginals. That is,

$$Q(A_1 \times \Omega_2) = \int_0^1 Q_1\{\omega_1 : 1_{A_1}(\omega_1 \geq t\} dt = Q_1(A_1)$$

and $Q(\Omega_1 \times A_2) = Q_2(A_2)$. The problem of defining $Q(X_2 \in A_2 | X_1 = \omega_1)$ when $Q(X_1 = \omega_1) = 0$ appeared in the context of Bayesian statistics concerning lower and upper probabilities. There, only a partial solution was given. An open problem is this: if Q is a joint fuzzy measure on $(\Omega_1 \times \Omega_2, \mathcal{A}_1 \otimes \mathcal{A}_2)$, under which conditions can a fuzzy kernel K from Ω_1 to Ω_2 be found such that

$$Q_2(A_2) = \int_0^1 Q_1\{\omega_1 : K(A_2, \omega_1) \geq t\} dt$$

where Q_i are marginal fuzzy measures of Q? The need for considering conditional fuzzy measures is apparent in situations such as fuzzy regression problems, interpretation and assignments of weights to fuzzy rules in control or expert systems, and so on.

The discussion above is concerned with the problem of existence of conditional fuzzy measures of the form $\mu(Y \in A : X = x)$ when $\mu(X = x) = 0$. It is a mathematical problem. Of course, from a subjective evaluation point of view, one can specify such measures for inference purposes. A mathematically difficult problem arises if we insist on a respectable level of the theory of fuzzy measures. In this regard, much research needs to be done. Fuzzy kernels can be used to define conditional fuzzy measures. (See, for example, [101, 17].)

Research in the field of fuzzy measures and integrals is currently very active, with many papers appearing. The reader should be aware that there is no consensus even in terminology. For example, a Radon-Nikodym theorem for fuzzy valued measures is not a Radon-Nikodym theorem for nonadditive set functions. While dropping σ-additivity of ordinary measures, most of the researchers add assumptions to fuzzy measures which make the latter almost similar to the former. Perhaps one should fix a precise case, say Choquet capacities, and see what fundamental new results will surface. The problem of conditional fuzzy measures $\mu(Y \in A : X \in B)$, where $\mu(X \in B) > 0$, does not cause technical problems. As in the case of belief conditioning of Shafer, the problem with these straightforward definitions is that it yields a whole host of choices. Without a clear logic or application-oriented motive, various definitions of conditional fuzzy measures look like mathematical games of extending conditional probability measures. If the root of fuzzy measures is in subjective evaluation processes, the conditioning problem should involve human factors, and any proposed conditioning procedure should be tested empirically.

11.5 Exercises

1. Verify that μ is monotone of order 2 when

$$\mu(A \cup B) + \mu(A \cap B) \geq \mu(A) + \mu(B)$$

2. Show that if μ is monotone of order $n \geq 2$, then it is monotone of order m for $2 \leq m \leq n$.

3. Show that μ in Example 11.2.3 of Section 11.2 is monotone of infinite order.

4. Let μ be a fuzzy measure. For $A, B, X \in \mathcal{A}$, define

$$\Delta_1(X, A) = \mu(X) - \mu(X \cap A)$$
$$\Delta_2(X, A, B) = \Delta_1(X, A) - \Delta_1(X \cap B, A)$$

Show that μ is monotone of order 2 if and only if $\Delta_2(X, A, B) \geq 0$ for all X, A, B.

5. *Show that if the Möbius inversion of a fuzzy measure μ is non-negative, then μ is a belief function, that is, is monotone of infinite order.

6. *Let Ω be finite. Let Q be a probability measure on Ω such that $Q(A) > 0$ for every subset $A \neq \varnothing$. Let μ be a fuzzy measure on Ω such that $\mu(\Omega) = 1$, and f its Möbius inversion.

   (a) Define $P : 2^\Omega \to \mathbb{R}$ by

$$P(A) = \sum_{\varnothing \neq B \subseteq \Omega} f(B)Q(A|B)$$

   where $Q(A|B) = \dfrac{Q(A \cap B)}{Q(B)}$. Show that P is a probability measure.

   (b) Show that P dominates μ if and only if μ is monotone of order 2.

7. Let $\mathcal{F}$ be a pavage of subsets of Ω such that if $F_n \in \mathcal{F}$ with F_n decreasing in n and $\neq \varnothing$, then $\cap F_n \neq \varnothing$. Define $I : 2^\Omega \to \{0, 1\}$ by $I(A) = 1$ if $A \neq \varnothing$ and 0 otherwise. Show that I is an $\mathcal{F}$-capacity.

8. Let $\mathcal{C}$ be a collection of subsets of Ω. Let $\mu : \mathcal{C} \to [0, \infty)$ be monotone.

(a) Verify that μ_* and μ^* defined in 11.1 are monotone.

(b) Verify that any monotone extension ν of μ to 2^Ω satisfies $\mu_* \le \nu \le \mu^*$.

9. Let $\mathcal{C}$ be a collection of subsets of Ω such that $\varnothing \in \mathcal{C}$ and $\mathcal{C}$ is closed under complementation. Let $\mu : \mathcal{C} \to [0, \infty)$ be monotone. Let $\nu(A) = \mu(\Omega) - \mu(A^c)$. Show that ν is monotone.

10. Let $(\Omega, \mathcal{A}, P)$ be a probability space. *Let $\lambda > 0$. Define $\mu_\lambda : \mathcal{A} \to [0, 1]$ by*

$$\mu_\lambda(A) = \frac{1}{\lambda}[(1 + \lambda)^{P(A)} - 1]$$

(a) Show that $\lim_{\lambda \to 0} \mu_\lambda(A) = P(A)$.

(b) Show that μ_λ is a fuzzy measure.

(c) Show that μ_λ is a Sugeno λ-measure.

11. Let $(\Omega, \mathcal{A}, P)$ be a fuzzy measure space, where $\mathcal{A}$ is an algebra of subsets of Ω. Assuming $\mu(\Omega) < \infty$, consider $\mu^*(A) = \mu(\Omega) - \mu(A')$, where $A' = \Omega - A$.

(a) Verify that μ^* is a fuzzy measure on $\mathcal{A}$.

(b) Show that μ is monotone of order 2 if and only if

$$\mu^*(A \cup B) + \mu^*(A \cap B) \le \mu^*(A) + \mu^*(B)$$

This condition is called **alternating of order** 2, or **submodular**.

12. Show that the formula

$$\mu(A|B) = \frac{\mu(A \cap B)}{\mu(A \cap B) + \mu^*(A' \cap B)}$$

for finite fuzzy measures μ reduces to the usual definition of conditional probability when μ is additive and $\mu(\Omega) = 1$.

13. Let $(\Omega, \mathcal{A}, P)$ be a finite fuzzy measure space. With the notation of Section 10.1, show that $\mathcal{D}(B) = \mathcal{A}$ when $\mu^*(B) > 0$, and μ is subadditive.

14. Let X be a nonnegative random variable defined on some probability space $(\Omega, \mathcal{A}, P)$. Consider the random set $S(\omega) = [0, X(\omega)]$. The random length $\mu(S)$ of S is X. Show that the expected value of $\mu(S)$ is $\int_0^\infty \pi(x)dx$, where $\pi(x) = P(\{\omega : x \in S(\omega)\})$ is the covering function of S.

15. Let S be a random closed set in $\mathbb{R}^k$. Let μ denote Lebesgue measure on $\mathbb{R}^k$. Then $E(\mu(S)) = \int_{\mathbb{R}^k} \pi(x)d\mu(x)$, where $\pi(x) = P(x \in S)$. Let $S_1, S_2, ..., S_n$ be a random sample from S. Define the empirical covering function to be $\pi_n(x) = \#\{j : x \in S_j\}/n$.

   (a) Show that $\pi_n(x) \to \pi(x)$ as $n \to \infty$, for each x, with probability one.

   (b) Show that $\int_{\mathbb{R}^k} \pi_n(x)d\mu(x) \to E(\mu(S))$ as $n \to \infty$, for each x, with probability one.

16. Let X be a nonnegative random variable. Consider the random closed set $S(\omega) = [0, X(\omega)]$ on $\mathbb{R}^+$. Let $\pi(x) = P(x \in S)$. Determine the capacity functional of S in terms of π.

17. Let $\pi : \mathbb{R} \to [0, 1]$ be upper semicontinuous. For $K \in \mathcal{K}(\mathbb{R})$, let $T(K) = \sup_{x \in K} \pi(x)$. Show that if $K_n \in \mathcal{K}(\mathbb{R})$ with $K_n \searrow$, then $T(\cap_n K_n) = \lim_{n \to \infty} T(K_n)$.

18. Let $(\Omega, \mathcal{A})$ be a measurable space. An **ideal** $\mathcal{J}$ of $\mathcal{A}$ is a subset of $\mathcal{A}$ such that $\varnothing \in \mathcal{J}$, if $A \in \mathcal{J}$ and $B \in \mathcal{A}$ with $B \subseteq B$, then $B \in \mathcal{J}$, and $A, B \in \mathcal{J}$ imply that $A \cup B \in \mathcal{J}$. A σ-ideal $\mathcal{J}$ of $\mathcal{A}$ is a ideal $\mathcal{J}$ of $\mathcal{A}$ such that if $A_n \in \mathcal{J}$, then $\cup_n A_n \in \mathcal{J}$.

   (a) If P is a probability measure on $\mathcal{A}$, verify that $\mathcal{J} = \{A \subseteq \mathcal{A} : P(A) = 0\}$ is a σ-ideal of $\mathcal{A}$.

   (b) Let $\mathcal{J}_t, t \geq 0$, be a family of σ-ideals of $\mathcal{A}$ such that $\mathcal{J}_s \subseteq \mathcal{J}_t$ whenever $s \leq t$. Show that $T(A) = \inf\{t \geq 0 : A \in \mathcal{J}_t\}$ is , that is, for any sequence $A_n \in \mathcal{A}$, $T(\cup_n A_n) = \sup_n T(A_n)$.

   (c) Suppose that Ω is a metric space, and $d(A)$ is the diameter of A. An **outer measure** on $\mathcal{P}(\Omega)$ is a function $\mu : \mathcal{P}(\Omega) \to [0, \infty]$ such that $\mu(\varnothing) = 0$, is monotone increasing, and is σ-subadditive, that is, if $A_n \in \mathcal{P}(\Omega)$, then $\mu(\cup_n A_n) \leq \sum_n \mu(A_n)$. For $\alpha \geq 0$, the **Hausdorff α-measure** μ_α is defined by

$$\mu_\alpha(A) = \lim_{\varepsilon \to 0}\left\{ \inf \sum_n (d(A_n))^\alpha \right\}$$

where the inf is taken over all countable coverings of A by closed balls A_n such that $d(A_n) < \epsilon$. Verify that each μ_α is an outer measure.

   (d) Define the **Hausdorff dimension** as $D(A) = \inf\{\alpha \geq 0 : \mu_\alpha(A) = 0\}$. Show that D is strongly maxitive.

   (e) Let P be a probability measure on $(\Omega, \mathcal{A})$, and $f : \Omega \to [0, \infty]$
       measurable and bounded. Show that $T(A) = \inf\{t \geq 0 : P(A \cap (f > t)) = 0\}$ is strongly maxitive.

19. (General construction of maxitive set functions) Let $(\Omega, \mathcal{A})$ be a
    measurable space.

   (a) Let μ be a maxitive set function defined on $\mathcal{A}$ with $\mu(\varnothing) = 0$.
       Show that for each $t > 0$, $\mathcal{J}_t = \{A \in \mathcal{A} : \mu(A) \leq t\}$ is an ideal
       in $\mathcal{A}$. Show also that the family of ideals $\{\mathcal{J}_t\}_{t \geq 0}$ is increasing
       in t. Verify that $\mu(A) = \inf\{t \geq 0 : A \in \mathcal{J}_t\}$.

   (b) Let $\{\mathcal{K}_t\}_{t \geq 0}$ be a family of ideals of $\mathcal{A}$ increasing in t. Let
       $\mu(A) = \inf\{t \geq 0 : A \in \mathcal{K}_t\}$. Show that μ is maxitive.

   (c) Let $\mathcal{J}$ be an ideal of $\mathcal{A}$ and $f : \Omega \to [0, \infty)$ be measurable. Let
       $\mathcal{J}_t = \{A \in \mathcal{A} : A \cap (f > t) \in \mathcal{J}\}$. Show that $\mu(A) = \inf\{t \geq 0 : A \in \mathcal{J}_t\}$ is maxitive.

Chapter 12

FUZZY INTEGRALS

In Chapter 11, we have introduced fuzzy measures in various situations involving general subjective evaluations. Pursuing this investigation, we address now the problem of integration of functions with respect to monotone set-functions.

12.1 Review of the Lebesgue integral

Here we will review the elements of integration with respect to σ-additive set functions. Let $(\Omega, \mathcal{A}, \mu)$ be a measure space. We are going to describe how to construct abstract integrals for a large class of numerical functions defined on Ω. This class of functions $\mathcal{M}$ consists of the *measurable* ones.

Definition 12.1.1 *Let $f : \Omega \to \mathbb{R}$. The function f is **measurable** if $f^{-1}(B) \in \mathcal{A}$ whenever $B \in \mathcal{B}(\mathbb{R})$.*

For any function $g : S \to T$, the function $g^{-1} : 2^T \to 2^S$ preserves set operations. For example $g^{-1}(\cup A_i) = \cup g^{-1}(A_i)$. Thus $\mu \circ f^{-1} : \mathcal{B} \to [0, \infty]$ is a measure on $(\mathbb{R}, \mathcal{B})$.

It turns out that measurability of numerical functions can be checked by using some special classes of subsets of $\mathbb{R}$. Specifically, f is measurable if and only if for $x \in \mathbb{R}$, $\{\omega : f(\omega) < x\} \in \mathcal{A}$, or equivalently if $\{\omega : f(\omega) > x\} \in \mathcal{A}$.

A function $f : \Omega \to \mathbb{R}$ can be written as

$$f(\omega) = f^+(\omega) - f^-(\omega)$$

where $f^+(\omega) = \max\{0, f(\omega)\}$ and $f^-(\omega) = \max\{0, -f(\omega)\}$. Note that f^+ and f^- are nonnegative. If f is measurable, then f^+ and f^- are also measurable.

Let $A_1, A_2, ..., A_n$ be a partition of the space Ω into measurable sets. A nonnegative **simple function** is a function φ of the form $\varphi(\omega) = \sum_{i=1}^{n} \alpha_i 1_{A_i}(\omega)$. Thus φ has value α_i on A_i. It should be clear that φ is measurable. It turns out that nonnegative measurable function can be approximated by simple functions. Let f be such a function and for $n \geq 1$ define

$$f_n(\omega) = \sum_{k=1}^{n2^n} \frac{k}{2^n} 1_{A_n^k}(\omega) + n 1_{A_n}(\omega)$$

where

$$A_n^k = \{\omega : \frac{k-1}{2^n} < f(\omega) \leq \frac{k}{2^n}\}$$

for $k = 1, 2, ..., n2^n$, and $A_n = \{\omega : f(\omega) > n\}$. Then f_n is an increasing sequence of nonnegative functions converging pointwise to f.

With the concepts above, we can describe abstract integrals. Let f be a nonnegative simple function, say, $f(\omega) = \sum_{i=1}^{n} \alpha_i 1_{A_i}(\omega)$. The integral of f with respect to μ is defined by

$$\int_\Omega f(\omega) d\mu(\omega) = \sum_{i=1}^{n} \alpha_i \mu(A_i)$$

Now let f be nonnegative and measurable. There is an increasing sequence of nonnegative simple functions f_n converging pointwise to f. Since $\int_\Omega f(\omega) d\mu(\omega)$ is an increasing sequence of numbers, it has a limit in $[0, \infty]$, and we define

$$\int_\Omega f(\omega) d\mu(\omega) = \lim_{n \to \infty} \int_\Omega f_n(\omega) d\mu(\omega)$$

Of course, it remains to be verified that this limit is independent of the choice of the simple functions f_n converging to f, but we omit those details.

As in the case of simple functions, the map $f \to \int_\Omega f(\omega) d\mu(\omega)$ is monotone increasing. Moreover, the following *monotone convergence theorem* is fundamental.

Theorem 12.1.2 *(Monotone convergence theorem) If f is an increasing sequence of nonnegative measurable functions, then*

$$\lim_{n \to \infty} \int_\Omega f_n(\omega) d\mu(\omega) = \int_\Omega \lim_{n \to \infty} f_n(\omega) d\mu(\omega)$$

Finally, for f measurable, we write $f = f^+ - f^-$ and define

$$\int_\Omega f(\omega)d\mu(\omega) = \int_\Omega f^+(\omega)d\mu(\omega) - \int_\Omega f^-(\omega)d\mu(\omega)$$

Of course this is defined when both f^+ and f^- are integrable, that is when both $\int_\Omega f^+(\omega)d\mu(\omega)$ and $\int_\Omega f^-(\omega)d\mu(\omega)$ are finite. More generally, this is defined if not both are infinite.

12.2 The Sugeno integral

As we have seen, *possibility distributions* are used to describe values of variables under fuzzy constraints. The typical situation is this: let X be a variable taking values in a set Ω. If we know that $X \in A$, where $A \subseteq \Omega$, then the set of possible values of X reduces to A. This reduction is expressed by viewing the function $\pi : \Omega \rightarrow [0,1] : \omega \rightarrow 1_A(\omega)$ as a possibility distribution of X in the sense that $\pi(\omega)$ is the degree of possibility for $X = \omega$. Given such a function π, Zadeh proposed assigning a possibility measure

$$P_\pi(B) = \sup_{\omega \in B} \pi(\omega)$$

to each subset B of Ω. Then P_π is a fuzzy measure, being 0 at $\varnothing$ and increasing. This fuzzy measure can be written

$$P_\pi(B) = \sup_{\omega \in \Omega}\{1_B(\omega) \wedge \pi(\omega)\}$$

Now in rule-based systems, for example, suppose that we only learn that X is $\tilde{A}$, where $\tilde{A}$ is a fuzzy subset of Ω with membership function $\tilde{A}(\cdot)$. This information induces a possibility distribution on X, namely $\pi(\cdot) = \tilde{A}(\cdot)$ as a natural generalization of the ordinary case. Moreover, if $\tilde{B}$ is a fuzzy subset of Ω, then one would like to quantify the statement "X is $\tilde{B}$" under $\pi(\cdot)$. A natural way for doing so is to extend the domain of $P_\pi(\cdot)$ to fuzzy subsets of Ω. Replacing $1_B(\cdot)$ by $\tilde{B}(\cdot)$ in the equation above leads to

$$P_\pi(\tilde{B}) = \sup_{\omega \in \Omega}\{\tilde{B}(\omega) \wedge \pi(\omega)\}$$

the sup of the function π over the fuzzy set $\tilde{B}$. Later in this section, we will formulate the general problem of optimization of bounded real-valued functions over fuzzy sets and discuss related issues in the context of integration.

If X is a random variable with probability measure μ on the measurable space $(\Omega, \mathcal{A})$, then one can quantify the statement "X is $\widetilde{B}$", where $\widetilde{B}$ is a measurable fuzzy subset of Ω, by

$$E_\mu\left(\widetilde{B}(X)\right) = \int_\Omega \widetilde{B}(\omega)d\mu(\omega)$$

Thus the domain of the probability measure μ can be extended via Lebesgue integration. One would like to know whether this can be obtained in a similar manner. The main issue is how to "integrate" the membership function $\widetilde{B}$ with respect to a fuzzy measure like P_π. We will show that

$$P_\pi(\widetilde{B}) = \sup_{\alpha \in [0,1]} \{\alpha \wedge P_\pi(\widetilde{B} \geq \alpha)\}$$

Indeed,

$$\sup_{\alpha \in [0,1]} \{\alpha \wedge P_\pi(\widetilde{B} \geq \alpha)\} = \sup_{\alpha \in [0,1]} \{\alpha \wedge \sup_{\widetilde{B}(\omega) \geq \alpha} \{\widetilde{B}(\omega) \wedge \pi(\omega)\}\}$$

$$\leq \sup_{\omega \in \Omega} \{\widetilde{B}(\omega) \wedge \pi(\omega)\}$$

Conversely, let $\varepsilon > 0$. For $\alpha \in [0,1]$ there is a $v \in (\widetilde{B} \geq \alpha)$ such that

$$\sup_{(\widetilde{B} \geq \alpha)} \pi(\omega) \leq \pi(v) + \varepsilon$$

Since $\pi(v) \geq \alpha$,

$$\sup_{(\widetilde{B} \geq \alpha)} \pi(\omega) \wedge \alpha \leq (\pi(v) + \varepsilon) \wedge \widetilde{B}(v)$$

$$\leq \sup_\Omega \{\widetilde{B}(\omega) \wedge \pi(\omega)\} + \varepsilon$$

so that

$$\sup_{\alpha \in [0,1]} \{\alpha \wedge \sup_{(\widetilde{B} \geq \alpha)} \pi(\omega)\} \leq \sup_\Omega \{\widetilde{B}(\omega) \wedge \pi(\omega)\} + \varepsilon$$

and the equality follows.

In the expression $\sup_{\alpha \in [0,1]}\{\alpha \wedge P_\pi(\widetilde{B} \geq \alpha)\}$ for $P_\pi(\widetilde{B})$, if we replace P_π by any normalized fuzzy measure μ on $(\Omega, \mathcal{A})$, and $\widetilde{B}$ by $f : \Omega \to [0,1]$, we get the quantity

$$I_\mu(f) = \sup_{\alpha \in [0,1]} \{\alpha \wedge \mu(f \geq \alpha)\}$$

This last quantity was proposed by Sugeno as the "fuzzy integral" of f with respect to μ. It is easily shown that

$$\sup_{\alpha \in [0,1]} \{\alpha \wedge \mu(f \geq \alpha)\} = \sup_{\alpha \in [0,1]} \{\alpha \wedge \mu(f > \alpha)\}$$

and this equality holds for any fuzzy measure μ with values in $[0, \infty)$ and f measurable with values in $[0, \infty)$. We use the notation $I_\mu(f)$ to denote

$$\sup_{\alpha \geq 0} \{\alpha \wedge \mu(f \geq \alpha)\} = \sup_{\alpha \geq 0} \{\alpha \wedge \mu(f > \alpha)\}$$

This definition of Sugeno's integral is obtained by replacing addition and multiplication of real numbers by sup and inf, respectively, in the formula relating abstract Lebesgue integral to Lebesgue integral on the real line. That is, if μ is a σ-additive measure on $(\Omega, \mathcal{A})$, then

$$\int_\Omega f(\omega) d\mu(\omega) = \int_0^\infty \mu(f > t) dt$$

However, $I_\mu(f)$ is constructed directly for any nonnegative measurable function f without considering first nonnegative simple functions, as is done in ordinary measure theory. The reason is that the operations of sup and inf are well defined and $I_\mu(f)$ involves only these, while in Lebesgue integration, the integral sign is a new operation which needs to be defined.

The functional I_μ is different from the Lebesgue integral functional when μ is additive. We have that $f \leq g$ implies $I_\mu(f) \leq I_\mu(g)$. Also, $I_\mu(1_A) = \mu(A)$. Thus I_μ is a generalization of Lebesgue measure, but is not a generalization of Lebesgue integration.

The quantity $I_\mu(f)$ can be constructed from simple functions. If φ is a nonnegative simple function with $\varphi(\omega) = \sum_{i=1}^n \alpha_i 1_{A_i}(\omega)$, where the A_i form a measurable partition of Ω, then define $Q(\varphi) = \vee_{i=1}^n (\alpha_i \wedge \mu(A_i))$, in analogy with ordinary integrals of simple functions. However, as we will see, $Q(\varphi)$ is not the fuzzy integral of φ as it is in the case of the Lebesgue integral.

Theorem 12.2.1 $I_\mu(f) = \sup Q(\varphi)$, *where the* sup *is taken over all simple functions* $\varphi \leq f$.

Proof. Let $J_\mu(f) = \sup Q(\varphi)$. Now J_μ is obviously an increasing functional. For $A \in \mathcal{A}$, the simple function

$$\varphi(x) = \left(\inf_{x \in A} f(x) \right) 1_A(x) \leq f(x)$$

which implies that

$$J_\mu(f) \geq J_\mu(\varphi) \geq Q(\varphi) = \alpha \wedge \mu(A) = \left(\inf_A f \right) \wedge \mu(A)$$

and thus that

$$J_\mu(f) \geq \sup_{A \in \mathcal{A}} [(\inf_A f) \wedge \mu(A)]$$

On the other hand, if $\varphi(x) = \sum_{i=1}^n \alpha_i 1_{A_i}(x)$ and $\varphi \leq f$, then

$$Q(\varphi) = \sup_i \{\alpha_i \wedge \mu(A_i(x))\} = \alpha_k \wedge \mu(A_k)$$

for some $k \in \{1, 2, ..., n\}$. Since $\varphi \leq f$ and $\varphi(x) = \alpha_k$ on A_k, we have $\alpha_k \leq \inf_{A_k} f(x)$. Thus

$$Q(\varphi) \leq \left(\inf_{A_k} f\right) \wedge \mu(A_k) \leq \sup_{A \in \mathcal{A}} [(\inf_A f) \wedge \mu(A)]$$

Hence

$$J_\mu(f) = \sup_{A \in \mathcal{A}} [(\inf_A f) \wedge \mu(A)]$$

It remains to show that this last equation holds when $J_\mu(f)$ is replaced by $I_\mu(f)$. For $A \in \mathcal{A}$ and $\alpha = \inf_A f$, we have $A \subseteq (f \geq \alpha)$, so that $\mu(A) \leq \mu(f \geq \alpha)$. Thus

$$\sup_{A \in \mathcal{A}} \left[\left(\inf_A f\right) \wedge \mu(A)\right] \leq I_\mu(f)$$

Conversely, for each $\alpha \geq 0$, let $A = (f \geq \alpha) \in \mathcal{A}$. Then

$$\alpha \wedge \mu(f \geq \alpha) \leq \left(\inf_A f\right) \wedge \mu(A) \leq \sup_{A \in \mathcal{A}} [(\inf_A f) \wedge \mu(A)]$$

so that

$$I_\mu(f) \leq \sup_{A \in \mathcal{A}} [(\inf_A f) \wedge \mu(A)]$$

■

The quantity

$$Q(\varphi) = \sup_{i=1}^n \{\alpha_i \wedge \mu(A_i)]\}$$

is not equal to $I_\mu(\mu)$ in general. For example, let $\Omega = [0, 1]$, $\mu = dx$, the Lebesgue measure on the Borel σ-field $\mathcal{B}$ of $[0, 1]$, and

$$\varphi(x) = (5/6)\, 1_{[0, \frac{1}{2}]}(x) + (7/8)\, 1_{[\frac{1}{2}, 1]}(x)$$

for $x \in [0,1]$. Then $Q(\varphi) = \frac{1}{2} < I_\mu(\varphi) = \frac{5}{6}$, as an easy computation shows. The reason for this phenomenon is that $I_\mu(\varphi) = \sup Q(\psi)$ where the sup is taken over all simple functions $\psi \le \varphi$, which is not, in general, equal to $Q(\varphi)$, since $Q(\varphi)$ is not an increasing functional. For example, let $\psi(x) \equiv \frac{3}{4}$ and

$$\varphi(x) = (5/6) \, 1_{[0,\frac{1}{2}]}(x) + (7/8) \, 1_{[\frac{1}{2},1]}(x)$$

Then $\psi \le \varphi$, but $Q(\psi) = \frac{3}{4} > Q(\varphi) = \frac{1}{2}$. Thus when computing $I_\mu(\varphi)$ for φ a simple function, one has to use the general definition

$$I_\mu(\varphi) = \sup_{\alpha \ge 0}\{\alpha \wedge \mu(\varphi > \alpha)\}$$

which does not reduce in general to $Q(\varphi)$. However, if

$$\alpha_1 \le \alpha_2 \le ... \le \alpha_n$$

and $B_i = \cup_{j=i}^i A_j$, then for

$$\varphi(\omega) = \sum_{i=1}^n \alpha_i 1_{A_i}(x)$$

we have

$$I_\mu(\varphi) = \sup_{i=1}^n\{\alpha_i \wedge \mu(B_i)\}$$

Indeed, for all i,

$$\sup_{\alpha \in [\alpha_{i-1}, \alpha_i)} \{\alpha \wedge \mu(\varphi > \alpha)\} = \alpha_i \wedge \mu(B_i)$$

The definition of $I_\mu(f)$ does not involve the monotone continuity of μ as opposed to a stronger condition of ordinary measures, namely, the σ-additivity in the construction of the Lebesgue integral. Thus, even in the case where quantities like $I_\mu(f)$ are to be taken as fuzzy integrals, the monotone continuity axiom for fuzzy measures is not needed. Adding this property to the set of axioms for fuzzy measures obscures the subjective evaluation processes. It is difficult to justify why human subjective evaluations should obey the monotone continuity axiom. On the technical side, the class of fuzzy measures with the monotone continuity property is almost that of Choquet's capacities. Zadeh's possibility measures do not possess the decreasing monotone continuity property, and yet they are some of the most plausible set functions for quantifying subjective evaluations.

Consider possibility measures again. If $\pi : \Omega \to [0,1]$, and $A \subseteq \Omega$, then

$$\sup_{\omega \in A} \pi(\omega) = \sup_{\Omega} \{\pi(\omega) \wedge 1_A(\omega)\}$$

When A is a fuzzy subset of Ω, $\sup_{\omega \in A} \pi(\omega)$ is taken to be $\sup_{\Omega} \{\pi(\omega) \wedge A(\omega)\}$. However, this expression does not lend itself to a generalization to $\pi : \Omega \to [0, \infty]$ to represent $\sup_A \pi$ for A fuzzy or not. For such π, $\sup_{\Omega} \{\pi(\omega) \wedge A(\omega)\} \leq 1$ for A fuzzy, while it is not bounded by 1 if A is not.

If we consider the fuzzy measure given by $P_\pi(A) = \sup_A \pi$ when $\pi : \Omega \to [0,1]$, then $I_{P_\pi}(1_A) = P_\pi(A)$ for subsets A of Ω, and

$$I_{P_\pi}(A) = \sup_{\Omega} \{\pi(\omega) \wedge A(\omega)\}$$

for A fuzzy. Thus we can use $I_{P_\pi}(A)$ to represent $\sup_A \pi$ whether or not A is fuzzy. When $\pi : \Omega \to [0,1]$,

$$I_{P_\pi}(A) = \sup_{\alpha \in [0,1]} \{\alpha \wedge P_\pi(A > \alpha)\}$$

and when $\pi : \Omega \to [0, \infty]$, we simply consider the fuzzy measure, still denoted by P_π, which is not a possibility measure since $P_\pi(\Omega)$ might be greater than 1. In that case,

$$I_{P_\pi}(A) = \sup_{\alpha \geq 0} \{\alpha \wedge P_\pi(A > \alpha)\}$$

can be used to represent $\sup_A \pi$ for fuzzy sets A.

Thus the Sugeno integral with respect to fuzzy measures, which are strong precapacities, provides a means for formulating the problem of maximizing functions over constraints expressed as fuzzy sets.

12.3   The Choquet integral

Standard measure theory is a theory of integrals motivated by the need to compute important quantities such as volumes, expected values, and so on. Motivated by subjective probabilities, we are led to consider fuzzy measures. Since additive measures are special cases of fuzzy measures, fuzzy integrals should be generalizations of Lebesgue integrals. Or, perhaps a radically different concept of integral should be developed specifically for nonadditive set functions. Whatever approach is taken, we need to consider carefully the meaning of the quantities fuzzy integrals are supposed to measure. Just what are those quantities of interest in reasoning with general knowledge that need to be measured? Guidelines and motivation from practical applications are needed to keep the theory on the right track.

The situations that we will consider have statistical flavor. For non-additive set functions like capacities, the difference between a statistical view and a subjective evaluation view is in the assignment of values to them. After that, they possess the same mathematical properties.

We first consider the following simple problem in decision making. Let X be a random variable with values in

$$\Theta = \{\theta_1, \theta_2, \theta_3, \theta_4\} \subseteq \mathbb{R}^+$$

and with density f_0. Suppose that f_0 is only known to satisfy

1. $f_0(\theta_1) \geq 0.4$,

2. $f_0(\theta_2) \geq 0.2$,

3. $f_0(\theta_3) \geq 0.2$, and

4. $f_0(\theta_4) \geq 0.1$.

Thus the expectation cannot be computed. Let $\mathcal{F}$ be the family of densities on Θ satisfying the inequalities above on its values. We might be interested in computing $\inf\{E_f(X) : f \in \mathcal{F}\}$. The situation can be described like this. Let P_f denote the probability on $\mathcal{P}(\Theta)$ generated by f. That is, for $A \subseteq \mathcal{P}(\Theta)$, $P_f(A) = \sum_{\theta \in A} f(\theta)$. Now let

$$F : \mathcal{P}(\Theta) \to [0,1] : A \to \inf\{P_f(A) : f \in \mathcal{F}\}$$

We will show that $\inf\{E_f(X) : f \in \mathcal{F}\}$ can be computed from the set function F which is clearly a fuzzy measure.

Suppose that $\theta_1 < \theta_2 < \theta_3 < \theta_4$, and let

$$g(\theta_1) = F\{\theta_1, \theta_2, \theta_3, \theta_4\} - F\{\theta_2, \theta_3, \theta_4\}$$

$$g(\theta_2) = F\{\theta_2, \theta_3, \theta_4\} - F\{\theta_3, \theta_4\}$$

$$g(\theta_3) = F\{\theta_3, \theta_4\} - F\{\theta_4\}$$

$$g(\theta_4) = F\{\theta_4\}$$

Then g is a density on Θ, and $g \in \mathcal{F}$, as is easily checked. Now

$$
\begin{aligned}
E_g(X) &= \int_{\Theta} X(\theta) dP_g(\theta) \\
&= \int_0^\infty P_g\{\theta_i : \theta_i > t\} dt \\
&= \int_0^\infty F\{\theta_i : \theta_i > t\} dt
\end{aligned}
$$

by construction of g in terms of F. This last integral is called the **Choquet integral** of $X(\theta) = \theta$ with respect to the increasing set function F.

It remains to show that $\inf\{E_f(X) : f \in \mathcal{F}\}$ is attained at g. Since

$$E_f(X) = \int_0^\infty P_f\{\theta_i : \theta_i > t\}dt$$

it suffices to show that for $t \in \mathbb{R}$,

$$P_g\{\theta_i : \theta_i > t\} \le P_f\{\theta_i : \theta_i > t\}$$

for all $f \in \mathcal{F}$. If $\{\theta : \theta > t\} = \{\theta_1, \theta_2, \theta_3, \theta_4\}$, then

$$P_g\{\theta > t\} = \sum g(\theta_i) = 1 = P_f\{\theta > t\}$$

For $\{\theta : \theta > t\} = \{\theta_2, \theta_3, \theta_4\}$,

$$P_g\{\theta > t\} = \sum_{j=2}^4 g(\theta_j) \le \sum_{j=2}^4 f(\theta_j) = P_f\{\theta > t\}$$

since by construction of g, we have $g(\theta_j) \le f(\theta_j)$ for $f \in \mathcal{F}$ and $j = 2, 3, 4$.

Thus the Choquet integral with respect to the nonadditivity set function F is the lower bound for expected values. This situation is general. If the fuzzy measure F is a normalized monotone capacity of infinite order, and $\mathcal{P} = \{P : F \le P\}$ is a class of probability measures on a measurable space $(\Omega, \mathcal{A})$, and $X : \Omega \to \mathbb{R}$ is a bounded random variable, then $\inf\{E_P(X) : P \in \mathcal{P}\}$ is the same as the Choquet integral of X with respect to F (although the inf may not be attained).

Consider now another situation with imprecise information. Let X be a random variable defined on a probability space $(\Omega, \mathcal{A}, P)$, and let $g : \mathbb{R} \to \mathbb{R}^+$ be a measurable function. Let $P_X = PX^{-1}$ be the probability law of X on $(\mathbb{R}, \mathcal{B})$. Then

$$Eg(X) = \int_\Omega g(X(\omega))dP(\omega) = \int_\mathbb{R} g(x)dP_X(x)$$

Now suppose that for each random experiment ω we cannot observe the exact outcome $X(\omega)$, but can locate $X(\omega)$ in some interval $[a, b]$. That is, $X(\omega) \in [a, b]$. Thus we have a mapping Γ defined on Ω with values in the class of nonempty closed subsets of $\mathbb{R}$ such that for each $\omega \in \Omega$, $X(\omega) \in \Gamma(\omega)$. The computation of $E_g(X)$ in this situation is carried out as follows: for each ω,

$$g(\Gamma(\omega)) = \{g(x) : x \in \Gamma(\omega)\}$$

Since $g(X(\omega)) \in g(\Gamma(\omega))$, we have

$$
\begin{aligned}
g_*(\omega) &= \inf\{g(x) : x \in \Gamma(\omega)\} \\
&\leq g(X(\omega)) \\
&\leq \sup\{g(x) : x \in \Gamma(\omega)\} \\
&= g^*(\omega)
\end{aligned}
$$

Thus the random variable $g \circ X$ of interest is bounded from below and above by the random variables g_* and g^*, respectively, which results in $E(g_*) \leq E(g(X)) \leq E(g^*)$. Thus we are led to compute $E(g_*) = \int_\Omega g_*(\omega)dP(\omega)$ and $E(g^*) = \int_\Omega g^*(\omega)dP(\omega)$. To remind us that these are bounds on $E(g(X))$, we write $E_*g(X) = Eg_*$, and $E^*g(X) = Eg^*$.

We will show that $E_*g(X)$ and $E^*g(X)$ are Choquet integrals of the function g with respect to some appropriate fuzzy measures on $(\mathbb{R}, \mathcal{B})$. First, we need to see when Eg_* and Eg^* exist. Indeed, even if g is measurable, g_* and g^* may not be. The following condition on the multi-valued mapping Γ will suffice.

We will say that Γ is **strongly measurable** if for all $B \in \mathcal{B}$,

$$
B_* = \{\omega : \Gamma(\omega) \subseteq B\} \in \mathcal{A}
$$
$$
B^* = \{\omega : \Gamma(\omega) \cap B\} \neq \varnothing\} \in \mathcal{A}
$$

This measurability reduces to ordinary measurability when Γ is single valued.

Lemma 12.3.1 *The following are equivalent.*

1. *Γ is strongly measurable.*

2. *If g is measurable, then g_* and g^* are measurable.*

Proof. Suppose that Γ is strongly measurable and g is measurable. To prove that g_* is measurable, it suffices to prove that $g_*^{-1}[c, \infty) \in \mathcal{A}$ for all $c \in \mathbb{R}$. Now $\omega \in g_*^{-1}[c, \infty)$ means that $\inf_{x \in \Gamma(\omega)} g(x) \geq c$, so that

$$
\Gamma(\omega) \subseteq \{x : g(x) \geq c\} = g^{-1}[c, \infty)
$$

and hence $\omega \in [g^{-1}[c, \infty)]_*$. If $\omega \in [g^{-1}[c, \infty)]_*$, then $\Gamma(\omega) \subseteq g^{-1}[c, \infty)$. That is, for all $x \in \Gamma(\omega)$, $g(x) \geq c$, implying that $\inf_{x \in \Gamma(\omega)} \geq c$, and thus $g_*(\omega) \in [c, \infty)$, or $\omega \in g_*^{-1}[c, \infty)$. Thus $g_*^{-1}[c, \infty) = [g^{-1}[c, \infty)]_*$. By assumption, $g^{-1}[c, \infty) \in \mathcal{B}$, the Borel sets of $\mathbb{R}$, and by the first condition, $[g^{-1}[c, \infty)]_* \in \mathcal{A}$. The measurability of g^* follows similarly.

For the converse, let $A \in \mathcal{B}$. Then $1_A = f$ is measurable and

$$
f_*(\omega) = \begin{cases} 1 \text{ if } \Gamma(\omega) \subseteq A \\ 0 \text{ otherwise} \end{cases}
$$

Hence $f_*^{-1}(\{1\}) = A_*$, and by hypothesis, $A_* \in \mathcal{A}$. Similarly, $A^* \in \mathcal{A}$. ∎

If we let $F_* : \mathcal{B} \to [0, 1]$ be defined by

$$F_*(B) = P\{\omega : \Gamma(\omega) \subseteq B\} = P(B_*)$$

then

$$
\begin{aligned}
E_* g(X) &= \int_\Omega g_*(\omega) dP(\omega) = \int_0^\infty P\{\omega : g_*(\omega) > t\} dt \\
&= \int_0^\infty P\{g_*^{-1}(t, \infty)\} dt = \int_0^\infty P[g^{-1}(t, \infty)]_* dt \\
&= \int_0^\infty P\{\omega : \Gamma(\omega) \subseteq g^{-1}(t, \infty)\} dt \\
&= \int_0^\infty F_*(g^{-1}(t, \infty)) dt
\end{aligned}
$$

Note that the multivalued mapping Γ is assumed to be strongly measurable, so that F_* is well defined on $\mathcal{B}$. Clearly F_* is a fuzzy measure. Since

$$F_*(g^{-1}(t, \infty)) = F_*\{x : g(x) > t\}$$

we have that

$$E_* g = \int_0^\infty F_*\{x : g(x) > t\} dt$$

Similarly, by setting

$$F^*(B) = P\{\omega : \Gamma(\omega) \cap B = \varnothing\} = P(B^*)$$

we get

$$E^* g = \int_0^\infty F^*\{x : g(x) > t\} dt$$

In the situation above, the set-function F_* is known theoretically (say, Γ is "observable" but X is not). Although F_* is not a probability measure, it can be used for approximate inference processes. Choquet integrals with respect to F_* represent some practical quantities of interest. Thus when subjective evaluations are quantified as fuzzy measures, a theory of fuzzy integrals, that is, of integration of functions with respect to monotone increasing set-functions, will be useful for inference purposes.

12.4   The Choquet integral as a fuzzy integral

Various concepts of fuzzy integrals are in the literature, resulting from diverse mathematical investigations into ways to extend the Sugeno or Lebesgue integrals. However, the main stream of thought seems to be to adopt the Choquet functional as the most reasonable way to define fuzzy integrals. The previous section gave the meaning of the Choquet integral in the context of capacities in some applications situations. The fuzzy integral is defined by replacing the capacity by a fuzzy measure in the original Choquet functional. In this process, only the monotone increasing property of fuzzy measures is used, allowing fuzzy integrals to be defined for the most general class of fuzzy measures. This is also consistent with the fact that if fuzzy measures are viewed as generalizations of Lebesgue measures, then the fuzzy integral should generalize the Lebesgue integral.

In all generality, a fuzzy measure μ on a set Ω is a real-valued set function defined on some class $\mathcal{A}$ of subsets of Ω containing the empty set $\varnothing$, such that $\mu(\varnothing) = 0$, and for $A, B \in \mathcal{A}$, with $A \subseteq B$, we have $\mu(A) \leq \mu(B)$. When the restriction is made to nonnegative values for μ, we allow ∞ as a possible value in order to cover classical cases such as Lebesgue measure on the real line. Thus $\mu : \mathcal{A} \to [0, \infty]$, and the structure of the collection of subsets $\mathcal{A}$ might be arbitrary. Of course, as a special case, when μ happens to be σ-additive, then $\mathcal{A}$ should be taken as a σ-field of subsets of Ω. It turns out that, in view of the integration procedure which will follow, $\mathcal{A}$ can be taken to be a σ-field in general.

We are going to consider the "integral" of real-valued functions defined on Ω. As in classical analysis, it is convenient to extend the possible values to $\pm\infty$, so that the range is the interval $[-\infty, \infty]$, and all sup such as $\sup_{\omega \in A} f(\omega)$ exist. ($\infty = +\infty$). Arithmetic operations are extended from $\mathbb{R}$ to $[-\infty, \infty]$ in the usual way.

1. $x(\pm\infty) = (\pm\infty)x = \pm\infty$ for $0 < x \leq \infty$, and $= \mp\infty$ for $-\infty \leq x < 0$.

2. $x + (\pm\infty) = (\pm\infty) + x = \pm\infty$ for $x \in \mathbb{R}$ and $(\pm\infty) + (\pm\infty) = (\pm\infty)$.

3. $\frac{x}{\pm\infty} = 0$ for $x \in \mathbb{R}$.

Other forms, such as $\infty + (-\infty)$ and $0(\infty)$ are undefined.

Let μ be a fuzzy measure on Ω, and $f : \Omega \to [-\infty, \infty]$. In the following, we will consider the function

$$[-\infty, \infty] \to [0, \infty] : t \to \mu(f > t)$$

which is well defined if for all $t \in [-\infty, \infty]$, the set

$$(f > t) = \{\omega \in \Omega : f(\omega) > t\} \in,$$

the domain of μ. Thus if we take $\mathcal{A}$ to be a σ-field of subsets of Ω, then this condition holds when f is $\mathcal{A}$ - $\mathcal{B}_{[-\infty,\infty]}$ measurable, where $\mathcal{B}_{[-\infty,\infty]}$ is the Borel σ-field of $[-\infty, \infty]$, defined as follows. The Borel σ-field $\mathcal{B}$ is the σ-field generated by the open intervals of $\mathbb{R}$. Now $\mathcal{B}_{[-\infty,\infty]}$ is the σ-field on $[-\infty, \infty]$ generated by $\mathcal{B}$ and $\{-\infty, \infty\}$. The Borel σ-field $\mathcal{B}_{[0,\infty]}$ is

$$\{A \cap [0, \infty] : A \in \mathcal{B}_{[-\infty,\infty]}\} = \{A : A \subseteq [0, \infty] \cap \mathcal{B}_{[-\infty,\infty]}\}$$

In the following, when the σ-fields are obvious, we will not mention them when speaking about the measurability of functions.

In ordinary integration, to avoid meaningless expressions like $-\infty + \infty$, one started out by considering nonnegative and nonpositive measurable functions separately. Let $(\Omega, \mathcal{A})$ be a measurable space, and let $f : \Omega \to [0, \infty]$ be a measurable function. Then there exists an increasing sequence $f_1, f_2, ...$ of simple functions $f_n : \Omega \to [0, \infty]$, that is, functions of the form $\sum_{j=1}^n a_j 1_{A_j}(\omega)$ with $a_j \in \mathbb{R}^+$ and the A_j pairwise disjoint elements of $\mathcal{A}$, such that for all $\omega \in \Omega$, $f(\omega) = \lim_{n \to \infty} f_n(\omega)$. If $f : \Omega \to [-\infty, \infty]$, then we write

$$
\begin{aligned}
f(\omega) \;&=\; f(\omega)1_{\{f \geq 0\}}(\omega) + f(\omega)1_{\{f < 0\}}(\omega) \\[2mm]
&=\; f(\omega)1_{\{f \geq 0\}}(\omega) - \left(-f(\omega)1_{\{f < 0\}}(\omega)\right) \\[2mm]
&=\; f^+(\omega) - f^-(\omega)
\end{aligned}
$$

Then f^+ and f^- are maps $\Omega \to [0, \infty]$, and f is measurable if and only if both f^+ and f^- are measurable. When μ is a nonnegative σ-additive measure on $(\Omega, \mathcal{A})$, the Lebesgue integral of a nonnegative f with respect to μ is defined to be

$$\int_\Omega f(\omega)d\mu(\omega) = \lim_{n \to \infty} \int_\Omega f_n(\omega)d\mu(\omega)$$

where f_n is simple and converges from below to f, and

$$\int_\Omega f_n(\omega)d\mu(\omega) = \sum_{j=1}^{k_n} a_j \mu(A_j)$$

when the A_j's form a partition. Because of the additivity of μ, the quantity $\int_\Omega f(\omega)d\mu(\omega)$ is well defined. It is independent of the particular choices of the f_n.

For $f : \Omega \to \mathbb{R}$ measurable, we define

$$\int_\Omega f(\omega) d\mu(\omega) = \int_\Omega f^+(\omega) d\mu(\omega) - \int_\Omega f^-(\omega) d\mu(\omega)$$

provided that not both terms in the right hand side are ∞. When $\mu(\Omega) < \infty$,

$$\int_\Omega f d\mu = \int_0^\infty \mu(f > t) dt + \int_{-\infty}^0 [\mu(f > t) - \mu(\Omega)] dt \qquad (12.1)$$

Indeed, since $f^+ \geq 0$, we have

$$\int_\Omega f^+ d\mu = \int_0^\infty \mu(f^+ > t) dt = \int_0^\infty \mu(f > t) dt$$

Similarly, $\int_\Omega f^- d\mu = \int_0^\infty \mu(f^- > t) dt$. For each $t > 0$, $(f^- > t)$ and $(f \geq -t)$ form a partition of Ω. By the additivity of μ,

$$\mu(f^- > t) = \mu(\Omega) - \mu(f \geq -t)$$

and 12.1 follows.

Still in the case $\mu(\Omega) < \infty$, for $A \in \mathcal{A}$, we get in a similar way that

$$
\begin{aligned}
\int_A f d\mu &= \int_\Omega (1_A f) d\mu \\
&= \int_0^\infty \mu((f > t) \cap A) dt + \int_{-\infty}^0 [\mu((f > t) \cap A) - \mu(A)] dt
\end{aligned}
$$

For $f \geq 0$, we have $\int_\Omega f d\mu = \int_\Omega \mu(f > t) dt$. The right-hand side is an integral of the numerical function $t \to \mu(f > t)$ with respect to the Lebesgue measure dt on $[0, \infty]$. It is well defined as long as the function is measurable. Thus if μ is a fuzzy measure on $(\Omega, \mathcal{A})$ and f is $\mathcal{A} - [0, \infty]$ measurable, then $(f > t) \in \mathcal{A}$ for all $t \in [0, \infty]$. Since μ is increasing, $t \to \mu(f > t)$ is $\mathcal{B}_{[0,\infty]} - \mathcal{B}_{[0,\infty]}$-measurable since it is a decreasing function. This is proved as follows: for $t \in [0, \infty]$, let $\varphi(t) = \mu(f > t)$. Then

$$(\varphi < t) = \begin{cases} [a, \infty] & \text{if } a = \inf\{\varphi < t\} \text{ is attained} \\ (a, \infty] & \text{if not} \end{cases}$$

In either case, $(\varphi < t) \in \mathcal{B}_{[0,\infty]}$. Indeed if $\inf\{\varphi < t\}$ is attained at a, then for $b \geq a$, we have $\varphi(b) \leq \varphi(a) < x$ so that $b \in (\varphi < x)$. Conversely, if $\varphi(c) < x$, then $a \leq x$ by the definition of a.

If $a = \inf\{\varphi < x\}$ is not attained, then if $\varphi(b) < x$, we must have $a < b$. Conversely, if $c > a$, then for $\varepsilon = c - a > 0$, there exists y such that $\varphi(y) < x$ and $a < y < c$. But φ is decreasing, so $\varphi(c) \leq \varphi(y) < x$. Thus $c \in (\varphi < x)$.

Thus for $f : \Omega \to [0, \infty]$, one can define the **fuzzy integral** of f with respect to a fuzzy measure μ by the Choquet integral $(C) \int_\Omega f d\mu = \int_0^\infty \mu(f > t) dt$. When $f : \Omega \to [-\infty, \infty]$ is measurable, one defines

$$(C) \int_\Omega f d\mu = (C) \int_\Omega f^+ d\mu - (C) \int_\Omega f^- d\mu$$

When $\mu(\Omega) < \infty$, consider

$$(C) \int_\Omega f d\mu = \int_0^\infty \mu(f > t) dt + \int_{-\infty}^0 [\mu(f \geq t) - \mu(\Omega)] dt$$

and for $A \in \mathcal{A}$,

$$(C) \int_A f d\mu = \int_0^\infty \mu[(f > t) \cap A] \, dt + \int_{-\infty}^0 [\mu((f \geq t) \cap A) - \mu(A)] dt$$

We say that f is μ-integrable on A when the right side is finite. When $\mu(\Omega) < \infty$, say $\mu(\Omega) = 1$, and μ represents subjective evaluations concerning variables, then the "fuzzy" variable f has the "distribution"

$$\varphi_f(t) = \begin{cases} \mu(f > t) & \text{if } t \geq 0 \\ \mu(f \geq t) - \mu(\Omega) & \text{if } t < 0 \end{cases}$$

The "expected value" of f under μ is the Lebesgue integral of the distribution φ_f, that is $\int_{-\infty}^\infty \varphi_f(t) dt$. Some remarks are in order.

When $f = 1_A$ with $A \in \Omega$, we have $\mu(f > t) = \mu(A) 1_{[0,1)}(t)$, so that

$$(C) \int_\Omega 1_A(\omega) d\mu(\omega) = \mu(A)$$

More generally, for $f(\omega) = \sum_{i=1}^n a_i 1_{A_i}(\omega)$ with the A_i's pairwise disjoint subsets of Ω and $a_0 = 0 < a_1 < \cdots < a_n$, we have

$$\mu(f > t) = \sum_{i=1}^n \mu \left(\cup_{j=i}^n A_j \right) 1_{[a_{i-1}, a_i)}$$

so that

$$(C) \int f(\omega) d\mu(\omega) = \sum_{i=1}^n (a_i - a_{i-1}) \mu \left(\cup_{j=i}^n A_j \right)$$

For an arbitrary simple function of the form $f = \sum_{i=1}^{n} a_i 1_{A_i}$, and with the A_i's forming a measurable partition of Ω and

$$a_1 < \cdots < a_k < 0 < a_{k+1} < \cdots < a_n$$

we have for $t \geq 0$

$$
(f > t) = \begin{cases} \varnothing \text{ if } t \in [a_n, \infty) \\ \cup_{i=j+1}^{n} A_i \text{ if } t \in [a_j, a_{j+1}) \\ \cup_{i=k+1}^{n} A_i \text{ if } t \in [0, a_{k+1}) \end{cases}
$$

and for $t < 0$

$$
(f \geq t) = \begin{cases} \cup_{i=k+1}^{n} A_i \text{ if } t \in (a_k, 0) \\ \cup_{i=j+1}^{n} A_i \text{ if } t \in (a_j, a_{j+1}] \\ \Omega = \cup_{i=1}^{n} A_i \text{ if } t \in (-\infty, a_1] \end{cases}
$$

Thus

$$\int_{0}^{\infty} \mu(f > t) dt = a_{k+1} \mu \left(\cup_{i=k+1}^{n} A_i \right)$$

$$+ \ldots + (a_{j+1} - a_j) \mu \left(\cup_{i=j+1}^{n} A_i \right) + \ldots + (a_n - a_{n-1}) \mu (A_n)$$

Also

$$\int_{-\infty}^{0} [\mu(f \geq t) - \mu(\Omega)] dt = (a_2 - a_1) [\mu (\cup_{i=2}^{n} A_i) - \mu (\cup_{i=1}^{n} A_i)] +$$

$$\ldots + (a_{j+1} - a_j) [\mu (\cup_{i=j+1}^{n} A_i) - \mu (\cup_{i=1}^{n} A_i)] +$$

$$\ldots + (-a_k) [\mu \cup_{i=k+1}^{n} A_i - \mu (\cup_{i=1}^{n} A_i)]$$

so that

$$(C) \int_{\Omega} f d\mu = \sum_{j=1}^{n} a_j [\mu (\cup_{i \geq j}^{n} A_i) - \mu (\cup_{i=j+1}^{n} A_i)]$$

While the Choquet integral is monotone and positively homogeneous of degree one, that is, $f \leq g$ implies that $(C) \int_{\Omega} f d\mu \leq (C) \int_{\Omega} g d\mu$ and for $\lambda > 0$, $(C) \int_\Omega \lambda f d\mu = \lambda(C) \int_\Omega f d\mu$, its additivity fails. For example, if $f = (1/4)\,1_A$ and $g = (1/2)\,1_B$, and $A \cap B = \varnothing$, then

$$(C) \int_\Omega (f+g)d\mu \neq (C) \int_\Omega f d\mu + (C) \int_\Omega g d\mu$$

as an easy calculation shows. However, if we consider two simple functions f and g of the form

$$f = a1_A + b1_B \text{ with } A \cap B = \varnothing,\, 0 \le a \le b$$

$$g = \alpha 1_A + \beta 1_B \text{ with } 0 \le \alpha \le \beta$$

then

$$(C) \int_\Omega (f+g)d\mu = (C) \int_\Omega f d\mu + (C) \int_\Omega g d\mu$$

More generally, this equality holds for $f = \sum_{j=1}^n a_j 1_{A_j}$ and $g = \sum_{j=1}^n b_j 1_{A_j}$, with the A_j pairwise disjoint and the a_i and b_i increasing and non-negative. Such pairs of functions satisfy the inequality

$$(f(\omega) - f(\omega'))(g(\omega) - g(\omega')) \ge 0$$

That is, the pair is **comonotonic,** or **similarly ordered.** It turns out that the concept of comonotonicity of real-valued bounded measurable functions is essential for the characterization of fuzzy integrals.

Definition 12.4.1 *Two real-valued functions f and g defined on Ω are* ***comonotonic*** *if for all ω and $\omega' \in \Omega$,*

$$(f(\omega) - f(\omega'))(g(\omega) - g(\omega')) \ge 0$$

Roughly speaking, this means that f and g have the same "tableau of variation". Here are a few elementary facts about comonotonic functions.

- The comonotonic relation is symmetric and reflexive, but not transitive.

- Any function is comonotonic with a constant function.

- If f and g are comonotonic and r and s are positive numbers, then rf and sg are comonotonic.

- As we saw above, two functions $f = \sum_{j=1}^n a_j 1_{A_j}$ and $g = \sum_{j=1}^n b_j 1_{A_j}$, with the A_j pairwise disjoint and the a_i and b_i increasing and non-negative are comonotonic.

Definition 12.4.2 *A functional H from the space* $\mathbb{B}$ *of bounded real-valued measurable functions on* $(\Omega, \mathcal{A})$ *to* $\mathbb{R}$ *is comonotonic additive if whenever f and g are comonotonic,* $H(f + g) = H(f) + H(g)$.

If $H(f) = \int_\Omega f d\mu$ with μ a Lebesgue measure, then H is additive, and in particular, comonotonic additive. Fuzzy integrals, as Choquet functionals, are comonotonic additive.

Here are some facts about comonotonic additivity.

- If H is comonotonic and additive, then $H(0) = 0$. This follows since 0 is comonotonic with itself, whence $H(0) = H(0 + 0) = H(0) + H(0)$.

- If H is comonotonic additive and $f \in \mathbb{B}$, then for positive integers n, $H(nf) = nH(f)$. This is an easy induction. It is clearly true for $n = 1$, and for $n > 1$ and using the induction hypothesis,

$$
\begin{aligned}
H(nf) &= H(f + (n-1)f) \\
&= H(f) + H((n-1)f) \\
&= H(f) + (n-1)H(f) \\
&= nH(f)
\end{aligned}
$$

- If H is comonotonic additive and $f \in \mathbb{B}$, then for positive integers m and n, $H((m/n)f) = (m/n)H(f)$. Indeed,

$$
\begin{aligned}
(m/n)H(f) &= (m/n)H(n\tfrac{f}{n}) \\
&= mH(f/n) \\
&= H((m/n)f)
\end{aligned}
$$

- If H is comonotonic additive and monotonic increasing, then $H(rf) = rH(f)$ for positive r and $f \in \mathbb{B}$. Just take an increasing sequence of positive rational numbers r_i converging to r. Then $H(r_i f) = r_i H(f)$ converges to $rH(f)$ and $r_i f$ converges to rf. Thus $H(r_i f)$ converges also to $H(rf)$.

We now state the result concerning the characterization of fuzzy integrals. Consider the case where $\mu(\Omega) = 1$. Let

$$
C_\mu(f) = \int_0^\infty \mu(f > t) dt + \int_{-\infty}^0 [\mu(f \geq t) - 1] dt
$$

Theorem 12.4.3 *The functional* C_μ *on* $\mathbb{B}$ *satisfies the following:*

1. $C_\mu(1_\Omega) = 1$.

2. C_μ *is monotone increasing.*

3. C_μ *is comonotonic additive.*

Conversely, if H is a functional on $\mathbb{B}$ satisfying these three conditions, then H is of the form C_μ for the fuzzy measure μ defined on $(\Omega, \mathcal{A})$ by $\mu(A) = H(1_A)$.

The first two parts are trivial. To get the third, it suffices to show that C_μ is comonotonic additive on simple functions. The proof is cumbersome and we omit the details.

For the converse, if $\mu(A) = H(1_A)$, then by the comonotonic additivity of H, $H(0) = 0$, so $\mu(\varnothing) = 0$. By the second condition, μ is a fuzzy measure. For the rest it is sufficient to consider nonnegative simple functions. We refer to [90] for the details.

12.5   Radon-Nikodym derivatives

Let μ and ν be two σ-additive set functions (measures) defined on a σ-algebra $\mathcal{U}$ of subsets of a set U. If there is a $\mathcal{U}$-measurable function $f : U \to [0, \infty)$ such that $\mu(A) = \int_A f d\nu$ for all $A \in \mathcal{U}$, then f is called (a version of) the **Radon-Nikodym derivative** of μ with respect to ν, and is written as $f = d\mu/d\nu$.

It is well known that the situation above happens if and only if μ is absolutely continuous with respect to ν, in symbols $\mu << \nu$, that is, if $\nu(A) = 0$, then $\mu(A) = 0$. When μ and ν are no longer additive, a similar situation still exists. Here is the motivating example. Let $f : U \to [0, \infty)$ be μ-measurable. Consider the (maxitive) fuzzy measure $\mu(A) = \sup_{u \in A} f(u)$. Also let $\nu_0 : \mathcal{U} \to [0, \infty)$ be defined by

$$\nu_0 = \begin{cases} 0 & \text{if} \quad A = \varnothing \\ 1 & \text{if} \quad A \neq \varnothing \end{cases}$$

Then μ can be written as the Choquet integral of f with respect to the fuzzy measure ν_0. Indeed

$$\nu_0(\{u : f(u) \geq t\} \cap A) = \begin{cases} 0 & \text{if} \quad f(u) < t \text{ for } u \in A \\ 1 & \text{if} \quad f(u) \geq t \text{ for some } u \in A \end{cases}$$

Thus

$$\int_0^\infty \nu_0(\{u : f(u) \geq t\} \cap A) dt = \int_0^{\sup_{u \in A} f(u)} dt = \mu(A)$$

By analogy with ordinary measure theory, we say that f is the Radon-Nikodym derivative of μ with respect to ν_0.

Definition 12.5.1 *Let μ, $\nu : U \to [0, \infty]$ be two fuzzy measures. If there exists a $\mathcal{U}$-measurable function $f : U \to [0, \infty)$ such that for $A \in \mathcal{U}$*

$$\mu(A) = \int_0^\infty \nu(\{u : f(u) \geq t\} \cap A)dt$$

then μ is said to have f as its Radon-Nikodym derivative with respect to ν, and we write $f = \frac{d\mu}{d\nu}$, or $d\mu = fd\nu$.

Here is another example. Let $U = \mathbb{R}$ and $\mathcal{U} = \mathcal{B}(\mathbb{R})$, the Borel σ-field of $\mathbb{R}$. Let $B = [0, 1)$, $A \in \mathcal{B}(\mathbb{R})$, $d(A, B) = \inf\{|x - y| : x \in A, y \in B\}$, and A' be the complement of A in $\mathcal{U}$. Define μ and ν by

$$\mu(A) = \begin{cases} 0 & \text{if} & A = \varnothing \\ \frac{1}{2}\sup\{|x| : x \in A\} & \text{if} & A \neq \varnothing \text{ and } d(A, B') > 0 \\ 1 & \text{if} & A \neq \varnothing \text{ and } d(A, B') = 0 \end{cases}$$

$$\nu(A) = \begin{cases} 0 & \text{if} & A = \varnothing \\ \frac{1}{2} & \text{if} & A \neq \varnothing \text{ and } d(A, B') > 0 \\ 1 & \text{if} & A \neq \varnothing \text{ and } d(A, B') = 0 \end{cases}$$

Then $f(x) = x$ if $x \in B$ and 1 otherwise is $\frac{d\mu}{d\nu}$.

Obviously, if $d\mu = fd\nu$, then $\mu << \nu$. But unlike the situation for ordinary measures, the absolute continuity is only a necessary condition for μ to admit a Radon-Nikodym derivative with respect to ν. For example, with $U = \mathbb{R}$ and $\mathcal{U} = \mathcal{B}(\mathbb{R})$ as before, and $\mathbb{N}$ the positive integers, let μ and ν be defined by

$$\mu(A) = \begin{cases} 0 & \text{if} & A \cap \mathbb{N} = \varnothing \\ 1 & \text{if} & A \cap \mathbb{N} \neq \varnothing \end{cases}$$

$$\nu(A) = \begin{cases} 0 & \text{if} & A \cap \mathbb{N} = \varnothing \\ \sup\{\frac{1}{x} : x \in A \cap \mathbb{N}\} & \text{if} & A \cap \mathbb{N} \neq \varnothing \end{cases}$$

By construction, $\mu << \nu$. Suppose that $d\mu = fd\nu$. Then $f : \mathbb{R} \to [0, \infty)$ is a $\mathcal{B}(\mathbb{R})$-measurable function such that for $A \in \mathcal{B}(\mathbb{R})$, $\mu(A) = \int_A fd\nu$

(in Choquet's sense). We then have for $n \in \mathbb{N}$,

$$
\begin{aligned}
\mu(\{n\}) &= \int_0^\infty \nu((f \geq t) \cap \{n\})dt \\
&= \int_0^{f(n)} \nu(\{n\})dt \\
&= \int_0^{f(n)} \frac{1}{n}dt \\
&= \frac{1}{n}f(n)
\end{aligned}
$$

But by construction of μ, $\mu(\{n\}) = 1$ so that $f(n) = n$. Now

$$
\begin{aligned}
\mu(\mathbb{N}) &= \int_\mathbb{N} fd\nu \\
&= \int_0^\infty \nu((f \geq t) \cap \mathbb{N})dt \\
&= \sum_{n=1}^\infty \int_{n-1}^n \nu((f \geq t) \cap \mathbb{N})dt \\
&\geq \sum_{n=1}^\infty \int_{n-1}^n \nu((f \geq n) \cap \mathbb{N})dt \\
&= \sum_{n=1}^\infty \int_{n-1}^n \nu(\{k : f(k) \geq n\})dt \\
&= \sum_{n=1}^\infty \int_{n-1}^n \nu(\{k : k \geq n\})dt \\
&= \sum_{n=1}^\infty \sup\{\frac{1}{k} : k \geq n\} \\
&= \sum_{n=1}^\infty \frac{1}{n} = \infty
\end{aligned}
$$

This contradicts the fact that $\mu(\mathbb{N}) = 1$. Thus, even though $\mu << \nu$, μ does not admit a Radon-Nikodym derivative with respect to ν. Depending on additive properties of μ and ν, sufficient conditions for μ to admit a Radon-Nikodym derivative with respect to ν can be found.

It is well known that probability measures (distribution of random variables) can be defined in terms of probability density functions via Lebesgue integration. For distributions of random sets, namely capacities alternating of infinite order, a similar situation occurs when the Radon-Nikodym property is satisfied. Specifically, suppose that $d\mu = fd\nu$. Then

μ is alternating of infinite order whenever ν is. We restrict the proof of this fact to the case where $\mathcal{U}$ is finite, in which case f can take only finitely many values, say $x_1 \le x_2 \le \cdots \le x_n$. with $x_0 = 0$, we can write

$$
\begin{aligned}
\mu(A) &= \int_A f d\nu \\
&= \sum_{k=1}^{n} \int_{x_{k-1}}^{x_k} \nu(\{x \in A : f(x) \ge t\}) dt \\
&= \sum_{k=1}^{n} (x_k - x_{k-1}) \nu(B_k \cap A)
\end{aligned}
$$

where $B_k = \{u \in U : f(u) \ge x_k\}$ for $k \le n$. Take $A = \cap_{i=1}^{m} A_i$. Then we have

$$
\mu(\bigcap_{i=1}^{m} A_i) = \sum_{k=1}^{n} (x_k - x_{k-1}) \nu(B_k \cap (\bigcap_{i=1}^{m} A_i))
$$

Since $B_k \cap (\cap_{i \in I} A_i) = \cap_{i \in I}(B_k \cap A_i)$ and ν is alternating of infinite order, we have

$$
\begin{aligned}
\nu(B_k \cap (\bigcap_{i=1}^{m} A_i)) &\le \sum_{\varnothing \ne I \subseteq \{1,2,\ldots,m\}} (-1)^{|I|+1} \nu(\cup_{i \in I}(B_k \cap A_i)) \\
&= \sum_{\varnothing \ne I \subseteq \{1,2,\ldots,m\}} (-1)^{|I|+1} \nu(B_k \cap (\cup_{i \in I} A_i))
\end{aligned}
$$

Thus

$$
\begin{aligned}
\mu(\bigcap_{i=1}^{m} A_i) &\le \sum_{k=1}^{n} (x_k - x_{k-1}) \sum_{\varnothing \ne I \subseteq \{1,2,\ldots,m\}} (-1)^{|I|+1} \nu(B_k \cap (\bigcap_{i \in I} A_i)) \\
&= \sum_{\varnothing \ne I \subseteq \{1,2,\ldots,m\}} (-1)^{|I|+1} \sum_{k=1}^{n} (x_k - x_{k-1}) \nu(B_k \cap (\bigcap_{i \in I} A_i)) \\
&= \sum_{\varnothing \ne I \subseteq \{1,2,\ldots,m\}} (-1)^{|I|+1} \mu(\bigcap_{i \in I} A_i)
\end{aligned}
$$

12.6   Exercises

1. Let $(\Omega, \mathcal{A})$ be a measurable space and $f : \Omega \to \mathbb{R}^+$. For $n \ge 1$ and $k \in \{1, 2, \ldots, n2^n\}$, let

$$
A_n^k = \{\omega : \frac{k-1}{2^n} \le f(\omega) < \frac{k}{2^n}\}
$$

and let $A_n = \{\omega : f(\omega) > n\}$.

(a) Show that for each $n \geq 1$, the sets A_n^k together with A_n form a measurable partition of Ω.

(b) For $n \geq 1$, let

$$f_n(\omega) = \sum_{k=1}^{n2^n} \frac{k}{2^n} 1_{A_n^k}(\omega) + n 1_{A_n}(\omega)$$

Show that f_n is an increasing sequence of simple functions which converges pointwise to f.

2. Let

$$f(\omega) = \sum_{i=1}^{m} \alpha_i \chi_{A_i}(\omega)$$

$$= \sum_{i=1}^{n} \beta_i \chi_{B_i}(\omega)$$

be two representations of the simple function f. Show that

$$\sum_{i=1}^{m} \alpha_i \mu(A_i) = \sum_{i=1}^{n} \beta_i \mu(B_i)$$

3. Let f and g be nonnegative simple functions, and $\alpha, \beta \geq 0$.

(a) Show that

$$\int_{\Omega} (\alpha f + \beta g)(\omega) d\mu(\omega) = \alpha \int_{\Omega} f(\omega) d\mu(\omega) + \beta \int_{\Omega} g(\omega) d\mu(\omega)$$

(b) If $f \leq g$ then

$$\int_{\Omega} f(\omega) d\mu(\omega) \leq \int_{\Omega} g(\omega) d\mu(\omega)$$

4. *Let $f \geq 0$ and measurable. Show that

$$\int_{\Omega} f(\omega) d\mu(\omega) = \sup\{ \int_{\Omega} \varphi(\omega) d\mu(\omega) : \varphi \text{ is simple and } \varphi \leq f\}$$

5. Let $(\Omega, \mathcal{A}, P)$ be a probability space, and $X : \Omega \to \mathbb{R}$ be a random variable. Write $X = X^+ - X^-$. Show that $P(X^- \geq t) = P(X \leq -t)$ and $P(X \leq t) = 1 - P(X > t)$.

6. Let $(\Omega, \mathcal{A}, \mu)$ be a fuzzy measure space with $\mu(\Omega) = 1$. For $h : \Omega \to [0, 1]$, define

$$\mathcal{S}_\mu(h) = \sup_{0 \le \alpha \le 1} (\alpha \wedge \mu(h_\alpha))$$

where $\wedge$ denotes minimum, and $h_\alpha = \{\omega : h(\omega) \ge \alpha\}$. Show that

(a) if $\alpha \in [0, 1]$ then $\mathcal{S}_\mu(\alpha) = \alpha$;

(b) for $A \in \mathcal{A}$, $\mathcal{S}_\mu(1_A) = \mu(A)$;

(c) for $f, g : \Omega \to [0, 1]$ with $f(\cdot) \le g(\cdot)$, one has $\mathcal{S}_\mu(f) \le \mathcal{S}_\mu(g)$.

7. Let $(\Omega, \mathcal{A}, \mu)$ be a fuzzy space. A set $N \in \mathcal{A}$ is called a μ-null set if

$$\mu(N \cup A) = \mu(A), \text{ for all } A \in \mathcal{A}$$

(a) Verify that if μ is additive then $\mu(N) = 0$.

(b) Define $H(A) = (0, \mu(A))$, for $A \in \mathcal{A}$. Verify that $\mu = L \circ H$ (i.e., $\mu(A) = L(H(A))$), where L denotes ordinary Lebesgue measure on $\mathbb{R}$.

(c) For a measurable function $f : \Omega \to \mathbb{R}^+$, define

$$C(f)(x) = \sup\{t : x \in H(\{f > t\})\}$$

for $x \in (0, \mu(\Omega))$. Show that

$$\mathcal{E}_\mu(f) = \int_0^{+\infty} \mu(f \ge t) dt = \int_0^{\mu(\Omega)} C(f)(x) dx$$

8. Let $(\Omega, \mathcal{A}, P)$ be a probability space, and $(\Theta, \mathcal{B})$ a measurable space. Let $\Gamma : \Omega \to \mathcal{B}$ such that $A_* = \{\omega : \Gamma(\omega) \subseteq A\} \in \mathcal{A}$ where $A \in \mathcal{A}$ and $P(\Upsilon = \varnothing) = 0$.

(a) Show that $\mu(A) = P(A_*)$ is a fuzzy measure.

(b) Let $f : \Theta \to \mathbb{R}^+$, and define

$$f_*(\omega) = \inf_{\theta \in \Gamma(\omega)} f(\theta)$$

Show that

$$\mathcal{E}_\mu(f) = \int_\Omega f_*(\omega) dP(\omega)$$

9. Let $f, g : \Omega \to \mathbb{R}$. Show that the following are equivalent.

(a) f and g are comonotonic.

(b) The collection $\{(f > t), (g > s), s, t \in \mathbb{R}\}$ of subsets is a chain with respect to set inclusion.

10. Let $U = \mathbb{R}$, $\mathcal{U} = \mathcal{B}(\mathbb{R})$, and $\mathbb{N}$ be the positive integers. Let $\mu, \nu,$ and $\gamma : \mathcal{B}(\mathbb{R}) \to [0, 1]$ be defined by

$$
\mu(A) = \begin{cases} 0 & \text{if } A \cap \mathbb{N} = \varnothing \\ \min\{1, \frac{1}{2}\sum_n\{\frac{1}{n} : n \in A \cap \mathbb{N}\} & \text{if } A \cap \mathbb{N} \neq \varnothing \end{cases}
$$

$$
\nu(A) = \begin{cases} 0 & \text{if } A \cap \mathbb{N} = \varnothing \\ \sup\{\frac{1}{n} : n \in A \cap \mathbb{N}\} & \text{if } A \cap \mathbb{N} \neq \varnothing \end{cases}
$$

$$
\gamma(A) = \begin{cases} 0 & \text{if } A \cap \mathbb{N} = \varnothing \\ 1 & \text{if } A \cap \mathbb{N} \neq \varnothing \end{cases}
$$

Show that

(a) $\mu << \nu << \gamma << \mu$.

(b) $\mu(\mathbb{R}) = \nu(\mathbb{R}) = \gamma(\mathbb{R}) = 1$.

(c) if $A_n \nearrow A$ (respectively $A_n \searrow A$), then $\nu(A_n) \nearrow \nu(A)$ (respectively $\nu(A_n) \searrow \nu(A)$).

(d) μ is alternating of order 2.

(e) ν and γ are alternating of infinite order.

Chapter 13

FUZZY MODELING AND CONTROL

In this last chapter, we discuss the use of fuzzy logic in an important area of applications, namely control engineering. However, we focus on aspects of modeling of domain knowledge and general principles of fuzzy control rather than on its practical designs and implementation. For readers who are interested in these aspects, [18] and [81] are good sources. The purpose here is to provide an understanding of the rationale behind fuzzy technology. No prior knowledge is needed. Since the methodology of fuzzy technology is general in the context of knowledge-based systems, we choose the topic of control to illustrate it. The reader can recognize easily that the discussions in this chapter apply to the field of expert systems as well.

13.1   Motivation for fuzzy control

The standard approach to designing controllers of dynamical systems relies on the availability of mathematical descriptions, that is, models, of these systems. In many situations, differential equations of dynamical systems can be specified from the laws of mechanics. For example, the analytic mathematical model of the motion of a car moving on a straight road is obtained as follows: in an ideal environment, suppose that the car is controlled by the throttle, producing an accelerating force $u_1(t)$, and by the brake, producing a retarding force $u_2(t)$. If we let $x_1(t)$ denote the distance of the car from a given initial position and $x_2(t)$ denote its velocity, then

$$\begin{aligned} \dot{x}_1 &= x_2 \\ \dot{x}_2 &= u_1 - u_2 \end{aligned} \tag{13.1}$$

where $\dot{x}_1$ is the derivative with respect to t. Consider the *state*, or *input*,

$$x = \begin{pmatrix} x_1 \\ x_2 \end{pmatrix}$$

and the *control*

$$u = \begin{pmatrix} u_1 \\ u_2 \end{pmatrix}$$

The system 13.1 of differential equations is written in the form of a time-invariant linear system

$$\dot{x} = Ax + Bu \tag{13.2}$$

where the matrices A and B are

$$A = \begin{pmatrix} 0 & 1 \\ 0 & 0 \end{pmatrix}$$

$$B = \begin{pmatrix} 0 & 0 \\ 1 & -1 \end{pmatrix}$$

In general, a dynamical system is represented by a nonlinear differential equation of the form

$$\dot{x} = f(x, u, t) \tag{13.3}$$

The aim of feedback control is to determine a *control law* $u = \varphi(x, t)$ in order to achieve some specific control objective.

For more complex problems, mathematical models might be hard to specify, or might be only partially known. On the other hand, if the control objectives are task-oriented, such as "park a car", then it is not clear how standard control theory can be immediately extended to cope with such new situations. A relevant question is whether it is possible to control plants without explicit knowledge of their mathematical models. A positive answer to this question will lead to a new approach to control engineering.

In the spirit of artificial intelligence, we observe that in everyday activities, humans through manual control strategies are quite capable of controlling complex systems without differential equations. They park cars and ride bicycles, for example. Doing so requires common sense reasoning, naive physics, or heuristic knowledge about the systems. This type of information is weaker than precise analytic descriptions of the dynamics by differential equations, and yet seems to be sufficient for control tasks. It is possible to acquire this control information, either by simulating the behavior of an expert who is capable of controlling a given system, or by asking him to describe his control strategy. A pioneering experiment by Mamdani and Assilian [60] showed that this is possible. In their experiment with a steam engine, the basic fact is that human operators express their control strategies linguistically rather than in precise mathematical terms. Thus in order to carry out a control synthesis, it is necessary to model linguistic information as well as the inference process. The use of fuzzy logic to achieve this goal leads to the field of fuzzy control. The adjective "fuzzy" is used to denote the mathematical way of modeling fuzziness in natural languages, analogous to "stochastic" control in which "noise" is modeled as randomness by using probability theory. Thus fuzzy control is a technique for deriving control laws when control information is expressed in linguistic terms.

It should be emphasized that from a general view of scientific investigations, each approach to a problem has its own domain of applicability. Standard control theory is efficient when precise mathematical models are available. The emergence of artificial intelligence technologies has suggested the additional use of manual control expertise in more complex problems. This is particularly appropriate when mathematical models are difficult to specify due, for example, to ill-defined problems or to the goal oriented objectives such as "park the car". The so-called intelligent control is born from the idea that it is possible to mimic human control strategies in designing automatic control laws. Of course, the modeling and synthesis of linguistic control rules of experts can be done in different ways. Fuzzy control denotes the approach to control engineering in which fuzzy logic is used to derive control laws.

13.2 The methodology of fuzzy control

We will use a very simple example to spell out the design methodology of fuzzy control. Note that this is an example of a "data driven" method: it is the data that dictates which appropriate method to use.

Suppose we wish to design an automatic device to keep the temperature of a room at a fixed value T_0. Let $T(t)$ denote the temperature of the room at time t. The purpose of the device is to bring $T(t)$ to T_0. By switching the heater or cooler on and off, we can control the rate of temperature changes. Let $x = T - T_0$ be the input variable and $u = \dot{x}$

the control variable. For an observed value x, it is required to find an appropriate value $u(x)$ for controlling the temperature. Now, suppose by "naive physics" or by "expert knowledge" we have the following information.

1. If x is "small positive" then u should be "small negative".

2. If x is "small negative" then u should be "small positive".

From the form of this knowledge, we see that each piece of information can be represented as an "If ... then ..." rule. This situation is similar to the general framework of knowledge-based systems. For example, in simple medical diagnostic systems, the input variables like x take only two possible values, 0 or 1. The relationships between variables are typically of causal nature and can be expressed in terms of implications. If we view variables as random variables and model strengths of variable relationships by conditional probabilities, then we can model the knowledge structure entirely within the framework of probability theory, namely by the joint probability distributions of the variables. In our actual example of automatic control, although the global knowledge structure is similar, there is a significant difference at the local level: the input variable x is a linguistic. That is it is a variable whose possible values are terms in a natural language rather than numerical ones. The output variables are of the same nature. This situation is general in real world complex problems where it is difficult to describe knowledge about control strategies in precise terms. For example, in describing car driving strategies, an expert driver will not be able to give a precise answer to a question such as "how many seconds should you apply the brakes if your car is going quite fast and there is an obstacle in front of you that seems quite near"?

Since the obvious task of a design engineer is to use these "If ... then ... " rules to build an automatic device behaving in a manner similar to a human expert, he has to translate first these rules into implementable mathematical representations. This amounts first to modeling fuzzy concepts in these rules, such as "positive small", "negative big", and so on. For any given control problem, after identifying input and output variables and their ranges, in order to search for a collection of "If...then..." rules in order to form a rule base, defining a control law, it is necessary to cover these ranges by appropriate fuzzy partitions, since essentially these rules deal with linguistic variables such as "small positive", "small negative", and so on, and hence need to be modeled by fuzzy sets.

Suppose the first rule R_1 is

"If x is A_1 then u is B_1"

where A_1 and B_1 stand for "positive small" and "negative small", respectively. To represent A_1 and B_1 mathematically, we use fuzzy sets. Let X and U denote the ranges of the temperature and of the rate of temperature changes, respectively. Then A_1 and B_1 can be modeled as fuzzy sets

$$A_1 \; : \; X \to [0,1]$$
$$B_1 \; : \; U \to [0,1]$$

In a more complicated rule such as "If x_1 is A_1 and ... x_n is A_n then u is B_1", we represent the input as a vector

$$x = \begin{bmatrix} x_1 \\ x_2 \\ \vdots \\ x_n \end{bmatrix}$$

and rewrite the rule as "If x is A then u is B_1" where A is the fuzzy Cartesian product of the $A_i's$ characterized by

$$A(x_1, x_2, ..., x_n) = \bigwedge \{A_i(x_i), i = 1, 2, ..., n\}$$

When u is also a vector, we have a multi-input multi-output rule.

In general scientific investigations, the emphasis on qualitative models has become routine. For example in mathematical psychology where the goal is to model behavior, one usually translates qualitative models into quantitative models since the latter provide representations necessary for testing. For this to be possible, one has to make several basic assumptions. For example, in providing qualitative models for association learning, one assumes, or considers, only the case that associations are in only two possible states: present in full, or completely absent. That is, partial associations are ignored. Furthermore, if one assumes that transitions from one state to another are quantified by probabilities, then under further plausible assumptions, Markov processes can be used as models. Of course, these probabilistic views of "all-or-none" models should be validated in specific situations. The modeling of knowledge using fuzzy sets is a further step in qualitative modeling. Basically, "all-or-none" models are extended to more flexible and realistic ones in which partial degrees of "truth" are incorporated. In each rule, besides the modeling of linguistic labels such as A_i and B_j, we need to model the connectives "and", "or", "not" and "if ... then ... ". It is precisely here that fuzzy logic is called upon. The standard logical connectives "and", "or", and "not" are modeled using t-norms, t-conorms, and negations, respectively. The logical connective "if ... then ... " is much much more subtle. A rule in our example such as "If x is positive small then u is negative small" does not represent a causal relationship between x and u, nor an implication in the usual sense. It simply conveys the idea that for every x, if x is positive and small, then there is a control value u which is negative and small, and which is a "reasonable" value to use. Thus in a collection of rules of the form

$$R_j : \text{"If } x \text{ is } A_j \text{ then } u \text{ is } B_j \text{"}, \ j =, 1, 2, ..., r \qquad (13.4)$$

we let $M(x, u)$ be the binary fuzzy predicate "u is a reasonable control for x" and consider a theory consisting of logical statements of the form

$$A_j(x) \rightarrow \exists u \ (M(x, u) \text{ and } B_j(u)), \ j = 1, 2, ..., r$$

Thus, for every x, if x satisfies the property A_j then there exists a value u which is "reasonable" for x, and for which B_j hold. It can be shown that (x, u) satisfies the formula "$A_j(x)$ and $B_j(u)$" if and only if $M(x, u)$ belongs to a minimal model of the theory above (in the context of non-monotonic logic instead of classical logic [62]). This observation can be used to justify the translation of a conditional statement of the form "if x is A then u is B" into a fuzzy binary relation $M(x, u) = A(x) \triangledown B(u)$, where $\triangledown$ is a t-norm, such as min. For a collection of rules R_j, the combined fuzzy relation M given by

$$M(x, u) = \bigvee \{ (A_j(x) \bigwedge B_j(u)), \ j = 1, 2, ..., n \}$$

is referred to as **Mamdani's rule**. The use of the t-conorm max reflects the fact that the connection between the rules is specified by the logical connective "or". Of course, in general, this connection depends on the meaning of the knowledge represented. For MIMO (multi-inputs multi-outputs) systems, some additional symbols, or metalanguage, are needed in the consequent part of each rule. For example, in

$$R_j : \text{"If } x \text{ is } A_j \text{ then } u_1 \text{ is } B_j^1; \ ... \ ; u_k \text{ is } B_j^k \text{"}$$

the symbol ";" is used to denote the noninteraction or independent control actions u_i. In this case, the rule R_j can be decomposed into k rules involving scalar control variables u_i with the same antecedent part, so that the MIMO system is equivalent to k MISO (multi-inputs single output)subsystems. Here, we focus only on MISO systems for simplicity.

The rule base 13.4 plays the role of a mathematical model. This is the heuristic approach to fuzzy control. The flexibility in describing a rule base in mathematical terms by using fuzzy logic comes from the fact that there are different ways to model fuzzy concepts by membership functions, as well as choices of logical connectives involved in rules. By a **fuzzy design**, we will mean a specific choice of such systems of parameters. We will elaborate on fuzzy design in the next section.

The following relational approach to deriving control laws from rule bases is popular in practice. First, a **fuzzy system** is a mapping M from fuzzy subsets of X to fuzzy subsets of U. Such a system can be defined by specifying a fuzzy relation on the product space $X \times U$. Thus for $A \in [0,1]^X$, we get $B = M \circ A \in [0,1]^U$, where $\circ$ denotes some composition operation among fuzzy relations. For example, the sup-$\triangle$composition, where $\triangle$ is a t-norm is defined by

$$B(u) = \bigvee_{x \in X} (M(x,u) \triangle A(x)) \tag{13.5}$$

When A is a singleton, that is, when $A(x) = 1_{\{x_0\}}(x)$ for some $x_0 \in X$, then by properties of t-norms, B is reduced to $B(u) = M(x_0, u)$. By a **fuzzy logic system** we mean a fuzzy system in which M is constructed from a rule base like 13.4 using fuzzy logic. If each rule R_j is translated into

$$M(x,u) = A_j(x) \triangle B_j(u)$$

then

$$M(x,u) = \triangledown[M_j(x,u), j = 1, 2, ..., r]$$

where $\triangledown$ is a t-conorm.

When $A(x) = 1_{\{x_0\}}(x)$, the **fuzzy output** B becomes

$$\begin{aligned} B(u) &= \triangledown[M_j(x,u), j = 1, 2, ..., r] \\ &= \triangledown[A_j(x_0) \triangle B_j(u), \ j = 1, 2, ..., r] \end{aligned}$$

In the context of control, the input variable takes values in the input space X. If the observed input value x^* is accurate, we take $A(x) = 1_{\{x^*\}}(x)$ as input to the fuzzy logic system, producing the fuzzy output of the rule R_j given by

$$u \to A_j(x^*) \triangle B_j(u)$$

Note that when $X = X_1 \times X_2 \times ... \times X_n$, then

$$\begin{aligned} A_j(x^*) &= A_j(x_1^*, x_2^*, ..., x_n^*) \\ &= \bigwedge\{A_j^i(x_i^*), i = 1, 2, ..., n\} \end{aligned}$$

where $A_j = A_j^1 \times A_j^2 \times ... \times A_j^n$ and represents the degree to which x^* satisfies the antecedent part of R_j. It can be interpreted as the **firing degree** of R_j when the input is x^*

If the observed input x^* contains some sort of error or imprecision, then we can "fuzzify" x^*. This amounts to building a fuzzy subset of X around x^* in a fashion appropriate to cope with the imprecision present. For example, if the basic statistics of the stochastic processes generating x^* are known, they can be used to construct a fuzzy subset A associated with x^*.

The general relational equation $B = M \circ A$ will produce the fuzzy output B from this fuzzy input A. In this case, the firing degree of R_j with respect to the input A is generalized to the **degree of matching** between A and the antecedent A_j of R_j, namely

$$\bigvee_{x \in X} \{A(x) \wedge A_j(x)\}$$

This quantity can also be interpreted as the possibility measure of A_j given A.

Now, for a crisp input A, the fuzzy logic system M produces a fuzzy output B. Referring back to our example of controlling a thermostat, we need a crisp value for the control action. Thus, the fuzzy output B should be summarized to obtain such a value. This procedure is termed **defuzzification**. For example, in the case where all spaces of interest are copies of the real line $\mathbb{R}$, the fuzzy set B can be summarized as

$$D(B) = \frac{\int_{\mathbb{R}} uB(u)du}{\int_{\mathbb{R}} B(u)du}$$

This defuzzification procedure is called the **centroid method**. For a crisp input x^*, if we view the quantity

$$\frac{B(\cdot | x^*)}{\int_{\mathbb{R}} B(u|x^*)du}$$

as a conditional probability density of the control variable given the input value x^*, then $D(B)$ has the form of a conditional mean. Other possible defuzzification procedures will be discussed in the next section.

We mention now another approach, one due to Sugeno, to construct fuzzy logic controllers where defuzzification is not needed. This approach also illustrates the obvious fact that if additional information about the dynamics of the plant under control is available, even in approximate form, it should be incorporated into the design of the controller. This approach is due to Sugeno and is essentially based on the possibility of describing locally the dynamics of a plant in approximate terms, in other words, by a fuzzy model. This is the case, for example, when for each member of a fuzzy partition of the input space X, the difference equation of the plant is linear to some degree. This suggests forming control rules as follows:

- R_j : "If x_1 is A_j^1 and ... x_n is A_j^n then $u = f_j(x_1, x_2, ..., x_n)$",
  $j = 1, 2, ..., r$

Here the x_i are actual observed values of the input variables and f_j is some specific linear function

$$f_j(x_1, x_2, ..., x_n) = \sum_{i=1}^{n} \alpha_{ij} x_i$$

and the fuzzy subsets A_j^i form a fuzzy partition of the input space. Note that the consequent part in each R_j is precise. The firing degree, or the degree of applicability, of R_j is $T(A_j^i(x_i), i = 1, 2, ..., n) = \tau_j$, where T is some t-norm such as $T(a, b) = ab$. The rule R_j will produce a crisp output given by $u_j = \tau_j f_j(x_1, x_2, ..., x_n)$. The **overall output control value** is taken as a weighted average

$$u(x_1, x_2, ..., x_n) = \frac{\sum_{j=1}^{r} \tau_j f_j(x_1, x_2, ..., x_n)}{\sum_{j=1}^{r} \tau_j}$$

The Sugeno approach, or more precisely, the Takagi-Sugeno model, is essentially a model-based fuzzy control.

To conclude this section, we mention some basic design issues.

- The rule base plays the role of control knowledge needed to derive control laws. As such, various factors in this knowledge base need to be examined to achieve a sufficiency of needed information—for example, the number of rules, the choice of parameters, membership functions of fuzzy concepts involved, and the logical connectives. As in standard control theory, any choice of these nominal parameters should be done with some robustness properties in mind. Also, in some cases, where imprecision arises in eliciting membership functions, interval-valued membership functions might be required.

- The description of control knowledge in linguistic rules involves linguistic variables whose values should form fuzzy partitions of associated spaces. The fuzzy partitions of both input and output spaces are necessary to insure that for any observed input values some rule in the rule base should "fire" with some positive firing degree.

- The principle of fuzzy control described in this section aims at producing a control law $u(x)$ from a rule base. Thus we can view a fuzzy logic system as a mechanism for producing an approximation to some ideal control law, or as a model-free regression type problem The property of being a "good approximator" can be used to justify its capability in approximating any continuous functions defined on

a compact domain. Note that such a theoretical result is only a
sort of existence theorem. Success of designs of fuzzy controllers
still depend on the skill of the knowledge engineers.

- Controllers are judged by their performance, such as robustness
  and stability. In this respect, it seems fruitful to have some form of
  knowledge about the dynamics of the plants, such as fuzzy models.
  Then an analysis can be done using both fuzzy models and control
  rules.

13.3   Optimal fuzzy control

As stated earlier, fuzzy control seems promising when the dynamics of
systems are not known with precision. Standard control relies on the
knowledge of a differential equation $\dot{x} = f(x, u)$ with $x(t_0) = x_0$ describing the dynamics. If such an equation is not available, then a weaker
form of knowledge might serve as a means to derive control laws. For
example, the knowledge about a plant might be expressed in the form of
a collection

$$R_j : \text{if } (x, u) \text{ is } A_j \text{ then } x \text{ is } B_j, \ j = 1, 2, ..., r$$

of rules. The standard approach to optimal control is this. Given the
differential equation together with a set of constraints specifying the goals
such as production planning, terminal control, tracking problems and so
on, an optimal control law is the one that maximized some objective
function $J(u)$ subject to the set of constraints. Suppose now that instead
of a differential equation with its constraints, we have a collection of rules
as above. Now instead of maximizing an objective function subject to a
differential equation, we face a maximizing problem over a set of rules.
The mathematical formulation of this optimization problem is not quite
clear. If the set of rules is combined into a single fuzzy relation on an
appropriate space then we have the problem of optimizing a function
over a fuzzy set. But what does this mean? One formulation is this. Let
J be a function defined on a set X and let A be a fuzzy subset of X.
Suppose that J takes real values in an interval $[a, b]$. Then we may as
well assume that $J : X \to [0, 1]$. If A is a crisp set so that $A(x) = 1$ for
all $x \in X$, then maximizing J over A should simply mean maximizing
J, which can be expressed as maximizing the function $J \wedge A$. So in the
fuzzy case, we could formulate the problem of maximizing a function J
over a fuzzy set A as that of maximizing the function $J \wedge A$. This is
not the only reasonable formulation. In the crisp case, maximizing the
product $J(x)A(x)$ is also the same as maximizing J, so one could state
the problem in general as one of maximizing the $J(x)A(x)$.

13.4 An analysis of fuzzy control techniques

Fuzzy control is a way to transform linguistic knowledge into control laws. A systematic way of doing so was spelled out in the previous section. It consisted of

- modeling fuzzy concepts in rules,

- a choice of fuzzy connectives in rules or combination of rules, and

- a choice of a defuzzification procedure.

We look at these steps in more detail. Consider a rule base of a MISO system, namely where the input variable is $x = (x_1, x_2 ..., x_n) \in X = X_1 \times X_2 \times .... \times X_n$, and the output variable is $u \in Y$. One needs first to specify fuzzy partitions of all spaces involved. Let $A'_j s$ and $B'_j s$ be fuzzy partitions of X and Y, respectively, corresponding to linguistic variables such as "small positive", and so on. Next linguistic "If ...then ..." rules corresponding to these fuzzy partition can be found either from experts or through training data. These rules form the knowledge base from which a control law can be derived. In fact, these rules are a weaker way of defining a function that can approximate the ideal control law. In this regard, see Chapter 8 on universal approximation.

Step 1. R_j : "If x is A_j then u is B_j", $j = 1, 2, ..., r$

The linguistic labels A^i_j and B^i_j are viewed as fuzzy subsets of appropriate spaces. The membership functions of these fuzzy concepts need to be specified. This can be done by asking experts or by using common sense. In asking experts to specify the membership function of a fuzzy concept A of X, we can only obtain a finite number of degrees of belief. For $x_i \in X$, $i = 1, 2, ..., m$, we can ask N experts whether they believe the statement "x_i is A" is true. The degree of x_i in A, that is, the value $A(x_i)$, is taken to be M/N where M is the number of experts who say yes. There are other ways to obtain a value of $A(x_i)$. Experts can use their subjective probabilities. In any case, from a finite number of values $A(x_i)$, one extrapolates to obtain a function A defined on all of X.

Control rules expressed in natural language might contain **fuzzy modifiers** such as "very", or "almost all". Modeling fuzzy quantifiers was discussed in the Chapter 9.

When it is not reasonable to give experts' degrees of belief as numbers in the unit interval, one has to allow membership functions to take values elsewhere, such as subintervals of $[0, 1]$. When this happens, we are in the realm of interval-valued fuzzy logic. An expert may specify a degree of belief of 8 on a scale from 0 to 10. But this means the expert's degree of belief is closer to 8 than to 7 or to 9, so belongs to the interval $[0.75, 0.85]$.

Membership functions of fuzzy concepts can be chosen in some simplified forms especially easy for computations. Some of these simple forms are parametric, that is, are determined by the choice of a finite number of parameters. Here are some examples of such fuzzy subsets of $\mathbb{R}$.

1. **Triangular membership functions** are those of the form

$$A(x) = \begin{cases} 0 & \text{if} \quad x < a - b \text{ or } x > a + b \\ 1 + \dfrac{x - a}{b} & \text{if} \quad a - b < x \leq a \\ 1 - \dfrac{x - a}{b} & \text{if} \quad a < x \leq a + b \end{cases}$$

   where a and b are any real numbers.

2. **Trapezoidal membership functions** are those of the form

$$A(x) = \begin{cases} 0 & \text{if} \quad x < a \text{ or } x > d \\ \dfrac{x - a}{b - a} & \text{if} \quad a \leq x < b \\ 1 & \text{if} \quad b \leq x < c \\ \dfrac{x - d}{c - d} & \text{if} \quad c \leq x \leq d \end{cases}$$

   where $a < b < c < d$.

3. **Gaussian type membership functions** are those of the form

$$A(x) = \exp\left(\frac{(x - \mu)^2}{\sigma^2}\right)$$

4. **Piecewise polynomial functions**, or **splines**, including piecewise linear ones are often used.

5. **Piecewise fractionally linear functions** are those with each piece a function of the form

$$A(x) = \frac{ax + b}{cx + d}$$

After the forms of the various membership functions have been decided, the parameters of the functions are specified in such a way that the fuzzy controller will behave in some optimal fashion. That part is called **tuning**. Those interested in tuning procedures should consult books on applications of fuzzy control.

Step 2. After modeling linguistic labels in rules as fuzzy subsets of appropriate sets, the next task is to decide on the choices of logical connectives involved. The theory of fuzzy logic provides classes of candidate function such as t-norms, t-conorms, and negations. Such choices should be guided by the universal approximation property of fuzzy systems and by sensitivity analysis. For example, popular t-norms are min and product. Popular t-conorms are max and the dual of product with respect to the negation $1 - x$. When the connection between rules is expressed as "or" (or "else"), and the combination of rules is max $-T$ composition, then when $T = $ min or product, we obtain max-min or max-product composition (inference), respectively.

Step 3. Having obtained an overall fuzzy output B as a fuzzy subset of the output space V, we need to summarize it into a single value u^* to be used as the actual control action. A **defuzzification procedure** is a map

$$D : [0, 1]^U \to U$$

Here are some examples of defuzzification procedures.

1. **Centroid (or center-of-gravity) defuzzification.** Here is a heuristic motivation of this method. Since $B(u)$ is the degree to which u is compatible with B, it is proportional to some probability density function $f(u)$. For example, $B(u)$ is proportional to the number of experts who believe that u is good, so that the more experts who believe in u, the greater the chance that u is actually good. Letting

$$f(u) = \frac{B(u)}{\int_U B(u)du}$$

we can choose u^* to minimize the average deviation $\int (u-u^*)^2 f(u)du$. Differentiation with respect to u^* yields

$$u^* = \frac{\int_U uB(u)du}{\int_U B(u)du}$$

which is called the **centroid value**.

2. **Mean-of-maxima defuzzification.** Somewhat similar to measures of location in statistics, and also because of the meaning of degrees of membership, values u^* which have the highest values $B(u^*)$ are also natural to consider. Let E be those elements of U which maximize B. If E is finite then put u^* to be the average of the elements of E. In particular, if E consists of just one element, u^* is taken to be that element. If E is infinite, then one takes u^* to be some sort of average of the elements of E.

3. **Center-of-area defuzzification.** This method is the counterpart of the concept of a median of a probability density function. The center of area of B is the value u^* splitting the area under B into two parts of equal size.

As for logical connectives, the choice of a defuzzification procedure should be guided by considerations such as robustness or sensitivity to errors in its argument B. It can be shown that the centroid defuzzification procedure is continuous with respect to a suitable topology on the space of membership functions, and hence tends to have good robustness properties.

13.5   Exercises.

1. Let I be a bounded interval of the real line $\mathbb{R}$. Give a fuzzy partition of I with triangular membership functions.

2. Let I and J be two bounded intervals of $\mathbb{R}$. Consider a collection of rules:

   $R_j :$ "If x is A_j then u is B_j", $j = 1, 2, 3$

   where the A_j and B_j are fuzzy subsets of I and J, respectively.

   (a) Specify A_j and B_j so that (A_1, A_2, A_3) and (B_1, B_2, B_3) are fuzzy partitions of I and J, respectively.

   (b) In view of the previous part, compute for a given x the fuzzy output of each rule R_j using min-inference.

   (c) Suppose the connection between the rules is expressed as the logical connective "or". Compute the overall fuzzy output by using the t-conorm $\max(a, b)$. Do the same for the t-conorm $x + y - xy$.

   (d) In each case of the previous part, compute the crisp output value using the centroid defuzzification procedure.

3. With an example specifying membership functions of the rule base of Exercise 2, using the t-norm product, t-conorm $x + y - xy$, and center-of-area defuzzification procedure, give an explicit formula for computing the overall crisp output.

4. Let f be a probability density function on $\mathbb{R}$, that is, $f \geq 0$ and $\int_{-\infty}^{\infty} f(x)dx = 1$. Let $J(u) = \int_{-\infty}^{\infty} (x - u)^2 f(x)dx$. Show that $u = \int_{-\infty}^{\infty} xf(x)dx$ minimizes J.

5. Let the overall fuzzy output B of a fuzzy logic system be a trape-zoidal function with parameters a, b, c, d. Compute the defuzzified value of B by using

   (a) the centroid method,

   (b) the mean-of-maxima method, and

   (c) the center-of-area method.

Bibliography

[1] Abel, N., Untersuchungen der Funktionen Zweier unabhängig veränderlichen Grössen x und y, welche Eigenschaft haben, dass $f(z, f(x, y))$ eine symmetrische Funktion von x, y und z ist, *J. Reine Angew. Math.* 1 (1826), 11-15.

[2] Aczél, J., Sur les opérations définies pour nombres réels, *Bull. Soc. Math. France* 76 (1949) 59-64.

[3] Aczél, J., *Lectures on Functional Equations and Their Applications*, Mathematics in Science and Engineering 19, Academic Press, New York-London, 1966.

[4] Aczél, J., On mean values, *Bull. Amer. Math. Soc.*, 54, 392-400 (1948).

[5] Aczél, J. and J. Dhombres, *Functional Equations in Several Variables*, Cambridge University Press, 1989.

[6] Alsina, C., E. Trillas, and L. Valverde, On some logical connectives for fuzzy sets, 71-76, in *Fuzzy Sets for Intelligent Systems*, D. Dubois, H. Prade, and R. Yager, eds., Morgan Kaufmann, San Mateo, 1993.

[7] Aumann, R. J. and L. S. Shapley, *Values of Non-atomic Games*, Princeton University Press, Princeton, 1974.

[8] Bacchus, F., *Representation and Reasoning with Probability Knowledge*, MIT Press, Cambridge, 1990.

[9] Barnett, S. and R. G. Cameron, *Introduction to Mathematical Control Theory*, Clarendon Press, 1985.

[10] Barr, M., **-Autonomous Categories*, Lecture Notes in Mathematics, #752, Springer-Verlag, 1979.

[11] Bertoluza, C. and Bodini, A., A new proof of Nguyen's compatibility theorem in a more general context, *J. Fuzzy Sets and Systems*, 95 (1998), 99-102.

[12] Bouchon-Meunier, B., *La Logique Floue*, Collection Que-Sais-Je, Presses Universitaires de France, Paris, 1992.

[13] Burris, S. and H. D. Sankappavaran, *A Course in Universal Algebra*, Springer-Verlag, New York, 1980.

[14] Chateauneuf, A. and J. Jaffray, Some characterizations of lower probabilities and other monotone capacities through the use of Möbius inversion, *Math. Soc. Sci.* 17 (1989), 263-283.

[15] Cignoli, R. and Mundici, D., An elementary proof of Chang's completeness theorem for the infinite valued-calculus of Lukasiewicz, *Studia Logica*, 1997.

[16] Comer, S. D., *On connections between information systems, rough sets, and algebraic logic*, preprint.

[17] Denneberg, D., *Non-additive Measure and Integral*, Kluwer Academic Press, Dordrecht, 1994.

[18] Driankov, D. H. Hellendoorn, and M. Reinfrank, *An Introduction to Fuzzy Control*, Springer-Verlag, 1996.

[19] Dombi, J., A general class of fuzzy operators, the De Morgan class of fuzzy operators and fuzziness measures induced by fuzzy operators, *J. Fuzzy Sets Syst.*, 8 (1982), 149-163.

[20] Dubois, D. and H. Prade, *Fuzzy Sets and Systems: Theory and Applications*, Academic Press, New York, 1980.

[21] Dubois, D. and H. Prade, *Possibility Theory: An Approach to Computerized Processing of Uncertainty*, Plenum Press, New York, 1988.

[22] Dubois, D., H. Prade. and R. Yager, *Readings in Fuzzy Sets for Intelligent Systems*, Morgan-Kaufmann, San Mateo, 1993.

[23] D. Dubois and H. Prade, A review of fuzzy sets aggregation connectives, *Information Sciences* 36(1985), 85-121.

[24] Falconer, K., *Fractal Geometry: Mathematical Foundations and Applications*, John Wiley, 1990.

[25] Foias, C., Commutant lifting techniques for computing optimal H_∞ controllers, H_∞-control, *Lecture Notes in Mathematics*, 1496(1991), 1-36, Springer-Verlag, 1991.

[26] Frank, C., On the simultaneous associativity of $F(x,y)$ and $x+y-F(x,y)$, *Acqua. Math.* 19 (1979), 194-226.

[27] Fuchs, L., On mean systems, *Acta Math. Acad. Sci. Hungar.*, 1(1950), 03-320.

[28] Fuller, R. and T. Keresztvalvi, On generalizations of Nguyen's theorem, *J. Fuzzy Sets Syst* 4 (1990), 371-374.

[29] Gardenfors, P., *Knowledge in Flux*, MIT Press, Cambridge, 1988.

[30] Gehrke, M. and E. Walker, Iterated conditionals and symmetric Stone algebras, *Discrete Mathematics* 148(1996), 49-61.

[31] Gehrke, M. and E. Walker, The structure of rough sets, *Bull. Pol. Acad. Sci. Math* 40(1992), 235-245.

[32] Gehrke, M., C. Walker, and E. Walker, De Morgan systems on the unit interval, *Int. J. Intelligent Syst.* 11(1996), 733-750.

[33] Gehrke, M., C. Walker, and E. Walker, Some comments on interval valued fuzzy sets, *Int. J. Intelligent Syst.* 11(1996), 751-759.

[34] Gehrke, M., C. Walker, and E. Walker, A mathematical setting for fuzzy logic, *Int. J. Uncertainty Fuzziness Knowledge-Based Syst.* 5(3)(1997), 223-238.

[35] Gehrke, M., C. Walker, and E. Walker, A Note on Negations and Nilpotent t-norms, *Int. J. of Approx. Reasoning*, 14(1999), to appear.

[36] Gehrke, M., C. Walker, and E. Walker, Stone Algebra Extensions with Bounded Dense Sets, *Algebra Univers.*, 37(1997), 1-23.

[37] Giarratano, J. and G. Riley, *Expert Systems: Principles and Programming*, PWS, Boston, 1994.

[38] Girard, J. Y., Linear logic, *Theo. Comp. Sci.* 50(1987), 1-12.

[39] Goodman, I. R. and H. T. Nguyen, *Uncertainty Models for Knowledge-Based Systems*, North-Holland, Amsterdam, 1985.

[40] Goodman, I. R., H. T. Nguyen, and G. S. Rogers, On the scoring approach to admissibilities of uncertainty measures in expert systems, *Jour. Math. Anal. and Appl.*, 159(2)(1991, 550-594.

[41] Goodman, I. R., H. T. Nguyen, and E. A. Walker, *Conditional Inference and Logic for Intelligent Systems: A Theory of Measure Free Conditioning*, North Holland, Amsterdam, 1991.

[42] Grabisch, M., H. Nguyen, and E. Walker, *Fundamentals of Uncertainty Calculi with Applications to Fuzzy Systems*, Kluwer Academic Press, Boston, 1994.

[43] Grätzer, G., *Lattice theory*, W. H. Freeman and Co., San Francisco 1971.

[44] Grätzer, G., *General Lattice Theory*, Academic Press, New York, 1978.

[45] Grätzer, G. and E. T. Schmidt, On a Problem of M. H. Stone, *Acta Math. Acad. Sci. Hungar* (1957), 455-460.

[46] Hamacher, H., Uber logische Verknupfungen unscharfer Aussagen und deren Augehörige Bewertungsfunctionen, *Progress in Cybernetics and Systems Research*, vol. 3, Klir, G. and I. Ricciardi, eds., Hemishphere, Washington, D.C., (1978), 276-288.

[47] Hamilton, A. G., *Logic for Mathematicians*, Cambridge University Press, Cambridge, 1987.

[48] Hardy, G. H., J. E. Littlewooe, and G. Polya, *Inequalities*, Cambridge University Press, Cambridge, 1934.

[49] Harding, J., M. Marinacci, N. T. Nguyen, and T. Wang, Local Radon-Nikodym derivatives of set functions, *Int J. Uncertainty, Fuzziness and Knowledge-Based Systems*, 5(3) (1997), 379-394.

[50] Haten, H. et al, *Le Raisonnement en Intelligence Artificielle*, Inter Editions Paris, Paris, 1991.

[51] Iwiński, T. B., *Algebraic approach to rough sets*, Bull. Polish Acad. Sci. Math., 35(1987), 673-683.

[52] Iwinski, T. B., *Rough orders and rough concepts*, Bull. Polish Acad. Sci. Math., 36(1988), 187-192.

[53] Kandel, A. and G. Longholz, *Architectures for Hybrid Intelligent Systems*, CRC Press, Boca Raton, 1992.

[54] Klir, G. and T. A. Folger, *Fuzzy Sets, Uncertainty and Information*, Prentice Hall, Englewood Cliffs, 1988.

[55] Kolmogorov, A. N., Sur la notion de la moyenne, *Atti Accad. Naz. Lincei Rend.*, 12(1930), 388-391.

[56] Kruse, R., J. Gebhard, and F. Klawonn, *Foundations of Fuzzy Systems*, John Wiley, New York, 1994.

[57] Kruse, R., E. Schweke and J. Heinsohn, *Uncertainty and Vagueness in Knowledge-Based Systems*, Springer-Verlag, New York, 1991.

[58] Levi, I., *The Enterprise of Knowledge*, MIT Press, Cambridge, 1980.

[59] Ling, C. H., Representation of associative functions, *Publ. Math. Debrecen* 12 (1965), 189-212.

[60] Mamdani, E. H. and S. Assilian, An experiment in linguistic synthesis with a fuzzy logic controller, *Int. J. Man-mach. Stud.* 7 (1975), 1-13.

[61] Manton, K. G., M. A. Woodbury, and H. D. Tolley, *Statistical Applications Using Fuzzy Sets*, John Wiley, New York, 1994.

[62] Marek, V. W. and M. Truczczyński, *Non-monotonic Logic*, Springer-Verlag, New York, 1993.

[63] Marr, D., *Vision*, Freedman, 1982.

[64] Matheron, G., *Random Sets and Integral Geometry*, John Wiley, New York, 1975.

[65] Mesiar, R. and M. Navara, Diagonals of continuous triangular norms, *Fuzzy Sets and Systems*, 104(1999, 35-41.

[66] Meyerowitz, A., F. Richman, and E. Walker, Calculating maximum-entropy densities for belief functions, *Int. J. Uncertainty Fuzziness Knowledge-Based Syst.* 2(4) (1994), 377-389.

[67] Moser, B., E. Tsiporkova, and E. P. Klement, Convex Combinations in Terms of Triangular Norms: A Characterization of Idempotent, Bisymmetrical and Self-Dual Compensatory Operators, preprint.

[68] Mosteller, F. and C. Youtz, Quantifying probability expressions, *Stat. Sci.* 1(15) (1990), 2-34.

[69] Navara, M., Characterization of measures based on strict triangular norms, preprint.

[70] Mukaidono, M. and H. Kikuchi, Proposal on Fuzzy Interval Logic, *Japan. Jour. of Fuzzy Theory and Systems*, 2(2) (1990), 99-117.

[71] Nagumo, M., Über eine Klasse der Mittelwerte, *Japan. J. Math.*, 7(1930), 71-79.

[72] Negoita, C. V. and D. A. Ralescu, *Applications of Fuzzy Sets to Systems Analysis*, Birkhauser-Verlag, Basel, 1975.

[73] Nguyen, H. T., A note on the extension principle for fuzzy sets, *J. Math. Anal. and Appl*, 64(1978), 369-380.

[74] Nguyen, H. T., Intervals in Boolean rings: approximation and logic, *J. Foundations of Computing and Decision Sciences*, 17(3) (1992), 131-138.

[75] Nguyen, H. T., V. Kreinovich, and O. Kosheleva, Is the success of fuzzy logic really paradoxical? *Int. J. Intelligent Syst.*,5(1996), 295-326.

[76] Nguyen, H. T., V. Kreinovich, and D. Tolbert, A measure of average sensitivity for fuzzy logics, *Int. J. Uncertainty Fuzziness Knowledge-Based Syst.* 2(4) (1994), 361-375.

[77] Nguyen, N. T., N. Nguyen, and T. Wang, On capacity functionals in interval probabilities, *Int. J. Uncertainty Fuzziness Knowledge-Based Syst.*, 5(3) (1997), 359-377.

[78] Nguyen, H. T., M. Sugeno, R. Tong, and R. Yager, eds, *Theoretical Aspects of Fuzzy Control*, John Wiley, New York, 1994.

[79] Nilsson, N., Probability logic, *Art. Intell.*, 28(1986), 71-87.

[80] Orlowski, S. A., *Calculus of Decomposable Properties, Fuzzy Sets and Decisions*, Allerton Press, 1994.

[81] Palmn R., D. Driankov and H. Hellendoorn, *Model-Based Fuzzy Control*, Springer-Verlag, 1997.

[82] Pawlak, Z., Rough sets, *Int. J. Comp. Inf. Sci.*, 11(5) (1982), 341-356.

[83] Pawlak, Z., *Rough sets*, Kluwer Academic Publishing, Boston 1992.

[84] Pearl, J., *Probabilistic Reasoning in Intelligent Systems*, Morgan Kaufmann, San Mateo, 1988.

[85] Pomykala, J. and J. A. Pomykala, The Stone algebra of rough sets, *Bull. Polish Acad. Sci. Math.* 36(1988) 495-508.

[86] Rescher, N., *Many Valued Logics*, McGraw Hill, New York, 1969.

[87] Richman, F. and E. Walker, Some group theoretic aspects of t-norms, preprint.

[88] Robbins, H. E., On the measure of a random set, *Ann. Math. Statist.*, 15(1944), 70.

[89] Rudin, W., *Principles of Mathematical Analysis*, 3rd edition, McGraw-Hill, New York, 1976.

[90] Schmeidler, D., Integral representation without additivity, *Proc. Am. Math. Soc.* 97 (1986), 255-261.

[91] Schweizer, B. and A. Sklar, Associative functions and statistical triangle inequalities, *Publ. Math. Debrecen* 8 (1961), 169-186.

[92] Schweizer, B. and A. Sklar, *Probabilistic Metric Spaces,* North-Holland, Amsterdam, 1983.

[93] Shafer, G., *A Mathematical Theory of Evidence*, Princeton University Press, Princeton, 1976.

[94] Skowron, A., The rough sets theory and evidence theory, *Fundamenta Informaticae* XIII(1990), 245-262.

[95] Smets, P., E. H. Mamdani, D. Dubois, and H. Prade, *Non-standard Logics for Automated Reasoning,* Academic Press, New York, 1988.

[96] Sundberg, C. and C. Wagner, Characterizations of monotone and 2-monotone capacities, *J. Theo. Probab.* 5 (1992), 150-167.

[97] Trillas, E., Sobre funciones de negación en la teoria de conjuntos difusos, *Stochastica* III-1 (1979), 47-60.

[98] Walker, C. and E. Walker, Powers of t-norms, to appear.

[99] Valverde, L., Construction of F-indistinguishibility operators, *Fuzzy Sets Syst.* 17 (1985), 313-328.

[100] Wakker, P. P., *Additive Representations of Preferences*, Kluwer Academic Press, Boston, 1989.

[101] Walley, P., *Statistical Reasoning with Imprecise Probabilities*, Chapman and Hall, London, 1991.

[102] Weber, S., A general concept of fuzzy connectives, negations, and implications based on t-norms and t-conorms, *Fuzzy Sets Syst.* 11 (1983), 115-134.

[103] Weiss, S. M. and C. A. Kulikowski, *Computer Systems That Learn,* Morgan Kaufmann, San Mateo, 1991.

[104] Yager, R., On a general class of fuzzy connectives, *Fuzzy Sets Syst.* 4(1980), 235-242.

[105] Yager, R., On ordered weighted averaging aggregation operators in multicriteria decision making, *Trans. Systems, Man & Cybern.* 18(1988), 183-190.

[106] Yager, R., S. Ovchinnoikov, R. Tong, and H. Nguyen (eds.), *Fuzzy Sets and Applications. Selected papers by L. A. Zadeh*, John Wiley, New York, 1987.

[107] Zadeh. L., Fuzzy Sets, *Information and Control*, 8(1965), 338-353.

[108] Zimmerman, H. J., *Fuzzy Sets, Decision Making and Expert Systems*, Kluwer Academic Press, Boston, 1986.

[109] Zimmerman, H. J., *Fuzzy Set Theory and its Applications*, Kluwer Academic Press, Boston, 1991.

Answers to Selected Exercises

The solutions to some of the exercises are given below. The exercises are chosen for various reasons. Generally, they are the more difficult ones. But some are chosen because they illustrate some technique or principle. Still others present material supplementary to that in the text. Complete details are not always given, but rather hints or an outline of the solutions are presented.

SOLUTIONS FOR CHAPTER 1

Exercise 6. $(A \vee B)(x) = A(x) \vee B(x) = 1$ if $X \geq 25$ and

$$\left(\frac{40 - x}{20}\right) \vee \left(1 + \left(\frac{x - 25}{5}\right)^2\right)^{-1}$$

if $25 \leq x$. See the plot on page 9. To determine when $(40 - x)/20$ is the larger involves solving a cubic polynomial. We leave the rest to the reader.

Exercise 9.

$$
\begin{aligned}
(A \vee B)'(x) &= 1 - (A \vee B)(x) \\
&= 1 - A(x) \vee B(x) \\
(A' \wedge B')(x) &= A'(x) \wedge B'(x) \\
&= (1 - A(x)) \wedge (1 - B(x)) \\
&= \begin{cases} 1 - A(x) & \text{if} \quad A(x) \geq B(x) \\ 1 - B(x) & \text{if} \quad B(x) \geq A(x) \end{cases} \\
&= 1 - A(x) \vee B(x)
\end{aligned}
$$

Exercise 10. (a) $S_f(\Omega) = \{A_t : 0 \leq t \leq 1\}$ where $A_t = \{u \in U : f(u) \geq t\}$. Thus $S_f(\Omega)$ is totally ordered.

(b) Let $A \subseteq \Omega$.

$$
\begin{aligned}
\{\omega \in \Omega : A \subseteq S_f(\omega)\} &= \{\omega : f(A) \subseteq [\alpha(\omega), 1]\} \\
&= \{\omega : \alpha(\omega) \leq \inf f(A)\} \in \mathcal{A}
\end{aligned}
$$

since α is a random variable.

(c) The equality holds when $A = \varnothing$, both sides being equal to 1. $P\{\omega : A \subseteq S_f(\omega)\} = P\{\omega : \alpha(\omega) \leq \inf f(A)\} = \inf f(A)$ since α is uniformly

distributed. Now, $P\{\omega : A \subseteq S(\omega)\}$ for $A \in S_f(\Omega)$ is well defined
by the definition of a nested random set. Also $P\{\omega : A \subseteq S(\omega)\} = $
$P\{\omega : y \in S(\omega)\}$ for all $y \in A\} \leq P\{\omega : z \in S(\omega)$ for some $z \in A\}$.
Thus $P\{\omega : A \subseteq S(\omega)\} \leq \inf_{y \in A} P\{\omega : y \in S(\omega)\} = \inf f(A)$, since by
hypothesis $f(y) = P\{\omega : y \in S(\omega)\}$.

Let $A \neq \varnothing$ and $A \in S_f(\Omega)$. Define $\mathcal{D}(A) = \{B \in S_f(\Omega) : \varnothing \neq B \subset$
$A\}$. If $\mathcal{D}(A) = \varnothing$, then for all $x \in A, P\{\omega : x \in S(\omega)\} = P\{\omega : A \subseteq S(\omega)\}$
since we always have $P\{\omega : x \in S(\omega)\} \geq P\{\omega : A \subseteq S(\omega)\}$. If the
inclusion is strict, then there exists ω such that $x \in B = S(\omega)$, so $B \neq \varnothing$,
and $B \subset A$, which is impossible.

If $\mathcal{D}(A) \neq \varnothing$, then let $x \in A \backslash B$ for all $B \in \mathcal{D}(A)$. For this x, using
the fact that $S_f(\Omega)$ is totally ordered. we get $\{\omega : x \in S(\omega)\} = \{\omega : A \subseteq$
$S(\omega)$. From all the above, we have for $A \in S_f(\Omega), P\{\omega : A \subseteq S(\omega)\} \leq$
$\inf f(A) = \inf_{y \in A} P\{\omega : y \in S(\omega)\} \leq P\{\omega : x \in S(\omega)\} = P\{\omega : A \subseteq$
$S(\omega)\}$, and therefore $P\{\omega : A \subseteq S(\omega)\} = \inf f(A) = P\{\omega : A \subseteq S_f(\omega)\}$.

SOLUTIONS FOR CHAPTER 2

Exercise 3. **(a)** That $\preceq$ is a preorder is very easy. It is linear since $\leq$
is linear.

(b) If $xy = 0$, then certainly $(x, y) \preceq (x, y)$. If $xy \neq 0$, then $\frac{x}{x} = \frac{y}{y} = 1$,
and hence $(x, y) \preceq (x, y)$, and we have reflexivity. Suppose that $(x, y) \preceq$
(u, v) and $(u, v) \preceq (s, t)$. If $xy = 0$, then $(x, y) \preceq (s, t)$ by definition. If
$xy \neq 0$, then $\frac{u}{x} = \frac{v}{y} \geq 1$, and $\frac{s}{u} = \frac{t}{v} \geq 1$, so $\frac{s}{x} = \frac{s}{u}\frac{u}{x} = \frac{t}{v}\frac{v}{y} = \frac{t}{y} \geq 1$.
Thus $(x, y) \preceq (s, t)$, and we have transitivity. This preorder is not linear
since neither $(1, 2) \preceq (1, 3)$ nor $(1, 3) \preceq (1, 2)$ holds.

(c) That the preorder in (a) is in Γ is clear. For the preorder in (b),
suppose $(x, y) \preceq (u, v)$. If $xy = 0$, then $axby = 0$ and $yx = 0$, so $(ax, by) \preceq$
(au, bv) and $(y, x) \preceq (v, u)$. If $xy \neq 0$, then $\frac{u}{x} = \frac{v}{y} \geq 1$. If $ab = 0$ then
certainly $(ax, by) \leq (au, bv)$. If $ab \neq 0$, then $\frac{au}{ax} = \frac{bv}{by} \geq 1$, and again
$(ax, by) \leq (au, bv)$. To show property (ii) is very easy. If $(x, x) \preceq (u, u)$,
and $xx = 0$, then $x = 0$ whence $x \leq u$. If $xx \neq 0$, then $\frac{u}{x} = \frac{u}{x} \geq 1$, and
so $x \leq u$. If $x \leq u$, and $x = 0$, then clearly $(x, x) \preceq (u, u)$. If $x \neq 0$, then
$u \neq 0$, and since $\frac{u}{x} \geq 1$, we get $(x, x) \preceq (u, u)$. Thus the preorder in (b)
is in Γ.

It remains to show that the preorder in (a) is the only linear one
in Γ. So suppose that $\preceq \in \Gamma$. We must show that $(x, y) \preceq (u, v)$ if
and only if $xy \leq uv$. Suppose that $(x, y) \preceq (u, v)$. Then $(xy, xy) \preceq$
(yu, xv), and since $(y, x) \preceq (v, u)$, we have $(yu, xv) \leq (uv, uv)$. By
transitivity, $(xy, xy) \leq (uv, uv)$, whence $xy \leq uv$. Now suppose that

$xy \leq uv$. We need that $(x,y) \leq (u,v)$. Suppose that $xy \neq 0$. Either $(x,y) \leq (\sqrt{xy}, \sqrt{xy})$ or $(\sqrt{xy}, \sqrt{xy}) \leq (x,y)$. If the former, then $(x\sqrt{\frac{y}{x}}, y\sqrt{\frac{x}{y}}) \leq (\sqrt{xy}\sqrt{\frac{y}{x}}, \sqrt{xy}\sqrt{\frac{x}{y}})$, or $(\sqrt{xy}, \sqrt{xy}) \leq (x,y)$. Similarly $(\sqrt{xy}, \sqrt{xy}) \leq (x,y)$ implies that $(x,y) \leq (\sqrt{xy}, \sqrt{xy})$, so that in any case, $(x,y) \leq (\sqrt{xy}, \sqrt{xy}) \leq (x,y)$. Since $xy \neq 0$, $uv \neq 0$, whence $(x,y) \leq (\sqrt{xy}, \sqrt{xy}) \leq (\sqrt{uv}, \sqrt{uv}) \leq (u,v)$. Now suppose that $xy = 0$. If both x and y are 0, then $(x,y) \leq (\sqrt{xy}, \sqrt{xy}) \leq (x,y)$. If $x = 0$ and $y > 0$, then the inequalities $(x,y) \leq (\sqrt{xy}, \sqrt{xy}) \leq (x,y)$ become $(0,y) \leq (0,0) \leq (0,y)$. One of $(0,1) \leq (1,0)$ and $(1,0) \leq (0,1)$ holds. If the former, then $(0,y) \leq (0,0)$ and $(0,0) \leq (y,0)$, using property (i). Therefore, in any case we have the inequalities $(x,y) \leq (\sqrt{xy}, \sqrt{xy}) \leq (x,y)$, and our result follows.

Exercise 5. We prove one of the absorption identities. The other properties are completely trivial. $a \vee (a \wedge b)$ is the sup of a and $a \wedge b$. Since $a \geq a$ and $a \geq a \wedge b$, $a \geq a \vee (a \wedge b)$. Clearly the other inequality holds.

Exercise 11. It is easy to see that $(u,v) \leq (x,y)$ if and only if $u \leq x$ and $v \leq y$ makes $X^{[2]}$ into a lattice. Suppose $'$ is an involution on L. Then

$$(x,y)'' = (y',x')' = (x,y)$$

If $(u,v) \leq (x,y)$, then $u \leq x$ and $v \leq y$, so

$$(u,v)' = (v',u') \leq y',x')$$

If X together with $'$ is De Morgan, then

$$
\begin{aligned}
((u \wedge v) \wedge (x,y))' &= (u \wedge x, v \wedge y)' \\
&= ((v \wedge y)', (u \wedge x)') \\
&= (v' \vee y', u' \vee x') \\
&= (v',u') \vee (y',x') \\
&= (u,v)' \vee (x,y)'
\end{aligned}
$$

Exercise 18. The Boolean algebra is not complete since U has a set of finite subsets whose union and complement are both infinite. Such a set of subsets of U has no sup in $\mathcal{F}$.

Exercise 24. Let R and S be two equivalence relations on U. The sup of these two equivalence relations is the set consisting of all pairs $(u,v) \in U \times U$ such there are elements $x_1, x_2, ..., x_n$ with $(u,x_1), (x_1,x_2), ...(x_{n-1}, x_n)$, (x_n, v) all in $R \cup S$. Note that $R \cup S$ is not an equivalence relation since it may not be transitive. It should be clear now what the sup of a set of equivalence relations on U is.

Exercise 28. $RS = \{(u, w) : \text{for some } v \in V, (u, v) \in R \text{ and } (v, w) \in S\}$. This is the definition of composition of relations. So

$$
\begin{aligned}
(RS)T &= \{(u, x) : \text{ for some } w \in W, (u, w) \in RS, \ (w, x) \in T\} \\
&= \{(u, x) : \text{ for some } v \in V, \ w \in W, \ (u, v) \in R , \\
&\quad\ (v, w) \in S, \ (w, x) \in T\} \\
R(ST) &= \{(u, x) : \text{ for some } v \in V, \ (u, v) \in R, \ (v, x) \in ST\} \\
&= \{(u, x) : \text{ for some } v \in V, \ w \in W, (u, v) \in R , \\
&\quad\ (v, w) \in S, \ (w, x) \in T\}
\end{aligned}
$$

Exercise 33. **(a)**

$$
\begin{aligned}
(A \vee B)_\alpha &= \{x \in U : A(x) \vee B(x) \geq \alpha\} \\
A_\alpha \cup B_\alpha &= \{x \in U : A(x) \geq \alpha \text{ or } B(x) \geq \alpha\}
\end{aligned}
$$

Since for each x, either $A(x) \leq B(x)$ or $B(x) \leq A(x)$, $A(x) \vee B(x)$ is either $A(x)$ or $B(x)$, and the desired equality follows. Similarly, $(A \wedge B)_\alpha = A_\alpha \cap B_\alpha$.

(b) No: $(A')_\alpha = \{x \in U : 1 - A(x) \geq \alpha\}$, whereas $(A_\alpha)' = \{x \in U : A(x) \geq\!\!< \alpha\}$.

(c) Obviously $\bigcap_{A \in \mathcal{S}} A_\alpha \leq (\bigwedge_{A \in \mathcal{S}} A)_\alpha$. If $\inf_{A \in \mathcal{S}} A(x) \geq \alpha$, then $A(x) \geq \alpha$ for all $A \in \mathcal{S}$.

(d) No in general. Obviously for all α, $\bigcup_{A \in \mathcal{S}} A_\alpha \subseteq (\bigvee_{A \in \mathcal{S}} A)_\alpha$. But $\sup_{A \in \mathcal{S}} A(x) \geq \alpha$ does not necessarily imply that there is some $A \in \mathcal{S}$ such that $A(x) \geq \alpha$, unless for each $x \in U$, $\sup_{A \in \mathcal{S}} A(x)$ is attained.

Exercise 39. **(a)** If f is one-to-one, then this follows from part 3 of Theorem 2.6.1.

(b) If $C = V$ and $f = A$, then this follows from part 2 of Theorem 2.6.1.

Exercise 41. **(a)** For $y < 0$, $f^{-1}(y) = \varnothing$, so that $f(A)(y) = 0$. For $y = 0$, $f^{-1}(0) = (-\infty, 0]$, so that $f(A)(0) = \sup_{-\infty < x \leq 0} A(x) = 1$. For $0 < y < 1$, $f^{-1}(y) = \{y\}$, so that $f(A)(y) = A(y)$.

(b) For $y = 1$, $f^{-1}(1) = [1, \infty)$, so that $f(A)(1) = \sup_{x \geq 1} e^{-\frac{1}{x}} = 1$. For $y > 1$, $f^{-1}(y) = \varnothing$, so that $f(A)(y) = 0$. In summary,

$$f(A)(y) = \begin{cases} 0 & \text{for } y < 0 \text{ or } y > 1 \\ 1 & \text{for } y = 0 \text{ or } y = 1 \\ e^{-\frac{1}{x}} & \text{for } 0 < y < 1 \end{cases}$$

(c) For $\alpha = 1$, we have $A_1 = \{x : A(x) \geq 1\} = \{0\}$. Thus $f(A) = \{f(x) : x \in A_1\} = \{f(0)\} = \{0\}$. Now $[f(A)]_1 = \{y : f(A)(y) \geq 1\} = \{0, 1\} \neq f(A_1)$.

Exercise 42. (a) $f^{-1}(0) = \{x, y) : x + y = 0\} = \{(x, -x) : x \in \mathbb{R}\}$
On $[-4, 4]$, the maximum value of $A(x) \wedge B(x)$ is $e^{-\frac{1}{x}} < 1$. And on $(-\infty, -4) \cup (4, \infty)$, $A(x) = B(-x) < 1$, and $\sup_{x > 4} A(x) = \sup e^{-\frac{1}{x}} = 1$.

(b) We have $A_1 = [3, 4]$, $B_1 = [-2, -1]$, and hence $f(A_1, B_1) = A_1 + B_1 = [1, 3]$. Note that $0 \notin f(A_1, B_1)$. But since $A(x) \wedge B(-x) < 1$ on $[0, 4]$, and $A(x) = B(x) = e^{-\frac{1}{x}}$ on $[0, 4]$,

$$f(A, B)(0) = \sup_{x \geq 0}(A(x) \wedge B(-x)) = \sup_{x > 4} e^{-\frac{1}{x}} = 1$$

Thus $0 \in [f(A, B)]_1$, showing that $[f(A, B)]_1 \neq f(A_1, B_1)$.

SOLUTIONS FOR CHAPTER 3

Exercise 2. These are all routine. We will prove 16. It is of some interest.

$$\begin{aligned} (A + (-B))(z) &= \bigvee_{x+y=z} (A(x) \wedge (-B)(y)) \\ &= \bigvee_{x+y=z} (A(x) \wedge B(-y)) \\ (A - B)(z) &= \bigvee_{x-y=z} (A(x) \wedge B(y)) \\ &= \bigvee_{x+y=z} (A(x) \wedge B(-y)) \end{aligned}$$

Exercise 4. Note that part a implies part b, and part c implies part d. We consider part a. $(A - A)(0) = \bigvee_{x-y=0} (A(x) \wedge A(y))$, and $\chi_0(0) = 1$. Any constant A will serve as an example.

Exercise 11. For a fuzzy quantity A to be convex means that its α-cuts are intervals. Equivalently, $A(y) \geq A(x) \wedge A(z)$ whenever $x \leq y \leq z$. So to show AB is convex, we need to show that $AB(y) \geq AB(x) \wedge AB(z)$. If $x \leq y \leq z < 0$ or $0 < x \leq y \leq z$, then $AB(y) \geq AB(x) \wedge AB(z)$. A proof of this can be modeled after the proof that $A + B$ is convex if A and B are convex. The other cases, for example, if $x < 0 \leq y \leq z$, are not clear.

Exercise 12. The only real problem is to show that two convex fuzzy quantities have a sup in C. But the inf of any set of convex fuzzy quantities is convex. This follows easily from the fact that a fuzzy quantity A is convex if and only if $A(y) \geq A(x) \wedge A(z)$ whenever $x \leq y \leq z$. The sup of two convex fuzzy quantities is the inf of all the convex fuzzy quantities $\geq$ both. It is very easy to find two convex fuzzy quantities whose sup as fuzzy quantities is not convex.

Exercise 13. The fuzzy quantity B is obviously convex, and it is a straightforward calculation to check that A is convex. $(A + B)(x) = \vee_{y+z=x}\{A(y) \wedge B(z)\} = \vee_{y\in\mathbb{R}} A(y) = \frac{1}{2}$. Now $(A + B)_{\frac{3}{4}} = \varnothing$, while $A_{\frac{3}{4}} = \varnothing$, $B_{\frac{3}{4}} = \mathbb{R}$, and so $A_{\frac{3}{4}} + B_{\frac{3}{4}} = \mathbb{R}$.

Exercise 14. Since A is continuous, the α-cuts are closed, and hence are closed intervals.

Exercise 15. Suppose that A is convex. Let $\lambda_1, \lambda, \lambda_2 \in [0, 1]$ with $\lambda_1 < \lambda < \lambda_2$. For $x, y \in \mathbb{R}$, let $x' = \lambda_1 x + (1 - \lambda_1 y)$ and $y' = \lambda_2 x + (1 - \lambda_2 y)$. Then $\lambda = \alpha\lambda_1 + (1 - \alpha)\lambda_2$ for some $\alpha \in [0, 1]$. A calculation shows that $\alpha x' + (1 - \alpha)y' = \lambda x + (1 - \lambda)y = z'$. Since A is convex, we have $A(z') \geq A(x') \wedge A(y')$, or that A is quasiconcave.

Suppose that A is quasiconcave and that $x < y < z$. Then $y = \lambda x + (1 - \lambda)z$ with $\lambda \in (0, 1)$. Since $0 < \lambda < 1$ and A is quasiconcave, $A(y) = A(\lambda x + (1 - \lambda)z) \geq A(z) \wedge A(x)$.

Exercise 16. By quasiconcavity of A, we have for $x, y \in \mathbb{R}$, $\lambda \in [0, 1]$ and $z = \lambda x + (1 - \lambda)y$ that $A(z) \geq A(x) \wedge A(y)$. We need strict inequality when $A(x) \neq A(y)$ and $\lambda \in (0, 1)$. But this is immediate from A being one-to-one on $\{x \in U : A(x) < 1\}$ since $A(z)$ can be neither $A(x)$ nor $A(y)$.

SOLUTIONS FOR CHAPTER 4

Exercise 6. Using the definitions of logical equivalence and tautologies, this problem is just a matter of checking the table for $\Longleftrightarrow$ in Section 4.2. In Lukasiewicz' three valued logic, implication is *not* material implication.

Exercise 11. $t(a \vee a') = t(a) \vee t(a') = t(a) \vee (1 - t(a))$ which cannot be 1 unless $t(a) \in \{0, 1\}$.

Exercise 13. In $([0, 1], \vee, \wedge,', 0, 1)$, $x \wedge x' = x \wedge (1 - x) \leq 1/2$, while $y \vee y' \geq 1/2$. It is easy to find examples in $([0, 1]^{[2]}, \vee, \wedge,', 0, 1)$ where the inequality does not hold.

Exercise 14. See the discussion in Section 4.5. Letting $A(x) = u$, $B(x) = v$, and $C(x) = 0$ shows the equality does not hold.

SOLUTIONS FOR CHAPTER 5

Exercise 2. $x \triangle_0 y = 0$ unless $x \vee y = 1$. Now $x \triangle y = x \wedge y$ if $x \vee y = 1$, so $\triangle_0 \leq \triangle$. Since $x \triangle y \leq x \triangle 1 = x$ and $x \triangle y \leq 1 \triangle y = y$, $x \triangle y \leq x \wedge y$, so $\triangle \leq \triangle_5$.

Exercise 8. Induct on n, starting at $n = 2$. For $n = 2$, the assertion is true by definition of generator. Assume true for $n - 1$. So for n, we get $a \triangle f^{-1}(f(a)^{n-1} \vee f(0)) = f^{-1} \left[f(a)(f(a)^{n-1} \vee f(0)) \vee f(0) \right]$. There are two cases: $f(a)^{n-1} \vee f(0) = f(0)$ and $f(a)^{n-1} \vee f(0) = f(a)^{n-1}$. Both are easy to check.

If $f(0) > 0$, then for sufficiently large n, $f(a)^n < f(0)$, so $\overbrace{a \triangle \cdots \triangle a}^{n \ times} = f^{-1}(f(a)^n \vee f(0)) = 0$. If $f(0) = 0$, then $f^{-1}(f(a)^n \vee f(0)) = f^{-1}(f(a)^n) > 0$ for $a > 0$.

Exercise 13. If $f(x \triangle y) = f(x) \wedge f(y)$, then $x \triangle y = f^{-1}(f(x) \wedge f(y)) = f^{-1}(f(x)) \wedge f^{-1}(f(y)) = x \wedge y$.

Exercise 15. We need f to be monotone increasing. Let $x \leq y$. Then $x = y \circ z$ for some z, and so $f(x) = f(y \circ z) = f(y) \circ f(z)$. Therefore $f(x) \leq f(y)$.

Exercise 20. The function $(f(x) + g(x))/2$ is one-to-one since f and g are one-to-one and monotone increasing. The rest is clear.

Exercise 25. (a)

$$
\begin{aligned}
f_r(f_s(x)) &= f_r\left(\frac{x}{s-(s-1)x}\right)\\[1mm]
&= \frac{\frac{x}{s-(s-1)x}}{r-(r-1)\frac{x}{s-(s-1)x}}\\[1mm]
&= \frac{x}{r(s-(s-1)x)-(r-1)x}\\[1mm]
&= \frac{x}{rs-rsx+rx-rx+x}\\[1mm]
&= \frac{x}{rs-(rs-1)x}
\end{aligned}
$$

so $f_r f_s = f_{rs}$ Since $f_r f_{r-1} = f_1$ is the identity automorphism, and so G is a subgroup.

(b) This follows from the proof of the first part.

(c) $x \blacktriangle_{f_r} y = (f_r(x) + f_r(y) - 1) \vee 0 = \left(\frac{x}{r-(r-1)x} + \frac{y}{r-(r-1)y} - 1\right) \vee 0$,
which may be written in various ways.

(d)

$$
\begin{aligned}
f_{r-1}\alpha f_r(x) &= f_{r-1}\left(1 - \frac{x}{r-(r-1)x}\right)\\[1mm]
&= f_{r-1}\left(\frac{r-rx+x-x}{r-(r-1)x}\right)\\[1mm]
&= f_{r-1}\left(\frac{r(1-x)}{r-(r-1)x}\right)\\[1mm]
&= f_{r-1}\left(\frac{1-x}{1-(\frac{r-1}{r})x}\right)\\[1mm]
&= \frac{\frac{1-x}{1-(\frac{r-1}{r})x}}{r^{-1}-(r^{-1}-1)\frac{1-x}{1-(\frac{r-1}{r})x}}\\[1mm]
&= r^2\frac{x-1}{-x+xr^2-r^2}
\end{aligned}
$$

which may be written in various forms.

Exercise 26. It is easy to check that $f(x) = \frac{x}{2-x}$ is an automorphism of $[0,1]$. Thus $x \triangle y = f^{-1}(f(x)f(y))$ is a strict t-norm. Now $f^{-1}(x) = \frac{2x}{1+x}$, so that

$$x \bigtriangleup y = \frac{2\left(\frac{x}{2-x}\frac{y}{2-y}\right)}{1 + \frac{x}{2-x}\frac{y}{2-y}}$$

$$= \frac{2xy}{4 - 2x - 2y + 2xy}$$

$$= \frac{xy}{2 + xy - (x+y)}$$

This shows that $x \bigtriangleup y = \frac{xy}{2+xy-(x+y)}$ is a strict t-norm and that f is a generator of it. Finally, for part (c) just check that $f(x) = e^{-g(x)}$.

Exercise 37. $g^{-1}f$ is an isomorphism from $(\mathbb{I}, \blacktriangle_f)$ to $(\mathbb{I}, \blacktriangle_g)$. Were h another such isomorphism, then $h^{-1}g^{-1}f$ would be a non-trivial automorphism of $(\mathbb{I}, \blacktriangle_f)$. Corollary 5.3.11 prevents that.

Exercise 39. If $\eta(x) = x < y = \eta(y)$, then $x = \eta(x) > \eta(y) = y$.

Exercise 50. They are clearly continuous. The associative law for $\bigtriangleup$ is a bit tedius, but follows readily using the distributive law $x(y\vee z) = xy\vee xz)$. The t-norm is not Archimedean since $a \bigtriangleup a = a$. The natural dual of $\bigtriangleup$ is $\bigtriangledown$.

Exercise 53. We need $f^{-1}rf \geq f^{-1}rf$ whenever $r \leq s$. Suppose that $r \leq s$. Now for $x \in [0,1]$, $x^{f^{-1}rf} = f^{-1}rf(x) = f^{-1}f(x)^r \leq f^{-1}f(x)^s = x^{f^{-1}sf}$ since $f(x)^r \geq f(x)^s$ and f^{-1} is an antiautomorphism. Thus $f^{-1}rf \geq f^{-1}rf$.

Exercise 56. If $\bigtriangledown$ is a conorm on $\mathbb{I}^{[2]}$, then $x \bigtriangledown y = \eta(\eta(x) \bigtriangleup \eta(y))$ for some t-norm $\bigtriangleup$ and negation η on $\mathbb{I}^{[2]}$. Let $x = (a,b)$ and $y = (c,d)$. Then

$$\begin{aligned}
x \bigtriangledown y &= \eta(\eta(x) \bigtriangleup \eta(y)) \\
&= \eta(\eta(a,b) \bigtriangleup \eta(c,d)) \\
&= \eta((\eta(b),\eta(a)) \bigtriangleup (\eta(d),\eta(c))) \\
&= \eta((\eta(b) \circ \eta(d)),(\eta(a) \circ \eta(c))) \\
&= ((\eta(\eta(a) \circ \eta(c)),\eta(\eta(b) \circ \eta(d)))
\end{aligned}$$

and $(\eta(\eta(a) \circ \eta(c))$ is a conorm $\diamond$ on $[0,1]$, being dual to the norm $\circ$ with respect to η. The symbol η is used to denote both a negation on $[0,1]$ and the negation it induces on $\mathbb{I}^{[2]}$.

SOLUTIONS FOR CHAPTER 6

Exercise 4. We do part a. $(1-x)\triangledown xy = 1 - x + xy - (1-x)xy = 1 - x + x^2 y$

Exercise 8. $(x \Rightarrow y) = \bigvee\{z \in [0,1] : x \triangle z \leq y\}$, which we need to be $\eta_\triangle[x \triangle \eta_\triangle(y)]$.

$$\eta_\triangle[x \triangle \eta_\triangle(y)] = \bigvee\{z : z \triangle (x \triangle \eta_\triangle(y)) = 0\}$$
$$= \bigvee\{z : (z \triangle x) \triangle \eta_\triangle(y) = 0\}$$

Now $(z \triangle x) \triangle \eta_\triangle(y) = 0$ if and only if $z \triangle x \leq y$, which is what we want.

Exercise 11. Part (a) is easy. For example,

$$(0 \Rightarrow 1) = \eta(\wedge z : z \triangledown 1 \geq 0\} = \eta(0) = 1$$
$$(1 \Rightarrow 0) = \eta(\wedge z : z \triangledown 0 \geq 1\} = \eta(1) = 0$$

For the first part of (b), if $x \neq 0 \neq y$, then $(x \Rightarrow y) = 1$. If $x = 0$, then $(x \Rightarrow y) = 1$. If $x \neq 0$ and $y = 0$, then $(x \Rightarrow y) = \eta(x)$. A computation shows that for the second part of (b), $(x \Rightarrow y) = 1 - 0 \vee (x - y)$.

In part (c), $(1 \Rightarrow 0) = f^{-1}\left(\frac{f(0)f(0)}{f(1)} \vee f(0)\right) = f^{-1}f(0) = 0$. The rest is similar.

Exercise 14. $f^{-1}(x) = \log_a((a-1)y + 1)$. The function f generates the averaging operator

$$x \dot{+} y = f^{-1}\left(\frac{f(x) + f(y)}{2}\right)$$
$$= f^{-1}\left(\frac{\frac{a^x - 1}{a-1} + \frac{a^y - 1}{a-1}}{2}\right)$$
$$= \log_a\left(\frac{a^x - 1 + a^y - 1}{2} + 1\right)$$
$$= \log_a\left(\frac{a^x + a^y}{2}\right)$$

Exercise 18. We do the verification for the first two averaging operators. The negation of the averaging operator $\frac{x+y}{2}$ is the solution to

$\frac{x+\eta(x)}{2} = \frac{1}{2}$, which is $\eta(x) = 1 - x$. The negation of the averaging operator $\left(\frac{x^a + y^a}{2}\right)^{1/a}$ is the solution to $\left(\frac{x^a + (\eta(x))^a}{2}\right)^{1/a} = 0 \dot{+} 1 = \left(\frac{1}{2}\right)^{1/a}$, which a simple calculation shows is $\eta(x) = (1 - x^a)^{1/a}$.

Exercise 22. This follows immediately from Exercise 21, which in turn is a simple induction on n.

Exercise 24.

$$
\begin{aligned}
\rho_\nabla(\delta) &= \sup\{|(x \triangledown y) - (x' \triangledown y')| : |x - x'|, |y - y'| \le \delta\} \\
&= \sup\{|1 - (1 - x) \triangledown (1 - y) - (1 - (1 - x') \triangledown (1 - y'))| \\
&: \quad |1 - x - (1 - x')|, |1 - y - (1 - y')| \le \delta\} \\
&= \sup\{|(x \triangle y) - (x' \triangle y')| : |x - x'|, |y - y'| \le \delta\} \\
&= \rho_\triangle(\delta)
\end{aligned}
$$

Exercise 26. **(a)** If $|x - x'|$ and $|y - y'| \le \delta$, then it is easy to check that $|xy - x'y'| \le 2\delta - \delta^2$, while if $x = y = 1$ and $x' = y' = 1 - \delta$, then $|xy - x'y'| = 2\delta - \delta^2$.

(b) If $\delta < \frac{1}{2}$, then $\min\{1, 2\delta\} = 2\delta$, and for $|x' - x'|$ and $|y' - y'| \le \delta$ we have

$$|(x + y) - (x' + y')| = |x - x' + y - y'| \le 2\delta < 1$$

Taking $x = y = 0$ and $x' = y' = \delta$, we get $|(x + y) - (x' + y')| = 2\delta$. If $\delta \ge \frac{1}{2}$, $\min\{2\delta, 1\} = 1$, and taking $x = y = 0$ and $x' = y' = \delta$, we have $\rho(\delta) = 1$.

(c) This one is dual to $f(x, y) = xy$, so has the same extreme sensitivity, namely $2\delta - \delta^2$.

(d) Since max is dual to min, their extreme sensitivities are the same, and the sensitivity for min is calculated in Example 6.4.2. $\rho_\triangle(\delta) = \delta$.

Exercise 27. **(a)** If $f(x, y) = \min\{x, y\}$, and $x < y$, then $\frac{\partial f}{\partial x} = 1$ and $\frac{\partial f}{\partial y} = 0$. Thus $\left(\frac{\partial f}{\partial x}\right)^2 + \left(\frac{\partial f}{\partial y}\right)^2 = 1$, and similarly for $x > y$. It follows that $S(f) = 1$.

(b) In either case, $\left(\frac{\partial f}{\partial x}\right)^2 + \left(\frac{\partial f}{\partial y}\right)^2 = x^2 + y^2$ and $S(f) = \frac{2}{3}$.

(c) For $f(x, y) = \min\{1, x+y\}$, we have $\left(\frac{\partial f}{\partial x}\right)^2 + \left(\frac{\partial f}{\partial y}\right)^2 = 2$ or 0 according to whether $x + y < 1$ or not.

(d) $\frac{\partial g}{\partial x} = -\frac{\partial f(1-x,1-y)}{\partial x}$ and $\frac{\partial g}{\partial y} = -\frac{\partial f(1-x,1-y)}{\partial y}$. Thus

$$S(g) = \int_0^1 \int_0^1 \left[\left(\frac{\partial f(1-x,1-y)}{\partial x} \right)^2 + \left(\frac{\partial f(1-x,1-y)}{\partial y} \right)^2 \right] dxdy$$

Now changing variables by letting $x' = 1 - x$ and $y' = 1 - y$ gets $S(f) = S(g)$.

Exercise 29. (a) Since $x_{(n)}$ is the min of the x_i and $x_{(1)}$ is the max, we have

$$x_{(n)} = \left(\sum w_j \right) x_{(n)} \leq \left(\sum w_j x_{(j)} \right) \leq \left(\sum w_j \right) x_{(1)} = x_{(1)}$$

(b) These choices of w give the max, the min, and the ordinary average, respectively.

Exercise 35. Let $x \leq x'$ and $y \in [0,1]$. The rectangle with vertices $(x,y), (x,0), (0,y), (x',y)$ has volume $C(x\prime,y) - C(x,y)$, using the grounded property of C.

Exercise 36. (a) Let $x = (x_1, x_2, ..., x_n) \in \{0,1\}^n$. Then for $B = \{u_j : x_j = 1\}$, $P(X_1 = x_1, ..., X_n = x_n) = F(B)$. Conversely, let $B \subseteq U$. Then $f(B) = P(\omega : S(\omega) = B) = P(\omega : u_j \in S(\omega)$ for all $u_j \in B) = P(X_1 = x_1, ..., X_n = x_n)$ where $x_j = 1$ if $u_j \in B$ and 0 otherwise.

(b) Note that if $S \in \mathcal{S}(A)$, then

$$F_j(x) = \begin{array}{lll} 0 & \text{if} & x < 0 \\ 1 - A(x) & \text{if} & 0 \leq x < 1 \\ 1 & \text{if} & x \geq 1 \end{array}$$

Thus the membership function A determines all one-dimensional CDF's F_j of the X_j. By Sklar's Theorem, the joint CDF F of $X = (x_1, x_2, ..., x_n)$ is of the form $F(x_1, x_2, ..., x_n) = C([F_1(x_1), ..., F_n(x_n)]$. In view of (a), each such F determines a distribution for an S in $\mathcal{S}(A)$.

Exercise 37. (a) From $k = 1$ and $A = (x, y]$, we have $\Delta_f(A) = F(y) - F(x)$.

(b) It is obvious that F satisfies (i), (ii), and (iii). Now, for $A = (\frac{1}{4}, 1] \times (\frac{1}{2}, 1]$, we have $\Delta_f(A) = F(1, 1) - F(\frac{1}{4}, 1) - F(\frac{1}{2}, 1) + F(\frac{1}{4}, \frac{1}{2}) = -1$, so that F does not satisfy (iv).

(c) For $a_1 = (x_1, y_1) \leq a = (x, y)$, we have, applying the hypothesis to the case $n = 1$, that $F(a) \geq F(a_1)$. Next, for $A = (x_1, y_1] \times ... \times (x_k, y_k]$

arbitrary, $\Delta_f(A) \geq 0$ since it is a consequence of the hypothesis when $a = (y_1, y_2, ..., y_k)$, and the $n(= 2^k - 1)$ a_j being the vertices of A.

(d) If $(x, y) \leq (x', y')$ then $\max(x, y) \leq \max(x', y')$, and hence $F(x, y) \leq F(x', y')$. For $a = (x, y)$, $a_1 = (x_1, y_1)$, $a_2 = (x_2, y_2)$, with $0 \leq x_1 < 1 < x < x_2$ and $0 \leq y_2 < 1 < y < y_1$, we have $F(a \wedge a_1) = F(a \wedge a_2) = 1$, and $F(a \wedge a_1 \wedge a_2) = F(x_1, y_2) = 0$. Thus

$$\sum_{\emptyset \neq I \subseteq \{1,1\}} (-1)^{|I|+1} F(\wedge_{b \in \{a, a_i, i \in I\}} b = 1 + 1 - 0 = 2 > 1 = F(a)$$

and hence F does not satisfy the desired inequality.

Exercise 39. A t-norm is a 2-copula if and only if for $a \leq c$, $c \triangle b - a \triangle b \leq c - a$. To check that $(a + b - 1) \leq a \triangle b$, take $c = 1$.

SOLUTIONS FOR CHAPTER 7

Exercise 1. Reflexivity and symmetry of W are clear. For $\wedge$-transitivity, we have for all $x \in U$,

$$
\begin{aligned}
W(u, v) &= R(u, v) \wedge S(u, v) \\
&\geq (R(u, x) \wedge R(x, v)) \wedge (S(u, x) \wedge S(x, v)) \\
&= (R(u, x) \wedge S(x, v)) \wedge (R(u, x) \wedge S(x, v)) \\
&= W(u, x) \wedge W(x, v)
\end{aligned}
$$

Exercise 4. Suppose R is transitive. Then $R(u, v) \wedge R(v, w) \leq R(u, w)$, and so

$$\bigvee_{v \in U} (R(u, v) \wedge R(v, w)) \leq R(u, w)$$

so that $R \circ R \leq R$.

Conversely, suppose that $R \circ R \leq R$. Then certainly

$$\bigvee_{v \in U} (R(u, v) \wedge R(v, w)) \leq R(u, w)$$

so that for all v, $R(u, v) \wedge R(v, w) \leq R(u, w)$, whence R is transitive.

Exercise 5. Part (a) is trivial.

For part (b), suppose that R is transitive. For $R(u,v) \geq \alpha$ and $R(v,w) \geq \alpha$, we have

$$
\begin{aligned}
R(v,w) &\geq \bigvee_{x \in U} (R(u,x) \wedge R(x,w)) \leq R(u,w) \\
&\geq R(u,v) \wedge R(v,w) \\
&\geq \alpha
\end{aligned}
$$

Conversely, suppose that R_α is transitive for all α. Let $\alpha = R(u,v) \wedge R(v,w)$. Then $R(u,w) \geq \alpha$, and this for any v. Thus

$$
R(u,w) \geq \bigvee_{v \in U} (R(u,v) \wedge R(v,w))
$$

Exercise 6. No. It is easy to get an example with U a three element set and R and S equivalence relations, and in particular fuzzy equivalence relations.

Exercise 8. Suppose that each R_α is an equivalence relation. We show that R is transitive, which means that $R(u,w) \geq R(u,v) \wedge R(v,w)$. If $R(u,v) \wedge R(v,w) = \alpha$, then (u,v) and $(v,w) \in R_\alpha$, and since R_α is an equivalence relation and hence transitive, $(v,w) \in R_\alpha$, so $R(v,w) \geq \alpha = R(u,v) \wedge R(v,w)$.

Exercise 10. Suppose that each R_α is a partial order. To show that R is transitive, see the proof of transitivity in Exercise 8.

Exercise 14. R is $\triangle$-transitive, so $R(u,v) \geq \sup_w (R(u,w) \triangle R(w,v)) = R \circ R(u,v)$. Now

$$
\begin{aligned}
R \circ R(u,v) &= \sup_w (R(u,w) \triangle R(w,v)) \\
&\geq \sup_w (R(u,u) \triangle R(w,v)) \\
&= \sup_w (1 \triangle R(w,v)) \\
&= \sup_w (R(w,v)) \\
&\geq R(u,v)
\end{aligned}
$$

Exercise 23. Let

$$
\begin{aligned}
(f_1, f_2) &: \quad \mathcal{C}_1 \to \mathcal{C}_2 \\
(g_1, g_2) &: \quad \mathcal{C}_2 \to \mathcal{C}_3 \\
(h_1, h_2) &: \quad \mathcal{C}_3 \to \mathcal{C}_4
\end{aligned}
$$

be morphisms between the Chu spaces $C_1, ..., C_4$. Then

$$
\begin{aligned}
(h_1, h_2) \circ ((g_1, g_2) \circ (f_1, f_2)) &= (h_1, h_2) \circ (g_1 \circ f_1, \ f_2 \circ g_2) \\
&= (h_1 \circ (g_1 \circ f_1), \ (f_2 \circ g_2) \circ h_2) \\
&= ((h_1 \circ g_1) \circ f_1, \ f_2 \circ (g_2 \circ h_2)) \\
&= ((h_1, h_2) \circ (g_1, g_2)) \circ (f_1, f_2)
\end{aligned}
$$

That the appropriate identity exists is obvious.

SOLUTIONS FOR CHAPTER 8

Exercise 2. Polynomials are continuous, so $\mathcal{F} \subseteq C([a, b])$. Addition and multiplication of polynomials result in polynomials, so $\mathcal{F}$ is a subalgebra. Let $c, d \in [a, b]$ with $c \neq d$. The polynomial x is c at c and d at d. The polynomial $x - c + 1$ is nonzero at c.

Exercise 3. $\mathcal{F}$ does not separate points on $[-2, 2]$ since for and $x \in [-2, 2]$ and $f \in \mathcal{F}$, we have $f(x) = f(-x)$.

Exercise 4. **(a)** There are functions g and h in $\mathcal{F}$ such that $g(x) \neq g(y)$ and $h(x) \neq 0$. Let $\varphi = g + \lambda h$ with λ chosen as indicated. Then $\varphi(x) = g(x)$ if $g(x) \neq 0$ and $\varphi(x) = \lambda h(x)$ if $g(x) = 0$. Thus $\varphi(x) \neq 0$, and obviously $\varphi \in \mathcal{F}$. Now $\varphi(x) - \varphi(y) = g(x) - g(y) + \lambda h(x) - h(y) \neq 0$.

(b) Let $\alpha = \varphi^2(x) - \varphi(x)\varphi(y)$, and $f_1(x) = \frac{1}{\alpha} \left[\varphi^2(x) - \varphi(x)\varphi(y) \right]$. Then $\alpha \neq 0$ since $\varphi(x) \neq \varphi(y)$ and $\varphi(x) \neq 0$. Obviously $f_1 \in \mathcal{F}$. Now $f_1(y) = 0$ and $f_1(x) = 1$.

(c) The construction of f_2 is similar, and $f_1 + f_2$ clearly has the desired properties. Note that if $1 \in \mathcal{F}$, then one could let $f_2(w) = 1 - f_1(w)$ for all w.

Exercise 8. **(a)** $f + g$, fg, and af are continuous.

(b) Let $\mathcal{F}$ be a subalgebra of $C([a, b])$ containing 1 and x. Then any polynomial $\sum a_i x^i$ is in $\mathcal{F}$, and certainly is a subalgebra.

(c) The closure $\overline{\mathbb{P}}$ of $\mathbb{P}$ in $C([a, b])$ consists of those functions on $[a, b]$ which are uniform limits of sequences of elements of $\mathbb{P}$. According to the Weierstrass theorem, $\overline{\mathbb{P}} = C([a, b])$. But $|x| \in \overline{\mathbb{P}}$ but not in $\mathbb{P}$, so $\overline{\mathbb{P}} \neq \mathbb{P}$.

SOLUTIONS FOR CHAPTER 9

Exercise 3. (a) For $A \subseteq B$, we have $B = A \cup (B \backslash A)$, so that

$$\mu(B) = \mu(A) + \mu(B \backslash A) \geq \mu(A)$$

(b) For $A, B \in \mathcal{A}$, we have

$$
\begin{aligned}
A \cup B &= A \cup (B \backslash A) \\
B &= (B \backslash A) \cup (A \cap B)
\end{aligned}
$$

Thus

$$
\begin{aligned}
\mu(A \cup B) &= \mu(A) + \mu(B \backslash A) \\
&= \mu(A) + \mu(B) - \mu(A \cap B)
\end{aligned}
$$

(c) Let $A_n \nearrow$, $n \geq 1$. Let $A_0 = \varnothing$. We have

$$
\begin{aligned}
\mu \left(\bigcup_{n=1}^{\infty} A_n \right) &= \mu \left(\bigcup_{n=1}^{\infty} (A_n \backslash A_{n-1}) \right) \\
&= \sum_{n=1}^{\infty} \mu (A_n \backslash A_{n-1}) \\
&= \lim_{m \to \infty} \sum_{n=1}^{m} \mu (A_n \backslash A_{n-1}) \\
&= \lim_{m \to \infty} \mu \left(\bigcup_{n=1}^{m} A_n \backslash A_{n-1} \right) \\
&= \lim_{m \to \infty} \mu (A_m)
\end{aligned}
$$

(d) Let $B \in \mathcal{A}$, and define $\nu : \mathcal{A} \to \mathbb{R}^+$ by $\nu(A) = \mu(A \cap B)$. The details are routine.

Exercise 4. Let $\pi : \Omega \to [0, 1]$. The associated possibility measure on 2^{Ω} is defined by $\pi(A) = \sup\{\pi(\omega) : \omega \in A\}$. Since

$$\sup\{\pi(\omega) : \omega \in A_i\} = \sup_i \{\sup\{\pi(\omega) : \omega \in A_i\}\}$$

it follows that the restriction of π to $\mathcal{F}$ is a space law, and hence the result follows from Matheron's theorem.

Exercise 5. Let $f : \mathbb{R}^n \to [0, 1]$ be upper semicontinuous. It suffices to verify that $\pi(A) = \sup\{f(x) : x \in A\}$ is a Choquet capacity, alternating of infinite order. The fact that π is alternating of infinite order is proved as in Exercise **4.** Let $\pi : \Omega \rightarrow [0,1]$. The associated possibility measure on 2^Ω is defined by $\pi(A) = \sup\{\pi(\omega) : \omega \in A\}$. Since 4, it remains to show that π *is* upper semicontinuous on the set $\mathcal{K}$ of compact sets of $\mathbb{R}^n$, that is, if $K_m \searrow K$ in $\mathcal{K}$, then $\pi(K_m) \searrow \pi(K)$. Let $\alpha = \inf\{\pi(K_m)\}$ and $\beta = \pi(K)$. Clearly $\beta \leq \alpha$. Let $\varepsilon > 0$ and take $\delta = \alpha - \varepsilon < \alpha$. We then have $\pi(K_m) > \delta$ for all m, that is, $\sup\{f(x) : x \in K_m\} > \delta$ for all $m \geq 1$. Hence for

$$A_m = \{x : f(x) \geq \delta\} \cap K_m \neq \varnothing$$

we have that $A_m \subseteq K_m \subseteq K_1$ which is compact, and the A_m's are closed since f is upper semicontinuous. Therefore, $A = \bigcap_{m=1}^\infty A_m \neq \varnothing$. Since $A \subseteq K$, we have $\pi(A) \leq \pi(K) = \beta$. But by the construction of A_n, $\pi(A) \geq \delta$, and thus $\delta \leq \beta$.

Exercise 6. We verify the formula in a special case. Let X be a non-negative random variable. Let $S(\omega) = [0, X(\omega)]$. We have $\mu \circ S = X$, and $\pi(x) = P\{\omega : x \in S(\omega)\} = P(X \geq x)$. Now it is well known that $E(X) = \int_0^\infty P(X \geq x)dx$. But $E(X) = (E \circ \mu)(S)$, so that $(E \circ \mu)(S) = \int_0^\infty \pi(x)dx$.

Exercise 7. We have $\pi(A \cup A') = \pi(U) = \pi(A) \wedge \pi(A') = 1$.

Exercise 8. Let π be a possibility measure on U. Let $T : 2^U \rightarrow [0,1] :$ $A \rightarrow 1 - \pi(A')$. Then for part a, $T(\varnothing) = 1 - \pi(U) = 0$ and $T(U) = 1 - \pi(\varnothing) = 1$.

For part b.

$$
\begin{aligned}
T(\bigcap_{i \in I} A_i) &= 1 - \pi\left(\bigcup_{i \in I} A'_i\right) \\
&= 1 - \bigvee_{i \in I} \pi\left(A'_i\right) \\
&= \bigwedge_{i \in I} \left(1 - \pi\left(A'_{(i)}\right)\right) \\
&= \bigwedge_{i \in I} T(A_i)
\end{aligned}
$$

Exercise 13. Part (a) is a direct calculation, and for part(b), use l'Hospital's rule. For part (c), suppose that $I(A \cup B) = \min\{(A),(B)\}$ for A, B with $AB = \varnothing$. For arbitrary A and B we have

$$
\begin{aligned}
I(A \cup B) &= I(A'B \cup AB' \cup AB) \\
&= \min\{I(A'B), I(AB'), I(AB)\}
\end{aligned}
$$

But

$$I(A) = \min\{I(AB'), I(AB)\}$$
$$I(B) = \min\{I(A'B), I(AB)\}$$

Hence $I(A \cup B) = \min\{I(A), I(B)\}$.

SOLUTIONS FOR CHAPTER 10

Exercise 3. Let Q be a probability measure on a finite set Ω and let $0 < \varepsilon < 1$. Let

$$\mathcal{P} = \{\varepsilon P + (1 - \varepsilon)Q : P \text{ is a probability measure on } \Omega\}$$

Define $F(A) = \inf\{P(A) : P \in \mathcal{P}\}$. Obviously, F is increasing, $F(\varnothing) = 0$ and $F(\Omega) = 1$. It remains to show that for all $n \geq 1$,

$$F(\bigcup_{i=1}^{n} A_i) \geq \sum_{\varnothing \neq J \subseteq \{1,2,\ldots,n\}} (-1)^{|J|+1} F(\bigcap_{J} A_j)$$

If $\cup A_i \neq \Omega$ then there is a probability measure P such that $P(\cup A_i) = 0$, so that

$$
\begin{aligned}
F(\bigcup A_i) &= (1 - \varepsilon)Q(\bigcup A_i) \\
&\geq \sum_{\varnothing \neq J \subseteq \{1,2,\ldots,n\}} (-1)^{|J|+1}(1 - \varepsilon)Q(\bigcap_{J} A_j) \\
&= \sum_{\varnothing \neq J \subseteq \{1,2,\ldots,n\}} (-1)^{|J|+1}(1 - \varepsilon)F(\bigcap_{J} A_j)
\end{aligned}
$$

If $\bigcup A_i = \Omega$, then $F(\bigcup A_i) = 1$ and

$$
\begin{aligned}
\sum_{\varnothing \neq J \subseteq \{1,2,\ldots,n\}} (-1)^{|J|+1}(1 - \varepsilon)F(\bigcap_{J} A_j) &= (1 - \varepsilon)Q(\bigcup A_i) \\
&\leq 1 - \varepsilon
\end{aligned}
$$

Exercise 4. (a)

$$
\begin{aligned}
\sum_{\omega \in \Omega} g(\omega) &= \sum_{A \subseteq \Omega} \sum_{\omega \in A} \alpha(\omega, A) \\
&= \sum_{A \subseteq \Omega} f(A) = 1.
\end{aligned}
$$

(b) Let $P_g(A) = \sum_{w \in A} g(w)$. We have

$$
\begin{aligned}
F(A) &= \sum_{B \subseteq A} f(B) \\
&= \sum_{B \subseteq A} \sum_{w \in B} \alpha(w, B) \\
&\leq \sum_{w \in A} \sum_{\{B : w \in B\}} \alpha(w, B) \\
&= P_g(A)
\end{aligned}
$$

Exercise 8. (a) $g(w_j) = \sum_{A \in T(w_j)} f(A)$, where

$$
T(w_j) = \{A : w_j \in A \subseteq \{w_j, w_{j+1}, ..., w_k\}\}
$$

Then $\sum_{j=1}^{k} g(w_j) = \sum_{A \subseteq \Omega} f(A) = 1$

(b) $F(B) = \sum_{A \subseteq B} f(A) \leq \sum_{w_j \in B} g(w_j)$ since $\{A : A \subseteq B\} \subseteq \{A : A \in T(w_j) \text{ for some } w_j \in B\}$

(c) See Exercise 4.

Exercise 11. (a) Let $F(A) = [P(A)]^n$ and define $f : 2^\Omega \to [0, 1]$ by

$$
f(A) = \begin{cases}
0 \text{ if } |A| > n \\
\sum_{\substack{n_a \geq 0 \\ \sum n_a = n}} \prod_{a \in A} \frac{n!}{\prod n_a!} [P(a)]^{n_a} \text{ otherwise}
\end{cases}
$$

Since

$$
1 = [P(\Omega)]^n = [\sum_{w \in A} P(w)]^n = \sum_{A \subseteq \Omega} f(A)
$$

the function f is a probability mass assignment. It is easy to check that $F(A) = \sum_{B \subseteq A} f(B)$, so that $A \to [P(A)]^n$ is a belief function.

(b) Let $F_B(A) = [P(A|B)]^2$. Then F_B is a belief function from part (a), and hence its Möbius inversion is given by f_B.

(c)

$$
\begin{aligned}
F_B(A) &= F_B(A|C) \\
&= [P_B(A|C)]^2 \\
&= [P_B(\frac{AC}{P_B(C)})]^2 \\
&= [\frac{P(AC|B)}{P(C)|B)}] \\
&= [\frac{P(ABC)}{P(BC)}]^2 \\
&= (F_C)_B(A)
\end{aligned}
$$

(d) The "sandwich principle" holds. If $F(A) < F(A|B)$ then

$$
\begin{aligned}
P(A) &= P(AB\prime) + P(AB) \\
&= P(AB') + P(A|B)P(B) \\
&\geq P(AB') + P(A)P(B)
\end{aligned}
$$

from which it follows that

$$
P(A)[1 - P(B) \geq P(ABB')
$$

and

$$
P(A) \geq \frac{P(AB')}{P(B')} = P(A|B')
$$

Hence

$$
\begin{aligned}
F(A) &= [P(A)]^2 \\
&\geq [P(A|B')]^2 \\
&= F(A|B')
\end{aligned}
$$

and so $F(A) \geq \min\{F(A|B), F(A|B')\}$.

Exercise 14 It is suggested that the discussion of algorithm 2 in [64] be consulted.

Exercise 16. It is easy to get examples such that $A \sim B$ and $C \sim D$, yet $A \cup C$ is not equivalent to $B \cup D$, in fact so that $\underline{A \cup C} \neq \underline{B \cup D}$. Let $C = D$, and pick A and B so that $\underline{A \cup C}$ contains an element of $\mathcal{E}$ that $\underline{B \cup C}$ does not.

Exercise 18

$$P(b \Rightarrow a) = P(b' \vee a)$$
$$= P(b' \vee ab)$$
$$= P(b') + P(ab)$$
$$= 1 - P(b) + P(ab)$$
$$= [P(a|b) + P(a'|b)] - P(b) + P(ab)$$
$$= P(a|b) + P(a'|b) - P(b) + P(a|b)P(b)$$
$$= P(a|b) + P(a'|b) - P(b) + [1 - P(a'|b)\backslash P(b)$$
$$= P(a|b) + P(a'|b)P(b')$$

Exercise 23. The center of $\mathcal{R}$ consists of the elements $(\underline{X}, \overline{X})^* = \left(\overline{X}', \underline{X}'\right)$ of $\mathcal{R}$. Any $\underline{Y}$ is of the form $\overline{X}'$.

Exercise 24. (a) If $ab = cd$ and $b = d$, then $a + Rb' = a + ab' + R = a(1 + b') + Rb' = ab + Rb'$.Similarly $c + Rd' = cd + Rd' = cd + Rb'$, .so $a + Rb' = c + Rb'$. Conversely, if $a + Rb' = c + Rd'$, then $a = c + rd'$, and hence $ad = cd$. Also, $a + b' = c + sd'$, and so $ad + b'd = cd = ad$, which gets $b'd = 0$. Similarly, $bd' = 0$, and we have $(1 + b)d = d + bd = 0 = bd' = b(1 + d) = b + bd$, so $b = d$.

(b) $(a + rb') \geq (a+rb')\wedge ab = (a+rb')ab = ab$. Now $(b' \vee a)(a + rb') = (a + b' + ab')(a + rb') = a+rb'$, so $ab \leq a+rb' \leq b'\vee a$. If $ab \leq x \leq b'\vee a$, then $abx = ab$, and $x(b' \vee a) = x(a + b' + ab') = x(a + b' + a(1 + b)) = x(b' + ab) = xb' + abx = xb' + ab = xb' + a + ab' \in a + rb'$.

(c) $[a, b] = [a(b' \vee a), (b' \vee a) \vee a] = a + R(b' \vee a)'$, using part b.

SOLUTIONS FOR CHAPTER 11

Exercise 4. Suppose that μ is monotone of order 2, that is, $\mu(A \cup B) \geq \mu(A) + \mu(B) - \mu(A \cap B)$. Then for all $X \in \mathcal{A}$,

$$\mu(X) \geq \mu(X \cap (A \cup B))$$
$$= \mu((X \cap A) \cup (X \cap B))$$
$$= \mu(X \cap A) + \mu(X \cap B) - \mu(X \cap A \cap B)$$

so that $\Delta_2(X, A, B) = \mu(X) - \mu(X \cap A) - \mu(X \cap B) + \mu(X \cap A \cap B) \geq 0$. Conversely, suppose that $\Delta_2(X, A, B) \geq 0$. Take $X = A \cup B$. Then $0 \leq \Delta_2(X, A, B) = \mu(A \cup B) - \mu(A) - \mu(B) + \mu(A \cap B)$.

Exercise 5. This problem is lengthy. See [7].

Exercise 6. This exercise is difficult. See [51].

Exercise 11. (a) $\mu^*(\varnothing) = \mu(\Omega) - \mu(\Omega) = 0$. If $A \subseteq B$, then $\mu(A') \geq \mu(B')$, and $\mu^*(A) = \mu(\Omega) - \mu(A') \leq \mu(\Omega) - \mu(B') = \mu^*(B)$.

(b) This is the same as exercise 4.

Exercise 14. $E(\mu(S)) = E(X) = \int_0^\infty P(X \geq x)dx = \int_0^\infty \pi(x)dx$ by observing that $\{\omega : X(\omega) \geq x\} = \{\omega : x \in S(\omega)\}$.

Exercise 16. Recall that the capacity functional T of the random closed set S on $\mathbb{R}^+$ is $T(K) = P\{\omega : S(\omega) \cap K \neq \varnothing\}$ for compact subsets K of $\mathbb{R}^+$. When we have $K = \{x\}$, then $\pi(x) = T(\{x\}) = P(x \in S)$. Here, when $S(\omega) = [0, X(\omega)]$, it turns out that T can be determined from π. Indeed, $T(K) = P([0, x]] \cap K \neq \varnothing) = P(\wedge K \leq x) = \pi(\wedge K)$.

SOLUTIONS FOR CHAPTER 12

Exercise 1. (a) Clearly for each $n \geq 1$, the sets A_n^k, $k = 1, 2, ..., n2^n$ and A_n are pairwise disjoint and $A_n \bigcup(\bigcup_{k=1}^{n2^n} A_n^k) = \Omega$.

(b) If $f(\omega) > n+1$ then $f(\omega) > n$, so that $f_n(\omega) = n < n+1 = f_{n+1}(\omega)$. Suppose that $n < f(\omega) \leq n+1$. Then

$$\frac{n2^{n+1}}{2^{n+1}} < f(\omega) \leq \frac{(n+1)2^{n+1}}{2^{n+1}}$$

and

$$\frac{n2^{n+1}}{2^{n+1}} < \frac{n2^{n+1}+1}{2^{n+1}} < \frac{(n+1)2^{n+1}}{2^{n+1}}$$

If

$$\frac{n2^{n+1}}{2^{n+1}} \leq f(\omega) \leq \frac{n2^{n+1}+1}{2^{n+1}}$$

then $f_n(\omega) = n$ and $f_{n+1}(\omega) = \dfrac{n2^{n+1}+1}{2^{n+1}} > n$. If

$$\frac{n2^{n+1}+1}{2^{n+1}} < f(\omega) \leq \frac{(n+1)2^{n+1}}{2^{n+1}}$$

then $f_n(\omega) = n$, and $f_{n+1}(\omega) = \dfrac{(n+1)2^{n+1}}{2^{n+1}} > n$. Finally, if $f_n(\omega) \le n$, then for $k = n2^{n+1} + 1$,

$$\frac{k-1}{2^n} \le f(\omega) < \frac{k}{2^n}$$

for some $k = 1, 2, ..., n2^n$. So either

$$\frac{2(k-1)}{2^{n+1}} \le f(\omega) < \frac{2k-1}{2^{n+1}}$$

or

$$\frac{2k-1}{2^{n+1}} \le f(\omega) < \frac{2k}{2^{n+1}}$$

Reasoning as before, we get $f_n(\omega) \le f_{n+1}(\omega)$.

It remains to show that $\lim_{n \to \infty} f_n(\omega) = f(\omega)$. If $f(\omega) = \infty$, then $f(\omega) > n$ for all n, so that $f_n(\omega) = n$. If $f(\omega) < \infty$, then there is an n_0 such that $f(\omega) \le n_0$ and $|f(\omega) - f_n(\omega)| \le \frac{1}{2^n}$ for all $n \ge n_0$. In either case, $\lim_{n \to \infty} f_n(\omega) = f(\omega)$.

Exercise 2. With $\Omega = \bigcup_{j=1}^n A_j = \bigcup_{j=1}^n B_j$, we have $A_j = A_j \cap \Omega = \bigcup_i (A_j \cap B_i)$ and $B_i = B_i \cap \Omega = \bigcup_j (A_j \cap B_i)$. Also $f(\omega) = \alpha_j - \beta_i$ when $\omega \in A_j \cap B_i$.

$$
\begin{aligned}
\sum_j \alpha_j \mu(A_i) &= \sum_j \alpha_j \mu \left(\bigcup_i (A_j \cap B_i) \right) \\
&= \sum_j \alpha_j \sum_i \mu(A_j \cap B_i) \\
&= \sum_j \sum_i \alpha_j \mu(Aj \cap B_i) \\
&= \sum_j \sum_i \beta_i \mu(A_j \cap B_i) \\
&= \sum_{ij} \beta_i \sum_j \mu(A_j \cap B_i) \\
&= \sum_i \beta_i \mu \left(\bigcup_j (A_j \cap B_i) \right) \\
&= \sum_i \beta_i \mu(B_i)
\end{aligned}
$$

Exercise 6. **(a)** Let $h : \Omega \to [0,1]$ be $h(\omega) = \alpha$ for all $\omega \in \Omega$. Then $h_\beta = \varnothing$ for all $\beta > \alpha$. Thus $S_\mu(\alpha) = \sup_{0 \le \beta \le \alpha}(\beta \wedge \mu(\Omega)) = \alpha$.

(b) For $h = \chi_A$, $h_\alpha = A$ for all $\alpha > 0$. Thus $S_\mu(\chi_A) = \sup_{\alpha > 0}(\alpha \wedge \mu(A)) = \mu(A)$.

(c) Let $f, g : \Omega \to [0, 1]$ with $f \le g$. Then for $\alpha \in [0, 1]$, $f_\alpha \subseteq g_\alpha$ and so $\mu(f_\alpha) \le \mu(g_\alpha)$, implying that $S_\mu(f) \le S_\mu(g)$.

Exercise 8. Part (a) is trivial. For part (b), let $f_*(\omega) = \inf_{\theta \in \Gamma(\omega)} f(\theta)$. Then

$$
\begin{aligned}
\int_\Omega f_*(\omega) dP(\omega) &= \int_\Omega P(\omega : f_*(\omega) > t) dt \\
&= \int_0^\infty P(\omega : f(\Gamma(\omega)) \subset (t, \infty)) dt \\
&= \int_0^\infty \mu(f^{-1}(t, \infty)) dt \\
&= \int_0^\infty \mu\{\theta : f(\theta) > t\} \\
&= \mathcal{E}_\mu(f)
\end{aligned}
$$

Exercise 9. Assume (a) and suppose that there exists $A = (f > t)$ and $B = (g > s)$ such that $A \not\subseteq B$ and $B \not\subseteq A$. Take $\omega \in A - B$ and $\omega' \in B - A$. Then $f(\omega) > t \ge f(\omega')$. Similarly, $g(\omega') > s \ge g(\omega)$, so that $(f(\omega) - f(\omega'))(g(\omega) - g(\omega')) < 0$, contradicting (b). Thus (a) implies (b).

Suppose f and g are not comonotonic. Then there exist a and b such that $f(a) - f(b) > 0$ and $g(a) - g(b) < 0$. Let $A = \{\omega : f(\omega) > g(b)\}$ and $B = \{\omega : g(\omega :> g(a)\}$. Since $f(b) < f(a)$, we have $a \in A$ and $a \notin B$, so that $a \in A - B$. Similarly, $g(a)g(b)$ implies that $b \in B - A$. Thus $A \not\subseteq B$ and $B \not\subseteq A$. Thus (b) implies (a).

Exercise 10. Parts (a), (b), and (c) are easy to check. For (d) let

$$
S(A) = \begin{cases} 0 & \text{if } A \cap \mathbb{N} = \varnothing \\ \frac{1}{2} \sum_n \{\frac{1}{n} : n \in A \cap \mathbb{N}\} & \text{if } A \cap \mathbb{N} \ne \varnothing \end{cases}
$$

Observe that S is additive. There are four cases.

1. $\mu(A \cup B) < 1$. In this case, we have

$$
\begin{aligned}
\mu(A \cap B) &= S(A \cap B) \\
&= S(A) + S(B) - S(A \cup B) \\
&= \mu(A) + \mu(B) - \mu(A \cup B)
\end{aligned}
$$

2. $\mu(A \cup B) = 1, \mu(A) < 1, \mu(B) < 1$. In this case, we have

$$
\begin{aligned}
\mu(A \cap B) &\leq S(A \cap B) \\
&= S(A) + S(B) - S(A \cup B) \\
&\leq \mu(A) + \mu(B) - \mu(A \cup B)
\end{aligned}
$$

3. $\mu(A \cup B) = 1, \mu(A) = 1, \mu(B) < 1$. In this case, we have

$$
\begin{aligned}
\mu(A \cap B) &\leq \mu(B) \\
&= \mu(A) + \mu(B) - \mu(A \cup B)
\end{aligned}
$$

4. $\mu(A \cup B) = \mu(A) = \mu(B) = 1$. In this case, we have

$$
\begin{aligned}
\mu(A \cap B) &\leq \mu(B) \\
&= \mu(A) + \mu(B) - \mu(A \cup B)
\end{aligned}
$$

For part (e), it can be checked that both ν and γ are maxitive, and hence alternating of infinite order.

SOLUTIONS FOR CHAPTER 13

Exercise 1.

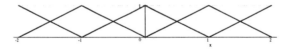

There are two fuzzy sets here, A and $1 - A$.

Exercise 4. Let $\mu = \int_{-\infty}^{\infty} x f(x) dx$. Since

$$
\begin{aligned}
(x - u)^2 &= (x - \mu + \mu - u)^2 \\
&= (x - u)^2 + (\mu - u)^2 + 2(x - \mu)(\mu - u)
\end{aligned}
$$

and $\int_{-\infty}^{\infty} f(x) dx = 1$, we have

$$
\int_{-\infty}^{\infty} (x - u)^2 f(x) dx = \int_{-\infty}^{\infty} (x - \mu)^2 f(x) dx + (\mu - u)^2
$$

Hence $u = \mu$ minimizes $J(u) = \int_{-\infty}^{\infty} (x - u)^2 f(x) dx$.

Index